OXFORD STATISTICAL SCIENCE SERIES

OXFORD STATISTICAL SCIENCE SERIES

Random Coefficient Models

NICHOLAS T. LONGFORD

Educational Testing Service
Princeton, New Jersey

CLARENDON PRESS · OXFORD
1993

Oxford University Press, Walton Street, Oxford OX2 6DP
Oxford   New York   Toronto
Delhi   Bombay   Calcutta   Madras   Karachi
Kuala Lumpur   Singapore   Hong Kong   Tokyo
Nairobi   Dar es Salaam   Cape Town
Melbourne   Auckland   Madrid
and associated companies in
Berlin   Ibadan

Oxford is a trade mark of Oxford University Press

Published in the United States
by Oxford University Press Inc., New York

A catalogue record for this book is available from the British Library

Library of Congress Cataloging in Publication Data
Longford, Nicholas T., 1955–
Random coefficient models / Nicholas T. Longford.
(Oxford statistical science series ; 11)
1. Regression analysis.   I. Title.   II. Series.
QA278.2.L66   1993   519.5'36–dc20   93-35629
ISBN 0 19 852264 9

Typeset by the author
Printed in Great Britain by
Bookcraft (Bath) Ltd
Midsomer Norton, Avon

To Elyse, Kalman,
and their 'Grandpa-Lady'

Preface

It is difficult to understand why statisticians commonly limit their inquiries to Averages, and do not revel in more comprehensive views. Their souls seem as dull to the charm of variety as that of the native of one of our flat English counties, whose retrospect of Switzerland was that, if its mountains could be thrown into its lakes, two nuisances would be got rid of at once.

Francis Galton, *Natural Inheritance.*

Although the development of statistics has been stimulated by a balanced mix of observational studies and experiments, some principles of statistical practice, such as adherence to experimental design, are motivated entirely by settings involving experiments. Experimental design has been instrumental in improvement, maintenance of quality, and efficient application of experiments in the entire spectrum of research activities.

Even when feasible, implementation of rigorous experimental design is often costly. The expenditures associated with its application are justified by yielding data that afford reliable inference. Rightly, statistical textbooks emphasize methods for the analysis of data obtained by designed experiments. However, experimental design is a very restricted framework for collecting data or information in general. Statistical practice and the bank of human knowledge would be extremely impoverished if we dismiss as a potential source of information any studies that do not involve experimental design.

A general approach to the analysis of data from an observational study relies on matching the context of the study with a hypothetical experimental design, and then proceeding with the analysis as if the data were the outcome of this 'experiment'. This approach is referred to as *statistical modelling*. In so far as the quality of matching can rarely be assessed by unambiguous criteria, statistical modelling involves subjective judgement. The position of the statistical 'modeller' can be enhanced by extending the range of models, so that they accomodate as many salient features of the modelled situation as possible.

An important element of experimental design is the independence of observations, which has implications for both economy of sample size and simplicity of optimal estimators. Time series and survey sampling methods provide important examples of data structures that involve dependent observations: temporal dependence in time series and symmetric patterns of dependence in sampling schemes without replacement. Structural equation modelling with latent variables is another generic example. In experimental design the subjects are randomly allocated to 'designed' (*a priori* set) values of the explanatory variables (e.g., treatment categories). Structural equation models assume that the explanatory variables are recorded subject to one or several sources of uncertainty.

Human subjects can rarely be coerced to comply with a rigorous experimental design. Ethical, legal, and practical considerations imply voluntary participation of human subjects in any experiment, and virtually bar manipulation of their background or of other variables associated with the subjects. The social sciences are an inexhaustible source of data describing situations in which several elements of experimental design (e.g., independence of observations, random allocation and assignment of explanatory variables, and missingness at random) cannot be adhered to.

Our society is organized in a complex network of structures, the simplest and most pervasive of which entails belonging to an aggregate unit, such as being a member of a familial, housing, educational, economic, geographical, political, or administrative unit. Such *clustering* leaves its mark on data that record phenomena of interest, on inferences desired, as well as on methods of data collection. Clustered sampling design is a case in point.

The need to define models that accomodate these features has been recognized only recently, and the methodological and technical know-how is slowly seeping through to the community of social scientists (statisticians). The purpose of this book is to make a modest contribution to this trend.

The principal aim of the book is an exposition of methods for the analysis of clustered observations; the secondary one is to provide substantive examples in which between-cluster variation is of substantive interest as a measure of uncertainty, quality, equity, or, generally, as a summary of differences among experimental or observational units. Another goal is to make a balanced presentation of the advantages and limitations of these methods.

The examples used for illustration of the methods are not drawn exclusively from the social sciences. Although the models are motivated mainly by social science problems, they are applicable in a variety of situations involving (imperfect) replication, such as repeated measurements, repeated experiments, longitudinal analysis, and analysis of covariance structures in general.

The methodological developments are by no means new. Random coefficient models have a distinguished record of applications in, among other areas, animal breeding experiments, longitudinal analysis, small-area estimation, and astronomy. However, their potential in the social sciences in particular is, in my view, not fully realized, or even appreciated.

The starting point of the book is ordinary linear regression. I assume that the reader is familiar with the basic concepts, as well as with basic linear algebra (including matrix operations), maximum likelihood, and the multivariate normal distribution. Chapter 7 requires familiarity with factor analysis and measurement error (latent variable) models, but the chapter is self-contained and can be skipped. Chapter 8 requires some background in generalized linear models.

Chapter 1 gives a brief summary of notation and of the prerequisities for the rest of the book, and motivates the development by two examples. Chapter 2 covers the simplest models that incorporate within-group correlation. Chapter 3 contains five illustrative examples. Chapter 4 presents a generalization of random-effects models which allow for a richer pattern of between-cluster variation. Applications of these models are illustrated in Section 5 for the same examples as given in Chapter 3. In Chapter 6 these models are extended to designs with multiple layers of nesting, such as individuals within families, families within towns, and so on. Chapters 2–6 present extensions of ordinary linear regression to correlated observations. Chapter 7 deals with extensions of the factor analysis and measurement error models for multilevel data. Chapter 8 describes a generalization of the random coefficient models to the class of generalized linear models.

Each chapter (except Chapters 3 and 5, which consist of examples) is concluded with bibliographical notes. These should in no way be regarded as a complete list of literature on the subject, but the references should be helpful in supplementing the text, enhancing the reader's interest in the subject, as well as for orientation in literature.

Acknowledgements

I am indebted to Martin Gilchrist, formerly of Oxford University Press, Oxford, England, for the original suggestion to write a book for the Statistics Series, and his persistence over several years to commit me to the project. Elizabeth Johnston and Richard Preston ably and enthusiastically followed up on his effort. My interest in the subject of the book has been greatly stimulated by Murray Aitkin and Harvey Goldstein. I want to thank my employer, Educational Testing Service, Princeton, New Jersey, the Institute of Electronic Systems, Aalborg University, Aalborg, Denmark, my host during most of 1991, and the Institute of Education, University of London, London, England, where I was a visiting ESRC Fellow in June and July 1991, for providing me with the environment and equipment and other support conducive to working on the manuscript. The following persons (in alphabetical order) have contributed to the book by providing me with data, and secretarial and technical assistance, by discussions, by commenting on drafts of chapters of the book, and by other means: Valerie Arneson, Liz Brophy, Martha Byelich, Adrian Davis, Eric Day, Jan DeLeeuw, Howard Everson, Susan Fieldsend, Mendel Fygenson, Karen Gold, Frank Jensen, Sung-Ho Kim, Lise-Lotte Kovsgaard, Kresten Krap Thorup, Steffen Lauritzen, Charles Lewis, Jo-ling Liang, Søren Lundbye-Christensen, Dirk Moore, Stuart McLeay, Bengt Muthén, Inge Novatkoski, Donald Powers, Jon Rasbash, Don Rock, Lisa Schneider, Gilg Seeber, Gerald Shure, Neal Thomas, Michael Wagner, Howard Wainer, and Minhwei Wang.

NICHOLAS T. LONGFORD

Princeton, NJ
May 1993

Contents

1
Introduction

1.1 Clustered observations

Clustering is an omnipresent feature in our everyday lives. We associate people with their families, neighbourhoods, companies, or classrooms; responses to the items of a questionnaire or test with the respondents; manufactured products with their batches, dates of manufacture, workshops, or factories, and so on. Familiar examples of clustering include:

- multiple measurements of a variable on a subject;
- observations on subjects in a geographical, political, or administrative unit;
- responses of members of a family or household to a survey question;
- responses of survey subjects contacted by an interviewer;
- experimental animals in a cage, or plants in a plot;
- animal litters;
- exam or test performance of students in a classroom or school;
- daily outputs of an industrial plant over a week.

The analyst's focus may be on *elementary units* (people, products, or questionnaire responses), or on the *aggregate units* (neighbourhoods, classrooms, factories, or interviewers). For inference about the population of elementary units it would seem sufficient to analyse a sample of these units. By symmetry, we should be able to make inferences about the population of aggregate units from a sample of aggregate units, even though each aggregate unit is represented not by a single observation but by a set of observations on a sample of its elementary units.

Data structures with categorized observations are routinely handled by analysis of variance (ANOVA) and analysis of covariance (ANCOVA). These methods are satisfactory when the number of categories is small and when specific pairwise comparisons of the clusters are of interest. The methods assume an *a priori* definition of a finite number of categories and representation of each of these categories in the sample by at least one observation. However, ANOVA and ANCOVA provide inference only about the *sampled* categories, not the entire population of categories (clusters). Also, they are unsatisfactory when the categories are represented by small numbers of elementary units. Random coefficient methods address these problems by providing summaries of the between-cluster differences which refer to the population of clusters. These summaries can be described as measures of *between-cluster variation*.

Often the number of clusters is large, though nevertheless finite. Then it is reasonable to assume that they comprise a sample from an infinite *population of clusters*, and that the elementary observations in each cluster comprise a sample from a population in that cluster. Such a 'working' assumption enables parsimonious description of between-cluster differences without representing each cluster in the data by one or several parameters.

The meaning of the terms 'observation' and 'cluster' may be unclear or misleading in some situations. For example, in longitudinal analysis with multiple observations on subjects, the set of observations on each subject constitutes a cluster; in multivariate analysis the vector of observations on each subject is a cluster, and its components are the elementary observations; in repeated experiments, data from each experiment form a cluster, and individual outcomes are the elementary observations. I will use the adjectives *elementary* and *elementary-level* for units, variables, variation, and the like, that refer to *atomic* observations, that is, observations that contain no elements. Cluster, or *second-level*, will refer to a set of elementary-level data that form a *cluster-level* unit.

The design involving elementary- and second-level units is sometimes referred to as *one-way layout*, or *two-level design*. Independent observations are contained within this design as a trivial case: either each cluster consists of a single observation or all the observations are contained in a single cluster. Multivariate data can be conceived as observations within vectors (subjects).

We will focus on situations where the aim is to explain or describe how an *outcome* variable is related to, or associated with, a set of *explanatory* variables. The relationship of the outcome variable may be perfectly replicated in each cluster, but most likely there will be some differences in this relationship. These differences, or between-cluster variation, could be ascribed to chance or to some observed or unobserved characteristics or attributes of the clusters.

I want to point out at the outset the essential difference between the terms *explanation* and *description* as used throughout the book. Explanation is an answer to the question *why* the variables are related in the way described by the adopted model, whereas description refers to an empirical finding of an association without any inference about the chain or network of causes that results in the inferred association. This is a recurrent issue in several examples in the book. It is essentially related to the differences between observational studies and experiments using a formal experimental design.

Table 1.1. Example of a one-way layout (two-level data). Each row represents a cluster; each unit is represented by its subscripts ij, where i is the order number within the cluster, and j is the order number of the cluster. The number of clusters is N_2 and cluster j contains n_j elementary units (for instance, $n_3 = 2$)

Cluster	Elementary units			
1	11	21	$\ldots$	$n_1 1$
2	12	22	$\ldots$	$n_2 2$
3	13	23		
$\vdots$				
N_2	$1N_2$	$2N_2$	$\ldots$	$n_{N_2} N_2$

In most examples discussed in the book the substantive problems refer to the explanation of phenomena, but, strictly speaking, the analysis carried out does not proceed any further than data description. The examples are left open-ended because their full treatment is beyond the scope of the book, but I indicate the line of inquiry that should be (was) followed.

We can think of data with one-way layout as

- a collection of datasets;
- multivariate data, with the dimension given by the size of the cluster with the largest number of observations (and other clusters possibly containing missing data);
- a 'ragged' matrix of data.

Table 1.1 depicts a dataset with one-way layout.

The order of the elementary units within a cluster is of no importance; the units within a cluster are *exchangeable*. Similarly, the order of the clusters is of no importance; the clusters as units are also exchangeable.

The rest of this chapter summarizes some of the statistical and analytical tools used in the book: matrix calculus in Section 1.2, the normal distribution in Section 1.3, ordinary regression in Section 1.4, maximum likelihood in Section 1.9, and the exponential family of distributions in Section 1.10. Simple variance component models are motivated by two examples introduced in Section 1.5, and discussed further in Sections 1.6, 1.7, and 1.8, with emphasis on the distinction between the classical ANOVA and variance component analysis as its random-effects counterpart.

1.1.1   SUBJECTS AND VARIABLES

We use the subscripts i and j to denote data or other quantities observation i in cluster j. Suppose cluster j is represented in the sample by n_j observations (the cluster may also contain subjects not included in the sample), and the sample consists of N_2 clusters, so that $N = \sum_{j=1}^{N_2} n_j$ is the total number of observations. We refer to the count n_j as the sample size of cluster j (cluster sample size), and to N_2 and N as the cluster-level and elementary-level sample sizes respectively. Unless stated otherwise, it is not assumed that the cluster sample sizes are identical. If they are identical we denote the common cluster sample size by n; then $N = N_2 n$.

In a regression framework, such as ordinary regression, each elementary observation is associated with an outcome y_{ij} and a $1 \times p$ (row) vector of explanatory variables $\mathbf{x}_{ij}$. The outcomes for cluster j form the $n_j \times 1$ (column) vector $\mathbf{y}_j$, $\mathbf{y}_j = (y_{1j}, \ldots, y_{n_j j})^\top$, and the explanatory variables for the cluster are stacked into the $n_j \times p$ matrix $\mathbf{X}_j = \left(\mathbf{x}_{1j}^\top, \ldots, \mathbf{x}_{n_j j}^\top \right)^\top$. We refer to $\mathbf{X}_j$ as the *design matrix* for cluster j. Further, we denote the vector of outcomes and the design matrix for the entire dataset as $\mathbf{y} = \left(\mathbf{y}_1^\top, \ldots, \mathbf{y}_{N_2}^\top \right)^\top$ and $\mathbf{X} = \left(\mathbf{X}_1^\top, \ldots, \mathbf{X}_{N_2}^\top \right)^\top$ respectively.

In the design matrices $\mathbf{X}$ and $\mathbf{X}_j$ the rows refer to subjects, and the columns to variables. We denote the respective kth columns of $\mathbf{X}$ and $\mathbf{X}_j$ by $\mathbf{x}^{(k)}$ and $\mathbf{x}_j^{(k)}$, and use the symbol $x^{(k)}$ for the kth variable, without specific reference to its values. Similarly, we use the symbol y for the outcome variable or for the random variable representing it. We make no notational distinction between the (realized) outcome y and the underlying random variable.

We assume throughout that the intercept is included in the design matrices $\mathbf{X}$ and $\{\mathbf{X}_j\}$ as the first column (variable), that is, $x_{ij}^{(1)} \equiv 1$. Each variable $x^{(k)}$, $k = 2, \ldots, p$, may be defined either for the elementary observations (*elementary-level variable*), or for the clusters (*cluster-level variable*). A cluster-level variable can be defined for the elementary observations, with constant values $x_j^{(k)}$ for each observation in cluster j. Note that an elementary-level variable may have constant values among the sampled units within each cluster, even though the variable is not constant among all the units within each cluster. We could make the fine distinction between *population* cluster-level variables and *sampling* cluster-level variables, but this is rarely necessary in practice.

1.2   Matrices

This section gives a brief overview of matrix notation, definitions, and properties of matrix operators. Some of the definitions are given in their least general form sufficient for all their uses in the book. Graybill (1969) and Searle (1982) are suitable references for a more thorough background.

1.2.1 MATRIX OPERATIONS AND PROPERTIES

We denote matrices by bold Latin or Greek capitals, such as $\mathbf{A}$, $\mathbf{B}$, and $\boldsymbol{\Sigma}$, and their elements by the corresponding character in plain italic type, subscripted with the row and column indices (e.g., A_{kh} and Σ_{kh}).

A column-vector of zeros of length n, $(0, \ldots, 0)^{\top}$, is denoted by $\mathbf{0}_n$. When the length of the vector is immaterial, or obvious from the context, we drop the subscript. Similarly, we denote by $\mathbf{1}$ ($\mathbf{1}_n$) the column-vector of ones (of length n). The $n \times n$ identity matrix is denoted by $\mathbf{I}_n$ or $\mathbf{I}$, and the square matrix of ones by $\mathbf{J}_n$ or $\mathbf{J}$. Note that $\mathbf{J}_n = \mathbf{1}_n \mathbf{1}_n^{\top}$ and $\mathbf{J}_n \mathbf{1}_n = n\mathbf{1}_n$.

We use the direct product notation. Let $\{\mathbf{A}_{kh}\}$ be a two-dimensional array of matrices, $k = 1, \ldots, K$, $h = 1, \ldots, H$, of dimensions $r_k \times s_h$, and let $\mathbf{B}$ be a $K \times H$ matrix. Then $\{\mathbf{A}_{kh}\} \otimes \mathbf{B}$ is defined as the $\sum_k r_k \times \sum_h s_h$ matrix consisting of the blocks $B_{kh} \mathbf{A}_{kh}$:

$$\mathbf{A} \otimes \mathbf{B} = \begin{pmatrix} B_{11}\mathbf{A}_{11} & B_{12}\mathbf{A}_{12} & \ldots & B_{1H}\mathbf{A}_{1H} \\ B_{21}\mathbf{A}_{21} & B_{22}\mathbf{A}_{22} & \ldots & B_{2H}\mathbf{A}_{2H} \\ \vdots & \vdots & \ddots & \vdots \\ B_{K1}\mathbf{A}_{K1} & B_{K2}\mathbf{A}_{K2} & \ldots & B_{KH}\mathbf{A}_{KH} \end{pmatrix}.$$

Direct product with vector $\mathbf{1}$ corresponds to stacking by columns. For example, the vector of outcomes $\mathbf{y}$, consisting of the N_2 subvectors $\mathbf{y}_j$, can be written as

$$\{\mathbf{y}_j\} \otimes \mathbf{1}_{N_2}.$$

Also, $\mathbf{X} = \{\mathbf{X}_j\} \otimes \mathbf{1}_{N_2}$ and $\mathbf{X}_j = \{\mathbf{x}_{ij}\} \otimes \mathbf{1}_{n_j}$.

Abusing the notation slightly, the block-diagonal matrix with blocks $\mathbf{V}_j$, $j = 1, \ldots, N$, can be written as

$$\{\mathbf{V}_j\} \otimes \mathbf{I}_N$$

(omitting the matrices for the off-diagonal blocks since they would be multiplied by 0). Also, for a $K \times H$ matrix $\mathbf{B}$, $\mathbf{B} \otimes \mathbf{J}_N$ will denote the $KN \times HN$ matrix with $N \times N$ identical blocks, each equal to $\mathbf{B}$.

A matrix with an equal number of rows and columns is called a square matrix. The *trace* of a square matrix is defined as the sum of its diagonal elements:

$$\mathrm{tr}(\mathbf{A}) = \sum_i A_{ii}.$$

The trace of a matrix is commutative, that is,

$$\mathrm{tr}(\mathbf{AB}) \;=\; \mathrm{tr}(\mathbf{BA}), \tag{1.1}$$

for any matrices $\mathbf{A}$ and $\mathbf{B}$ of respective dimensions (k, h) and (h, k), $k > 0$, $h > 0$, i.e., such that their products are well-defined square matrices. The transpose of a $K \times H$ matrix $\mathbf{A}$, denoted by $\mathbf{A}^\top$, is the $H \times K$ matrix with elements $(\mathbf{A}^\top)_{hk} = A_{kh}$. A square matrix such that $\mathbf{A} = \mathbf{A}^\top$ is called a *symmetric* matrix. A square matrix with zero off-diagonal elements ($A_{ij} = 0$ whenever $i \neq j$) is called a *diagonal* matrix.

The *rank* of a $K \times H$ matrix $\mathbf{A}$, denoted by $r(\mathbf{A})$, is defined as the maximum number of its linearly independent rows. It is also equal to the maximum number of linearly independent columns, and therefore $r(\mathbf{A}) = r(\mathbf{A}^\top)$. The largest possible rank of $\mathbf{A}$ is $\min(K, H)$. We say that $\mathbf{A}$ is *non-singular*, or that it is of *full rank*, if $r(\mathbf{A}) = \min(K, H)$. Otherwise, the matrix is *singular*. The rank of a diagonal matrix is equal to the number of its non-zero elements.

The *eigenvalues* of a square matrix $\mathbf{A}$ are defined as the scalar solutions c of the linear system of equations

$$\mathbf{Au} \;=\; c\mathbf{u},$$

where $\mathbf{u}$ is a vector of unit norm (i.e., $\mathbf{u}^\top\mathbf{u} = 1$). The *eigenvectors* are the corresponding vector solutions $\mathbf{u}$. The eigenvector corresponding to an eigenvalue is not unique; the set of all eigenvectors for an eigenvalue spans a linear space. The *multiplicity* of an eigenvalue is defined as the dimension of this linear space. By convention, an arbitrary orthonormal basis of this linear space is declared as the set of eigenvectors corresponding to the eigenvalue. Then all the eigenvectors of a matrix are mutually orthogonal; $\mathbf{u}_k^\top \mathbf{u}_h = 0$ whenever $k \neq h$. The total of multiplicities of the eigenvalues is equal to the size of the matrix.

A square matrix $\mathbf{A}$ is said to be *positive definite* (non-negative definite) if all its eigenvalues are positive (non-negative). Similarly, we define the terms negative definite and non-positive definite. A matrix $\mathbf{A}$ is positive definite (non-negative definite) if and only if $-\mathbf{A}$ is negative definite (non-positive definite). A square matrix is positive definite if and only if $\mathbf{a}^\top\mathbf{Aa} > 0$ for any $1 \times p$ non-zero vector $\mathbf{a}$; non-negative definiteness corresponds to $\mathbf{a}^\top\mathbf{Aa} \geq 0$.

The rank of a square matrix is equal to the number of its non-zero eigenvalues. Any symmetric matrix $\mathbf{A}$ can be reconstructed from its eigenvalues and eigenvectors as

$$\mathbf{A} \;=\; \sum_h c_h \mathbf{u}_h \mathbf{u}_h^\top . \tag{1.2}$$

This identity is referred to as the *eigenvalue decomposition* of $\mathbf{A}$. The trace of a square matrix is equal to the sum of its eigenvalues.

The *determinant* of a square matrix $\mathbf{A}$, denoted by $\det(\mathbf{A})$, is defined as the product of its eigenvalues. For any two square matrices $\mathbf{A}$ and $\mathbf{B}$ of the same size

$$\det(\mathbf{A}\,\mathbf{B}) = \det(\mathbf{A})\det(\mathbf{B}).$$

The *inverse* of a square matrix $\mathbf{A}$, denoted by $\mathbf{A}^{-1}$, is a matrix $\mathbf{B}$ such that their product is the identity matrix, $\mathbf{A}\,\mathbf{B} = \mathbf{I}$. A square matrix has an inverse if and only if it is non-singular, and then the inverse is unique. If $\mathbf{A}$ is symmetric, and (1.2) is its eigenvalue decomposition, then

$$\mathbf{A}^{-1} = \sum_h c_h^{-1}\mathbf{u}_h\mathbf{u}_h^{\mathsf{T}}.$$

1.2.2 MATRIX DIFFERENTIATION

Let $\mathbf{V}$ be a matrix, the elements of which are real functions of a variable θ (a matrix function of θ). We define the derivative $\partial\mathbf{V}/\partial\theta$ as the matrix of elementwise derivatives $\{dV_{kh}/d\theta\}$ if all these derivatives exist. Thus

$$\frac{\partial\mathbf{V}}{\partial\theta} = \lim_{\Delta\to 0}\frac{\mathbf{V}(\theta+\Delta) - \mathbf{V}(\theta)}{\Delta}, \tag{1.3}$$

where the limit is defined elementwise. The derivative $\partial\mathbf{V}/\partial\theta$ is a function of θ. Familiar rules for differentiation of univariate functions can be extended for matrices. For example,

$$
\begin{aligned}
\frac{\partial(\mathbf{V}_1 + \mathbf{V}_2)}{\partial\theta} &= \frac{\partial\mathbf{V}_1}{\partial\theta} + \frac{\partial\mathbf{V}_2}{\partial\theta}, \\[2mm]
\frac{\partial(\mathbf{A}\mathbf{V}\mathbf{B})}{\partial\theta} &= \mathbf{A}\frac{\partial\mathbf{V}}{\partial\theta}\mathbf{B}, \\[2mm]
\frac{\partial(\mathbf{V}_1\mathbf{V}_2)}{\partial\theta} &= \frac{\partial\mathbf{V}_1}{\partial\theta}\mathbf{V}_2 + \mathbf{V}_1\frac{\partial\mathbf{V}_2}{\partial\theta}, \\[2mm]
\frac{\partial\mathbf{V}^{-1}}{\partial\theta} &= -\mathbf{V}^{-1}\frac{\partial\mathbf{V}}{\partial\theta}\mathbf{V}^{-1},
\end{aligned}
\tag{1.4}
$$

where the matrices $\mathbf{V}_1$, $\mathbf{V}_2$, and $\mathbf{V}$ are functions of θ, $\mathbf{A}$ and $\mathbf{B}$ are matrices of constants, and all the matrices have such dimensions that the arithmetic operations on the left-hand sides of (1.4) are well-defined. The last identity in (1.4) assumes that $\mathbf{V}$ is invertible (square and non-singular) in the neighbourhood of θ. Note that attention has to be paid to the order of factors in matrix multiplication. The proofs of these identities are similar to their univariate analogues. For example, the derivative of the inverse is

$$\frac{\partial \mathbf{V}^{-1}}{\partial \theta} = \lim_{\Delta \to 0} \frac{\mathbf{V}^{-1}(\theta + \Delta) - \mathbf{V}^{-1}(\theta)}{\Delta}$$

$$= \lim_{\Delta \to 0} \frac{\mathbf{V}^{-1}(\theta + \Delta)\{\mathbf{V}(\theta) - \mathbf{V}(\theta + \Delta)\}\mathbf{V}^{-1}(\theta)}{\Delta}$$

$$= -\mathbf{V}^{-1} \frac{\partial \mathbf{V}}{\partial \theta} \mathbf{V}^{-1}.$$

The proof relies on the continuity of $\mathbf{V}^{-1}$ at θ, which can be ascertained by reference to the eigenvalue decomposition (1.2).

Further, for any non-singular matrix $\mathbf{V}$,

$$\frac{\partial \log(\det \mathbf{V})}{\partial \theta} = \mathrm{tr}\left(\mathbf{V}^{-1} \frac{\mathbf{V}^{-1}}{\partial \theta}\right). \tag{1.5}$$

A simple proof of (1.5), valid only for positive definite matrices, is given in Section 2.3 where the identity is used for the first time. See Graybill (1969) for an alternative proof and further background.

For a function f of a $K \times 1$ parameter vector $\boldsymbol{\theta}$ we define the partial derivative $\partial f / \partial \boldsymbol{\theta}$ as the vector of partial derivatives, $\{\partial f / \partial \theta_k\} \otimes \mathbf{1}_K$ (if all the K partial derivatives exist). Similarly, we define the derivative of a function with respect to a matrix, and the second-order partial derivative with respect to a pair of vectors. Magnus and Neudecker (1986) provide a comprehensive background to matrix differentiation, as well as to other matrix calculus.

1.3   Normal distribution

This section defines the univariate and the multivariate normal distributions and summarizes their most important properties.

The univariate normal distribution with mean μ and positive variance σ^2 is defined by the absolutely continuous density

$$\phi(y; \mu, \sigma^2) = \frac{1}{\sqrt{2\pi\sigma^2}} \exp\left\{-\frac{(y-\mu)^2}{2\sigma^2}\right\}. \tag{1.6}$$

The corresponding distribution function is

$$\Phi(y; \mu, \sigma^2) = \int_{-\infty}^{y} \phi(x; \mu, \sigma^2)dx. \tag{1.7}$$

The normal distribution with mean μ and variance σ^2 is denoted by the symbol $\mathcal{N}(\mu, \sigma^2)$. The notation $\phi(y)$ and $\Phi(y)$, with omitted mean and variance, is reserved for the *standard normal distribution*, $\mathcal{N}(0, 1)$.

The multivariate normal distribution has a number of equivalent definitions. Most familiar of them are:

1. The random vector $\mathbf{y}$ has a p-variate normal distribution if and only if for any non-zero p-variate vector of constants $\mathbf{a}$ the linear combination $\mathbf{a}^\top\mathbf{y}$ has a univariate normal distribution.

2. The absolutely continuous density of a p-variate normal vector $\mathbf{y}$ is

$$\phi_p(\mathbf{y};\boldsymbol{\mu},\boldsymbol{\Sigma}) \;=\; (2\pi\ \det\boldsymbol{\Sigma})^{-\frac{p}{2}}\exp\left\{-\frac{1}{2}(\mathbf{y}-\boldsymbol{\mu})^\top\boldsymbol{\Sigma}^{-1}(\mathbf{y}-\boldsymbol{\mu})\right\},$$
(1.8)

where $\boldsymbol{\mu}$ is a vector and $\boldsymbol{\Sigma}$ is a positive definite matrix; $\boldsymbol{\mu}$ and $\boldsymbol{\Sigma}$ are the mean and the variance matrix of the distribution respectively.

3. Let $\mathbf{y}$ be a $p \times 1$ vector of independent random variables, each with the standard normal distribution, $\mathcal{N}(0,1)$, $\boldsymbol{\mu}$ a $p \times 1$ vector, and $\mathbf{A}$ a non-singular $p \times p$ matrix. Then $\boldsymbol{\mu}+\mathbf{Ay}$ has the multivariate normal distribution with mean vector $\boldsymbol{\mu}$ and variance matrix $\mathbf{AA}^\top$.

We use the notation $\mathcal{N}_p(\boldsymbol{\mu},\boldsymbol{\Sigma})$ for the p-variate normal distribution with mean $\boldsymbol{\mu}$ and variance matrix $\boldsymbol{\Sigma}$. The subscript p indicating the dimension will be omitted whenever its value is immaterial or obvious from the context.

Any subvector and any permutation of a random vector with a multivariate normal distribution are multivariate normally distributed. Any non-singular linear transformation of a normal random vector $\mathbf{y}$, $\mathbf{Ay}$, where $\mathbf{A}$ is a non-singular matrix, is also normally distributed. If $\mathbf{y} \sim \mathcal{N}(\boldsymbol{\mu},\boldsymbol{\Sigma})$, then

$$\mathbf{Ay} \;\sim\; \mathcal{N}(\mathbf{A}\boldsymbol{\mu},\mathbf{A}\boldsymbol{\Sigma}\mathbf{A}^\top).$$

The definition of the multivariate normal distribution can be extended to non-negative definite variance matrices, e.g. by supplementing the components of the random vector with their linear combinations. Then any linear combination of two multivariate normal random vectors of the same length, $a_1\mathbf{y}_1 + a_2\mathbf{y}_2$, is normally distributed.

1.3.1 CONDITIONAL DISTRIBUTIONS

The multivariate normal distribution has the following property of 'closure under conditioning'. Let $\mathbf{y} = (\mathbf{y}_1^\top,\mathbf{y}_2^\top)^\top$ be a partitioning of the random vector $\mathbf{y}$ which has the distribution

$$\mathcal{N}\left\{\begin{pmatrix}\boldsymbol{\mu}_1\\\boldsymbol{\mu}_2\end{pmatrix},\begin{pmatrix}\boldsymbol{\Sigma}_{11} & \boldsymbol{\Sigma}_{12}\\\boldsymbol{\Sigma}_{21} & \boldsymbol{\Sigma}_{22}\end{pmatrix}\right\},$$
(1.9)

where the partitioning of the mean and of the variance matrix are compatible with the partitioning of $\mathbf{y}$. Obviously, the subvectors $\mathbf{y}_1$ and $\mathbf{y}_2$ have respective distributions $\mathcal{N}(\boldsymbol{\mu}_1, \boldsymbol{\Sigma}_{11})$ and $\mathcal{N}(\boldsymbol{\mu}_2, \boldsymbol{\Sigma}_{22})$. Suppose $\boldsymbol{\Sigma}_{22}$ is positive definite. Then the conditional distribution of the vector $\mathbf{y}_1$ given $\mathbf{y}_2$ is

$$\mathcal{N}\left\{ \boldsymbol{\mu}_1 + \boldsymbol{\Sigma}_{12}\boldsymbol{\Sigma}_{22}^{-1}(\mathbf{y}_2 - \boldsymbol{\mu}_2), \ \boldsymbol{\Sigma}_{11} - \boldsymbol{\Sigma}_{12}\boldsymbol{\Sigma}_{22}^{-1}\boldsymbol{\Sigma}_{21} \right\}. \tag{1.10}$$

An elementary proof of this statement can be conducted by simplifying the ratio of the joint densities of $\mathbf{y}$ and $\mathbf{y}_2$, but a more elegant proof by construction can be found in standard textbooks on multivariate analysis (e.g., Rao 1965, Morrison 1967, or Mardia *et al.* 1979).

1.4   Ordinary regression

Chapters 2–7 assume that the outcomes y are normally distributed. More precisely, we assume that each outcome y_i, $i = 1, 2, \ldots, N$, has a normal distribution with its expectation and variance related to the explanatory variables,

$$y_i \ \sim \ \mathcal{N}\left\{ f(\mathbf{x}_i; \boldsymbol{\beta}), \ \sigma^2(\mathbf{x}_i, \boldsymbol{\theta}) \right\}, \tag{1.11}$$

where f and σ^2 are smooth functions (e.g., infinitely differentiable) and σ^2 is non-negative. Note that the model specification by (1.11) is incomplete because it does not identify the joint distribution of $\mathbf{y}$, or the covariance structure of $\{y_i\}$, $\mathrm{var}(\mathbf{y})$. Ordinary regression corresponds to independent outcomes y_i, with a linear (regression) function f and constant variance σ^2:

$$y_i \ = \ \mathbf{x}_i\boldsymbol{\beta} + \varepsilon_i, \qquad \varepsilon_i \ \sim \ \mathcal{N}(0, \sigma^2), \quad \text{i.i.d.} \tag{1.12}$$

The *design matrix* is defined as $\mathbf{X} = \{\mathbf{x}_i\} \otimes \mathbf{1}_N$. Its dimensions are $N \times p$; we assume that $p < N$ and that the rank of $\mathbf{X}$ is p (that is, full rank). Then $\mathbf{X}^\top\mathbf{X}$ is also of full rank.

The ordinary least squares (OLS) estimator of the vector of regression parameters $\boldsymbol{\beta}$ is

$$\hat{\boldsymbol{\beta}} \ = \ \left(\mathbf{X}^\top\mathbf{X} \right)^{-1}\mathbf{X}^\top\mathbf{y}, \tag{1.13}$$

and the residual variance σ^2 is commonly estimated as

$$\hat{\sigma}^2 \ = \ \frac{\mathbf{y}^\top\left\{ \mathbf{I} - \mathbf{X}\left(\mathbf{X}^\top\mathbf{X} \right)^{-1}\mathbf{X}^\top \right\}\mathbf{y}}{N - p}. \tag{1.14}$$

The vector of residuals $\hat{\boldsymbol{\varepsilon}} = \mathbf{y} - \mathbf{X}\hat{\boldsymbol{\beta}}$ is important for model checking. If the assumptions of normality, independence, and of the functional form of the regression are satisfied and $\hat{\boldsymbol{\beta}}$ is close to $\boldsymbol{\beta}$, then the distribution of the values of $\hat{\varepsilon}$ resembles the normal distribution (e.g., symmetry, unimodality, and thin tails), and the values display no systematic pattern vis-à-vis the explanatory variables or the order of the observations (if a meaningful ordering is defined).

The dilemma that is often faced in analysis of clustered data is which set of outcomes should the regression model (1.11) or (1.12) be applied to. Essentially, there are three choices:

- the entire dataset of N observations;
- each cluster separately;
- the within-cluster means of the outcomes.

The first option amounts to ignoring clustering of the observations and the second option regards the clusters as unrelated datasets. The third option may at first appear attractive because the dataset of N observations has been essentially reduced to a smaller set of N_2 observations. However, we will see that the bias and loss of information incurred as a result of any form of averaging (*aggregating*) within clusters in most cases outweigh this trivial advantage. None of the options can be used for inference about how different/similar the clusters are.

1.4.1 ANALYSIS OF VARIANCE

Analysis of variance (ANOVA) is often considered for data with a one-way layout. The difference between the assumptions for ANOVA and the situation we focus on is that in ANOVA there is a small number of categories defined prior to and independently of the sampling and data collection procedures, whereas in our setup the clusters themselves are sampled. In ANOVA, inference is concerned with specific categories (treatments, states, regions, and the like). In our setup the focus is on a *sample* of clusters; the clusters can be thought of as anonymously labelled units, in the same way as can the elementary observations. Of interest is inference about the entire population of clusters, not only those that happen to be represented in the sample.

Analysis of variance is a special case of ordinary regression in which the categories (levels of a factor) are an explanatory variable. For K categories we can define $K - 1$ 'dummy' variables that identify the category of the observation. There is no unique choice for these variables. A standard choice is the following: the kth dummy variable is set to 1 if the observation belongs to the $(k + 1)$st category, and is set to 0 otherwise. Thus the set of dummy variables is equal to the $(K - 1)$-vector of zeros, $\mathbf{0}_{K-1}$, for an observation in the first category, and to the vector $(0, \ldots, 1, 0, \ldots, 0)$, with entry 1 in the $(k - 1)$st position, for an observation in category k.

Table 1.2. Commuting to work by bike, bus, and car. Commuting times data

Means	Time (in seconds)							
Bike	1020	1110	1110	1090	1040	1080	1060	1090
	1110	1090	1120	1060	1070	1070	1060	1170
Bus	1100	1140	1090	1130	1090	1110	1090	1130
	1130	1130	1120					
Car	1100	1080	1070	1120	1000	1010	1050	1060
	1090	1040	1080	1050	1060	1060	1040	1040
	1090	990	1070	1080	1110	1010		

1.5 Examples

1.5.1 COMMUTING TIMES

I measured the time it takes me to get to work by bike (in good weather), by bus (in bad weather), and by car, sharing the ride with a colleague who lives nearby. I would like to provide evidence that commuting to work is no slower by bike than by car or bus. Certainly, it is cheaper, has some environmental advantages, but it involves physical effort and greater exposure to risk of injury. If I wanted to be finicky I would add for the car ride a fraction of the time spent on maintenance of the car, dealing with the bills, purchasing petrol, and similar. For illustration, we restrict our attention to the skeletal version of the problem:

- Do all three means of commuting to work take the same amount of time?

Table 1.2 contains records of commuting times by the three means of transport.

The analysis of variance should start by checking whether the assumption of equal variance is acceptable. After all, the occasional traffic jams affect cyclists much less than car drivers. On the other hand, weather conditions are likely to affect car drivers much less than cyclists. The sample means and variances of the commuting times are (1084.4, 1266.2) for the bike, (1099.1, 1029.1) for the bus, and (1059.1, 1227.7) for the car-pool. The assumption of equal variances is acceptable. The analysis of variance is summarized in Table 1.3. The F-ratio statistic for the hypothesis of equal means is equal to $6.6/1.2 = 5.5$, higher than the 95 per cent critical value for the F-distribution with 2 and 46 degrees of freedom ($F_{2,46,.95} = 3.2$). We conclude that the three means of commuting are not equally fast. In fact, we could have guessed that the car ride is faster than the bus or bike.

Table 1.3. ANOVA for the commuting times data

	Sum of squares	Degrees of freedom	Mean sum of squares
Total	68 300	48	
Between	13 200	2	6600
Error	55 100	46	1200

Note that the comparisons of the commuting times may be biased because the data were not collected using a formal experimental design. I should have drawn a lot for each morning as to whether to take a bus, join my colleague, or cycle to work. But my colleague was not available every day (he had to run errands, travel away on business, and occasionally rode to work on a bike himself); in bad weather not only myself, but many other 'nice-weather' cyclists, took the bus to work, slowing the bus ride somewhat. The experimental design, which would have helped to arbitrate about the time taken by the three means of transport with more confidence, would have antagonized both my colleague and myself.

1.5.2 BLOOD PRESSURES

How high is the systolic blood pressure of heart-failure patients? Data from a hospital could be collected to answer the question. Of course, the question should be 'tightened' considerably: of interest is the blood pressure before surgery, immediately after surgery, while convalescing, 14 days after the surgery, or before leaving the hospital? The blood pressure fluctuates during the day, and so the time and conditions in which the measurement is to be taken have to be specified. And who should we regard as heart-failure patients? These and a host of similar questions have to be answered.

There are many reasons why heart-failure patients have varying blood pressures: first and foremost, all physiological variables exhibit natural variation among human subjects. There are various kinds of heart-failure, with a range of severity and possible complications. Next, there may be regional differences in the human population in general, and potential heart-failure victims in particular (for example, differences attributable to diet, environment, preventive medical care, prevalent employment, and life-style). Further, some hospitals may treat more acute cases than others (for example, depending on their reputation or location). And, of course, some hospitals may provide better treatment and care than others, and this results in additional reduction of blood pressure. But it is important to realize that these causes cannot easily be distinguished. For example, without additional information a hospital treating patients with relatively

Table 1.4.   Systolic blood pressures of patients in six hospitals. Each hospital H1–H6 is represented by a row of data

H					Blood pressures (in mm Hg)							
1	142	137	131	98	144	120	136	138	143	137	155	147
2	120	125	160	99	115	125	125	141	120	126	130	122
3	133	95	180	120	140	109	125	120	135	105	103	155
4	107	124	183	145	141	122	90	138	143	118	85	131
5	94	133	137	115	109	108	128	153	119	100	117	101
6	93	134	105	140	114	102	155	109	139	119	120	113

low blood pressures cannot unequivocally attribute its 'success' to any one of these (or some other) causes.

In order to proceed to the subject of this section we ignore these and related issues, and consider the dataset of systolic blood pressures of 12 patients each from six hospitals, H1–H6, given in Table 1.4. The units of measurement are millimetres of a column of mercury, mm Hg. The measurements were taken briefly after surgery. The within-hospital means are

$$135.67, \quad 125.67, \quad 126.67, \quad 127.25, \quad 117.83, \quad 120.25.$$

The familiar ANOVA procedure is summarized in Table 1.5. The F-ratio statistic for the comparison of the hospital means is equal to $465.9/389.7 = 1.20$, much smaller than the 95 per cent critical value of the F-distribution with 5 and 66 degrees of freedom ($F_{5,66,.95} = 2.35$).

We have found no evidence of systematic differences among the six hospitals, and so we have no evidence of systematic differences among the hospitals in general (e.g., in the country) either. The dataset is too small to provide evidence of small systematic differences among the hospitals. It is reasonable to regard the difference of, say, 10 mm Hg as substantial. The observed differences of the means for hospital H1, on the one hand, and hospitals H5 and H6, on the other, are greater than 15 mm.

Table 1.5.   ANOVA for the blood pressures data

	Sum of squares	Degrees of freedom	Mean sum of squares
Total	28 052	71	
Between	2330	5	465.9
Error	25 722	66	389.7

If we knew the within-hospital blood pressure means (say, for all the patients treated during a year), then the mean and variance of these averages would be a suitable summary of them. The sample mean is a suitable estimator for the former (some care would be required if the hospitals were represented by unequal numbers of patients), and the variance of the (population) within-hospital means can be estimated directly from ANOVA, Table 1.5, as shown below.

We assume that the blood pressure data contain two distinct additive influences, that of the hospital and that of the patient,

$$y_{ij} = \mu_j + \varepsilon_{ij}, \tag{1.15}$$

where $\{\mu_j\}$ and $\{\varepsilon_{ij}\}$ are two mutually independent random samples (i.i.d.) from $\mathcal{N}(0, \sigma_B^2)$ and $\mathcal{N}(0, \sigma_W^2)$. Then the expectations of the sums of squares in Table 1.5 are:

$$\mathbf{E}(SS_W) = (N - N_2)\sigma_W^2,$$

$$\mathbf{E}(SS_B) = (N_2 - 1)\sigma_W^2 + \left(N - \frac{1}{N}\sum_{j=1}^{N_2} n_j^2\right)\sigma_B^2,$$

for the within- ('error') and between-hospital sums of squares respectively. Here $N = 72$, $N_2 = 6$, and $n_j \equiv n = 12$, $j = 1, \ldots, 6$. After substitution we obtain $\mathbf{E}(SS_W) = 66\sigma_W^2$ and $\mathbf{E}(SS_B) = 5\sigma_W^2 + 60\sigma_B^2$. By matching the expected means $\mathbf{E}(SS_W)$ and $\mathbf{E}(SS_B)$ with their observed counterparts we obtain the estimates $\hat{\sigma}_W^2 = 389.7$ and $\hat{\sigma}_W^2 = 6.36$.

At first, this result appears to be contradictory. The estimated standard deviation of the (long-term) hospital means is only $\sqrt{6.36} = 2.52$, and so the standard deviation of the difference of two hospital means is $\sqrt{12.72} = 3.57$. However, two of the sample (observed) differences exceed 15 mm. The explanation rests on the large within-hospital variance and small samples within hospitals, as well as on the choice of comparison (*maximum* difference among six hospitals). The estimated standard deviation of the mean for a hospital is about $\sqrt{396/12} = 5.75$ mm; therefore the estimated standard deviation of the difference of two means is $5.75\sqrt{2} = 8.1$. Hence, the difference of 15 mm between two hospital means (based on 12 patients each) can arise purely by chance even when the two hospitals have identical long-term expectations.

1.6   Fixed versus random

Both examples in Section 1.5 involve one-way layout. In the commuting times example (Section 1.5.1) there are three possible means of commuting (bike, bus, and car), and we are interested in the comparison of these three *specific* means. Conceivably, I could walk, hitchhike, or even hangglide, but we are not interested in how long it would take me to get to work by these less traditional means. In the blood pressures example (Section 1.5.2) the hospitals are identified only by anonymous labels H1–H6; we do not know which one is the 'St XZ' hospital. The hospitals are *exchangeable*; the inferences we wish to draw do not depend on their identification. The patients within a hospital are also exchangeable; their identities are irrelevant for the purpose of comparing the hospitals. If the six hospitals are the only ones in the specific geographical region or city to which the study refers then we may want to identify the hospitals by their proper names. Even then we could consider the hospitals as a random sample from a hypothetical population of hospitals representing the surveyed medical care system. A hospital is not just the building, but also the medical and administrative management and staff, facilities, suppliers, and the like, and the hiring, procurement and maintenance of these can be regarded as a complex random process.

In the commuting times example the categories (means of transport) are defined unambiguously. We assume that each means of commuting is associated with a long-term average time which is a *fixed* constant. In the blood pressures example the hospitals are a *random* sample, and the mean blood pressures of their patients over a long period of time are regarded as independent realizations of a random variable. If we had the blood pressure data for a large number of patients from a hospital, their mean would be a precise estimate of *one realization* of this random variable.

Whether the categories in the one-way layout are 'fixed' or 'random' depends on the context of the problem. In general, when there are a large number of categories, all of them of comparable or equal importance, 'random' is a more appropriate choice. A strong argument for this choice is parsimony of the associated model description. Whereas in the 'fixed' case, each category of the factor is represented by a parameter (a degree of freedom), in the 'random' case a single parameter, the between-cluster variance, often provides an adequate description for the differences among the clusters. To distinguish between the corresponding models and their assumptions we refer to the former as *fixed-effects* ANOVA, and to the latter as *random-effects* ANOVA (RANOVA).

1.7 Shrinkage estimation of cluster means

Estimation of the bike, bus, and car-pool mean times in the commuting times example (Section 1.5.1) is straightforward. Each sample mean $\bar{y}_j$, $j = 1, 2, 3$, is normally distributed with expectation μ_j and variance σ^2/n_j. In the blood pressures example we ought to do better than just calculate the sample means within the hospitals. If the between-hospital variance were equal to zero, that is, there were no systematic (long-term) differences among the hospital means, the overall mean for the $N = 72$ patients would be a suitable estimate for each of the six hospitals, or, indeed, for each hospital in the population from which the six hospitals were drawn. This would represent a dramatic gain in efficiency. If σ^2 is the variance of the patients' blood pressures in a hospital, then the within-hospital sample mean has the variance $\sigma^2/12$, whereas the dataset sample mean has six times smaller variance, $\sigma^2/72$. The analysis in Section 1.5.2 suggests that the between-hospital differences are small relative to within-hospital (between-patient) differences.

One way of improving estimation of the average blood pressure among the patients in a hospital is to combine the two estimators: the within-hospital sample mean, $\bar{y}_j$, and the overall sample mean $\bar{y}$. Typically, the former is unbiased but has a large variance, and the latter is biased but has a small variance. When the between-cluster variance vanishes, the pooled estimator $\bar{y}$ is appropriate, and when the between-cluster variance is very large, the within-cluster mean $\bar{y}_j$ cannot be improved upon as an estimator of μ_j. In intermediate cases these two estimators can be combined linearly,

$$\bar{y}_{j,s} = w_j \bar{y}_j + s_j \bar{y}, \tag{1.16}$$

with weights w_j and $s_j = 1 - w_j$ inversely proportional to their mean squared errors. These are equal to

$$\mathbf{E}(\bar{y}_j - \mu_j)^2 = \sigma_W^2 / n_j$$

and

$$\mathbf{E}(\bar{y} - \mu_j)^2 = \mathbf{E}(\mu_j - \mu)^2 + \mathbf{E}(\bar{y} - \mu)^2$$

$$= \sigma_B^2 + \frac{\sigma_W^2}{N} + \sigma_B^2 \sum_j \frac{n_j^2}{N^2}.$$

If the terms involving negative powers of N are ignored then $\mathbf{E}(\bar{y} - \mu_j)^2 \approx \sigma_B^2$. This is equivalent to assuming that the sample mean $\bar{y}$ is equal to the expectation μ, an acceptable assumption for a large dataset (perhaps not for the blood pressures example). The estimator (1.16) is then

$$\bar{y}_{j,s} = \bar{y}_j \frac{g_j - 1}{g_j} + \frac{\bar{y}}{g_j}, \tag{1.17}$$

where $g_j = (1 + n_j \sigma_B^2 / \sigma_W^2)$. This is an example of the *shrinkage* estimator: the unbiased but inefficient estimator $\bar{y}_j$ is 'shrunk' towards the biased but stable estimator $\bar{y}$. The result is a biased estimator of the realized value of μ_j with a mean squared error smaller than that of $\bar{y}_j$; we have

$$\mathbf{E}(\bar{y}_{j,s} - \mu_j)^2 = \mathbf{E}\{w_j(\bar{y}_j - \mu_j) + s_j(\bar{y} - \mu_j)\}^2$$

$$\approx \frac{(g_j - 1)^2\sigma_W^2 + n_j\sigma_B^2}{n_j g_j^2}$$

$$= \frac{1}{\sigma_B^{-2} + n_j\sigma_W^{-2}}, \tag{1.18}$$

and so it is smaller than both σ_W^2/n_j and σ_B^2 whenever these quantities are positive. It approaches σ_W^2/n_j as σ_B^2 becomes very large, and σ_B^2 as σ_W^2 becomes very large. For relatively large elementary-level variance the shrinkage estimator $\bar{y}_{j,s}$ is much more efficient than $\bar{y}_j$ or $\bar{y}$, especially when n_j is small.

Usually, the variance components σ_B^2 and σ_W^2 are not known, and so the coefficients g_j in (1.17) have to be estimated. This may erode the efficiency of $\bar{y}_{j,s}$ somewhat, especially when the variances are estimated from small samples.

The estimator (1.17) is the conditional expectation of the cluster mean μ_j in the variance component model (1.15), given the observed mean $\bar{y}_j$. The variance matrix of $(\mu_j, \bar{y}_j)$ is

$$\begin{pmatrix} \sigma_B^2 & \sigma_B^2 \\ \sigma_B^2 & \sigma_B^2 + \sigma_W^2/n_j \end{pmatrix},$$

and, according to (1.10),

$$\mathbf{E}(\mu_j \,|\, \bar{y}_j; \mu, \sigma_W^2, \sigma_B^2) = \mu + \frac{\sigma_B^2}{\sigma_B^2 + \sigma_W^2/n_j}(\bar{y}_j - \mu), \tag{1.19}$$

which coincides with (1.17) when μ is replaced by $\bar{y}$.

1.8   Sources of variation

In the blood pressures example (Section 1.5.2) we identified two sources of variation: within hospitals and between hospitals. If we ignored the association of each patient with a hospital, then only one variance could be estimated and we would be unable to apportion it to patients and hospitals. But even with the existing one-way layout, and even if the data were much more extensive, we would have an incomplete picture. Maybe the hospitals do not matter, but the operating surgeons do. Without identifying each patient's surgeon and sampling surgeons within hospitals, we are unable to test this hypothesis.

Each of the *levels* (factors), such as hospital, surgeon, and patient, is considered as a *source of variation*. Each level is associated with the variance of the outcomes in the hypothetical experiment in which all the circumstances of the study are held constant but where the unit (patient, surgeon, or hospital) is randomly selected. A source of variation identified in the data often corresponds to a number of sources (causes), but the data contain information only about the aggregate of these sources of variation. For example, medical facilities and after-surgery care are components of between-hospital variation. Surgeon as a factor is a component of within-hospital variation.

A hospital may register a high mean blood pressure because it tends to have more severe cases. Without adjustment for the characteristics of the admitted cases (such as severity), an equitable comparison of hospitals is possible only if each hospital has the same distribution of patients' characteristics. This could be achieved by random allocation of patients to hospitals. In reality such a randomization is not possible; a number of other considerations prevail when a person suffers heart failure.

A source of variation identified in the data layout often has no unambiguous interpretation. For example, it may reflect the variation in the quality of the units or it may more appropriately be ascribed to the processes of selection, such as how a patient ends up in one hospital (or being treated by a given surgeon) rather than another. Even though we have indicated that the methods discussed are applicable for observational studies, they nevertheless rely on assumptions of random allocation, that is:

- The allocation of the elementary observations into clusters is non-informative.
- Elementary observations are a random sample from the within-cluster populations.
- The clusters are a random sample from the population of clusters.

In practice these three assumptions are rarely satisfied. A general strategy for analysis of observational studies is to assess the extent to which the analysed situation differs from this ideal, and to speculate about the extent and the direction in which the results would be altered had the study adhered to this scheme.

1.9 Maximum likelihood

The maximum likelihood is a general method for estimating parameters from realizations of random variables. In a general formulation, the data $\mathbf{y}$ is assumed to have a joint distribution given by the (discrete or absolutely continuous) density $L(\mathbf{y}; \boldsymbol{\theta})$, where $\boldsymbol{\theta}$ is a vector of parameters. The possible values of $\boldsymbol{\theta}$ form the *parameter space*, denoted by $\boldsymbol{\Theta}$. The density L, as a function of $\boldsymbol{\theta}$, and for a given vector of outcomes $\mathbf{y}$, is called the *likelihood*. The *maximum likelihood estimator* (MLE) of $\boldsymbol{\theta}$ is the value(s) of $\boldsymbol{\theta}$ for which the likelihood L attains its maximum. Note that the MLE may not exist, and, when it does, it may not be unique.

It is usually easier to work with the logarithm of the likelihood, the *log-likelihood*. For example, the log-likelihood for the ordinary regression, (1.12), is

$$l(\mathbf{y}; \boldsymbol{\beta}, \sigma) = -\frac{N}{2}\log(2\pi\sigma^2) - \frac{(\mathbf{y} - \mathbf{X}\boldsymbol{\beta})^\top(\mathbf{y} - \mathbf{X}\boldsymbol{\beta})}{2\sigma^2}. \qquad (1.20)$$

It is a quadratic function of $\boldsymbol{\beta}$, and so its maximization is straightforward. When $\mathbf{X}$ is of full rank its maximum is given by the OLS estimator, (1.13). The MLE of σ^2 is

$$\hat{\sigma}^2 = \frac{\hat{\varepsilon}^\top\hat{\varepsilon}}{N}, \qquad (1.21)$$

where $\hat{\varepsilon} = y - \mathbf{X}\hat{\boldsymbol{\beta}}$ is the vector of residuals. The familiar unbiased estimator of σ^2 is obtained by replacing the divisor N in (1.21) by $N - p$, where p is the number of estimated regression parameters, or *degrees of freedom*.

The vector of first-order partial derivatives of the log-likelihood is called the *scoring vector*. If a maximum of the likelihood occurs at an interior point $\hat{\boldsymbol{\theta}}$ of the parameter space, and the scoring vector is defined in the neighbourhood of this point, then the scoring vector has a root at $\hat{\boldsymbol{\theta}}$,

$$\frac{\partial l}{\partial\hat{\boldsymbol{\theta}}}\bigg|_{\boldsymbol{\theta}=\hat{\boldsymbol{\theta}}} = \mathbf{0}.$$

A standard approach to maximum likelihood estimation is to find all the roots of the scoring vector and to explore the behaviour of the log-likelihood on the boundary of the parameter space and at the points where the scoring vector is not defined. These are the only possible locations of the MLE.

1.9.1 NEWTON–RAPHSON METHOD

The Newton–Raphson algorithm is a general method for finding maximum likelihood estimates. Suppose the (log-)likelihood l is a twice continuously differentiable function of the parameters $\boldsymbol{\theta}$, the matrix of second-order partial derivatives is negative definite for each $\boldsymbol{\Theta}$, and the parameter space is an open set. Suppose we have a 'current' solution $\hat{\boldsymbol{\theta}}_{old}$; then the updated solution is defined as

$$\hat{\boldsymbol{\theta}}_{new} = \hat{\boldsymbol{\theta}}_{old} - \left(\frac{\partial l^2}{\partial\boldsymbol{\theta}\partial\boldsymbol{\theta}^\top}\right)^{-1}\frac{\partial l}{\partial\boldsymbol{\theta}}, \qquad (1.22)$$

where all the partial derivatives are calculated at $\boldsymbol{\theta} = \hat{\boldsymbol{\theta}}_{old}$. The Newton–Raphson algorithm consists of iterations of (1.22), with the 'new' solution from one iteration becoming the 'old' solution of the next one. Under certain regularity conditions (see, e.g., Gill *et al.* 1981) the Newton–Raphson iterations converge to the unique maximum of the likelihood function. We can apply the Newton–Raphson algorithm even in cases when the maximized function is not defined in an open parameter space, or the function's second-order partial derivative matrix is not negative definite throughout the parameter space. In that case, even if the Newton–Raphson iterations converge to a point, there is no guarantee that this point is a global maximum. Further exploration of the (log-)likelihood function is required to establish whether this point of convergence is a minimum, a maximum, or a point of inflexion, and if it is a maximum, whether it is unique. In many settings the matrix of second-order partial derivatives is negative definite and the likelihood has a unique maximum at the unique root of the scoring function.

A problem arises when the parameter space is not open, and the solution lies on the boundary. Estimation of a variance, or, more generally, of a parameter known to be non-negative is an example of this problem. There are several general approaches to alleviate such a problem:

- Do away with the boundary by a suitable reparametrization, e.g., by estimating the standard deviation σ instead of the variance σ^2.
- If the definition of the log-likelihood can be extended for negative values of σ^2 find the maximum in this *extended* parameter space. If the maximum is attained for a negative value of σ^2 then the maximum within the original parameter space is attained for $\sigma^2 = 0$.
- Constrained maximization by the method of Lagrange multipliers. Let $l(\boldsymbol{\theta})$ be a continuously differentiable function and $\mathbf{g}(\boldsymbol{\theta})$ a continuously differentiable vector function of $\boldsymbol{\theta}$. Then l has an extreme $\hat{\boldsymbol{\theta}}$ subject to the condition $\mathbf{g}(\boldsymbol{\theta}) = \mathbf{0}$ only if the function

$$L(\boldsymbol{\theta}) = l(\boldsymbol{\theta}) + \boldsymbol{\lambda}^\top \mathbf{g}(\boldsymbol{\theta})$$

satisfies $\partial L/\partial \boldsymbol{\theta}\big|_{\boldsymbol{\theta}=\hat{\boldsymbol{\theta}}} = \mathbf{0}$ and $\mathbf{g}(\hat{\boldsymbol{\theta}}) = \mathbf{0}$. See, e.g., Gill *et al.* (1981) for details.

The negative of the matrix of second-order partial derivatives is called the *sample information matrix*. Its expectation,

$$\mathbf{H}(\boldsymbol{\theta}) \;=\; -\mathbf{E}\left(\frac{\partial l^2}{\partial\boldsymbol{\theta}\partial\boldsymbol{\theta}^\top}\right), \tag{1.23}$$

is called the *expected* information matrix. The Fisher scoring algorithm differs from the Newton–Raphson one by using the expected information matrix instead of the sample one;

$$\hat{\boldsymbol{\theta}}_{new} \;=\; \hat{\boldsymbol{\theta}}_{old} \;+\; \left\{\mathbf{H}(\hat{\boldsymbol{\theta}}_{old})\right\}^{-1} \frac{\partial l}{\partial\boldsymbol{\theta}}\Big|_{\boldsymbol{\theta}=\hat{\boldsymbol{\theta}}_{old}}. \tag{1.24}$$

In the ordinary regression, with residual variance σ^2 assumed known, the Newton–Raphson and Fisher scoring algorithms coincide with OLS because the matrix of second-order partial derivatives, $-\sigma^{-2}\mathbf{X}^\top\mathbf{X}$, is not random.

In many settings it is simpler to evaluate $\mathbf{H}(\boldsymbol{\theta})$ than $\partial l^2/(\partial\boldsymbol{\theta}\partial\boldsymbol{\theta}^\top)$ and properties of $\mathbf{H}$ are easier to explore. Often, $\mathbf{H}(\hat{\boldsymbol{\theta}}) = -\partial l^2/(\partial\boldsymbol{\theta}\partial\boldsymbol{\theta}^\top)\big|_{\boldsymbol{\theta}=\hat{\boldsymbol{\theta}}}$ at the maximum likelihood solution $\hat{\boldsymbol{\theta}}$, and both matrix functions are continuous. As a consequence, Newton–Raphson and Fisher scoring methods have similar convergence properties.

1.10   The exponential family of distributions

A univariate distribution is said to belong to the exponential family if its (absolutely continuous or discrete) density has the form

$$f(y; \theta, \tau) = \exp\left[a(\tau)\{y\theta - b(\theta)\} + c(y, \tau)\right] \tag{1.25}$$

for some functions a, b, and c, and a given (scale) parameter τ. Several well-known distributions, such as the normal, binomial, gamma, and beta distributions, belong to the exponential family; for example the normal distribution $\mathcal{N}(\theta, \tau^2)$ corresponds to

$$a(\tau) = \tau^{-2}, \qquad b(\theta) = \frac{\theta^2}{2}, \qquad c(y, \tau) = \log(2\pi\tau^2) - \frac{y^2}{2\tau^2}.$$

The definition of the multivariate exponential family of distributions is analogous:

$$f(\mathbf{y}; \boldsymbol{\theta}, \boldsymbol{\tau}) = \exp\left[a(\boldsymbol{\tau})\{\mathbf{y}^\top\boldsymbol{\theta} - b(\boldsymbol{\theta})\} + c(\mathbf{y}, \boldsymbol{\tau})\right], \tag{1.26}$$

where $\boldsymbol{\theta}$ and $\boldsymbol{\tau}$ are vectors of parameters. The functions b in (1.25) and (1.26) are assumed to be twice differentiable.

The expectation and the variance matrix of a distribution in the exponential family are

$$\mathbf{E}(\mathbf{y}) = \frac{\partial b}{\partial\boldsymbol{\theta}},$$

$$\text{var}(\mathbf{y}) = a(\boldsymbol{\tau})^{-1}\frac{\partial^2 b}{\partial\boldsymbol{\theta}\partial\boldsymbol{\theta}^\top}. \tag{1.27}$$

We give a proof of these identities for absolutely continuous distributions. For discrete distributions the proof is analogous. The density in (1.26) integrates to unity:

$$\int f(\mathbf{y}; \boldsymbol{\theta}, \boldsymbol{\tau})d\mathbf{y} = 1, \tag{1.28}$$

and so its partial derivative with respect to $\boldsymbol{\theta}$ vanishes:

$$0 = \frac{\partial}{\partial\boldsymbol{\theta}} \int f(\mathbf{y}; \boldsymbol{\theta}, \boldsymbol{\tau}) d\mathbf{y} = \int \frac{\partial f(\mathbf{y}; \boldsymbol{\theta}, \boldsymbol{\tau})}{\partial\boldsymbol{\theta}} d\mathbf{y}$$

$$= \mathbf{E}\left\{ \frac{\partial \log f(\mathbf{y}; \boldsymbol{\theta}, \boldsymbol{\tau})}{\partial\boldsymbol{\theta}} \right\}.$$

Exchanging the order of differentiation and integration above is justified because the integral, as a function of $\boldsymbol{\theta}$, is bounded. By direct substitution we obtain

$$a(\boldsymbol{\tau})\left\{ \mathbf{E}(\mathbf{y}) - \frac{\partial b(\boldsymbol{\theta})}{\partial\boldsymbol{\theta}} \right\} = 0,$$

which implies the first identity in (1.27).

Taking the second-order partial derivatives of (1.28) yields

$$\mathbf{E}\left\{ \frac{\partial^2 \log f(\mathbf{y}; \boldsymbol{\theta}, \boldsymbol{\tau})}{\partial\boldsymbol{\theta}\partial\boldsymbol{\theta}^\top} \right\} + \mathbf{E}\left\{ \frac{\partial \log f(\mathbf{y}; \boldsymbol{\theta}, \boldsymbol{\tau})}{\partial\boldsymbol{\theta}} \frac{\partial \log f(\mathbf{y}; \boldsymbol{\theta}, \boldsymbol{\tau})}{\partial\boldsymbol{\theta}^\top} \right\} = 0,$$

and (1.27) follows by substitution of (1.26).

1.11 Bibliographical notes

Until the 1970s, development of variance component methods was motivated principally by agricultural and animal breeding experiments. Two of the earliest references to variance components, Airy (1861) and Galton (1886), are notable exceptions. Airy considered a standard problem in astronomy. Measurements of a phenomenon are taken several times each night. Since the process of measurement is not perfect (owing to imperfections of the apparatus and to the observer's imprecision) it is meaningful to consider two sources of variation: measurement (within a night) and time (between nights). The former is a nuisance factor; if it were eliminated, one observation per night would suffice. The variation in time (over occasions) is an inherent characteristic of the observed phenomenon and is therefore of considerable interest.

Galton studied the stature (height) of adult men and its association with the stature of their fathers. He identified between- and within-father sources of variation and proposed methods for their estimation. For the study Galton selected sets of brothers. Taller fathers tend to have taller sons, but brothers also vary in their stature, though to a lesser extent than adult men in general. Stigler (1986) gives a detailed description of the study from a historical perspective.

Eisenhart (1947) and Henderson (1953) presented important developments of variance component methods, originally targeted for animal breeding experiments. Recalling the principles of experimental design formulated by Fisher (1925), Eisenhart gave a comprehensive discussion of the assumptions underlying the analysis of variance with fixed and random effects. Interestingly, this article is followed by Cochran (1947) discussing examples in which some of these assumptions are violated.

Wilm (1945) discussed problems with the analysis of forestry experiments which are replicated in time and take as long as several years to realize. Marcuse (1949) analysed experiments with clustered sampling design; moisture content of cheese is measured for small quantities selected at random from each of a number of sampled lots. Crump (1951) contains a review of the state-of-the-art in the early postwar period. Most methods relied on balanced design and did not allow for any explanatory variables. Henderson (1953) defined three computational methods for estimation with variance component models which were universally adopted until the emergence of computers was anticipated in the late 1960s. These methods were a compromise of computational simplicity (they were intended for paper-and-pencil work or for calculators) and statistical efficiency. They catered for multiple regression with unbalanced design.

Maximum likelihood estimation with random effects models was described for the first time by Hartley and Rao (1967), using the Newton–Raphson and Fisher scoring methods. Subsequently, it was 're-invented' by several authors with only insubstantial improvements and in slightly different contexts (e.g., Jennrich and Sampson 1976, Jennrich and Schluchter 1986, Longford 1987, and others). In the 1980s variance component estimation came to be almost universally regarded as a standard application of Dempster et al.'s (1977) EM algorithm.

Searle (1971) reviews variance component methods prior to the introduction of computationally extensive methods. The review discusses a number of applications in biological sciences, and places equal emphasis on models, methods of estimation, and substantive interpretation. Most examples involve small sets of data, and so issues of bias, finiteness of the sampled population, negative estimated variances, as well as distributional assumptions of the outcomes, are emphasized. The computational methods rely on the classical analysis of variance which in some simple cases corresponds to maximum likelihood. Minimum quadratic unbiased estimation (MINQUE) is also reviewed.

Searle sets an intuitively appealing criterion for deciding whether the effects associated with a factor are fixed or random. The realization of an experiment is considered as a result of a random process. If a different realization of the same experiment has the same levels of the factor (e.g., specific diets or specific varieties of a plant), the factor is regarded as *fixed*. If the levels of the factor are selected for each realization, and thus vary from experiment to experiment (e.g., experimental animals, classrooms, or sites), then the factor is regarded as *random*. This rule is usually straightforward to apply in experimental design; in observational studies it is sometimes ambiguous since we may not have a clear conception of an alternative realization of the study.

Rao (1971a and b) constructed algorithms for minimum norm and minimum variance quadratic unbiased estimation.

Lindley and Smith (1972) made an important contribution to the discussion of alternatives to the classical analysis of (co-)variance methods, and to the efficient estimation of sets of parameters.

2
Analysis of covariance with random effects

2.1 Models

This chapter deals with models which combine ordinary regression and within-cluster correlation. The development closely parallels the combination of linear regression and ANOVA into the analysis of covariance (ANCOVA). Our focus is on settings with large numbers of clusters, or clusters being a random sample from a population, and so parsimonious description of (adjusted) between-cluster differences is a relevant issue. The principal difference between ANCOVA and our approach is in the role of the levels of the factor (clusters). In ANCOVA the effects associated with each cluster are unknown constants (they are *fixed*), in the random-effects approach, which we refer to as random-effects analysis of covariance, they are *random*.

We assume the one-way layout of elementary units i within clusters j. The outcome y_{ij} and a vector of *elementary-level* explanatory variables $\mathbf{x}_{ij}^{(1)}$ are recorded for each elementary unit. Additionally, a vector of *cluster-level* explanatory variables, $\mathbf{x}_j^{(2)}$, defined for clusters, may be available. The cluster-level variables can be redefined as elementary-level ones by assigning the values $\mathbf{x}_j^{(2)}$ to each elementary unit in cluster j. Conversely, if an elementary-level variable happens to be constant within each cluster, then it can be redefined as a cluster-level variable in the obvious way. We denote by $\mathbf{x}$ the vector of all explanatory variables (regressors) associated with an elementary observation;

$$\mathbf{x}_{ij} = \left(1, \ \mathbf{x}_{ij}^{(1)}, \ \mathbf{x}_{ij}^{(2)}\right).$$

We assume that the intercept term 1 is always included in the vector or regressors $\mathbf{x}$ as the first variable.

In many settings it is natural to assume that the outcomes within a cluster are correlated. The size of the correlation, or of the covariance, is often of considerable interest. In general, there may be an explanation for (positive) within-cluster correlation, such as similar or common circumstances of the observations within each cluster. These features of the data could be captured by one or several explanatory variables. When we take account of the available explanatory variables, it is of interest whether the adjusted outcomes remain correlated, or by how much the correlation has been reduced (increased). This concern is similar to that of attempting to explain the residual variance by additional explanatory variables in ordinary regression models.

We assume the linear regression model

$$\mathbf{y}_j = \mathbf{X}_j\boldsymbol{\beta} + \delta_j + \boldsymbol{\varepsilon}_j, \tag{2.1}$$

where $\boldsymbol{\varepsilon}_j = (\varepsilon_{1j}, \ldots, \varepsilon_{n_j j})^\top$, and the sets of random variables $\{\delta_j\}$ and $\{\varepsilon_{ij}\}$ are mutually independent random samples from $\mathcal{N}(0, \tau^2)$ and $\mathcal{N}(0, \sigma^2)$ respectively. This is the random-effects analysis of covariance model (RANCOVA). The only difference from the fixed effects (classical) ANCOVA is that the cluster 'effects' δ_j (i.e., the adjusted between-cluster differences) are a random sample rather then a set of unknown parameters.

Two observations in the same cluster are correlated; we have

$$\mathrm{var}(y_{ij}) = \sigma^2 + \tau^2,$$
$$\mathrm{cov}(y_{ij}, y_{i'j}) = \tau^2 \quad (i \neq i'), \tag{2.2}$$

or, in matrix notation,

$$(\mathbf{V}_j =) \quad \mathrm{var}(\mathbf{y}_j) = \sigma^2 \mathbf{I}_{n_j} + \tau^2 \mathbf{J}_{n_j}. \tag{2.3}$$

The variances σ^2 and τ^2 are called elementary- and cluster-level variance components. The correlation of two outcomes from the same cluster (the *within-cluster correlation*),

$$\rho = \frac{\tau^2}{\sigma^2 + \tau^2}, \tag{2.4}$$

is also referred to as the *variance component ratio*; it is the fraction of the residual variance attributed to between-cluster variation.

The *variance ratio* is defined as

$$\omega = \frac{\tau^2}{\sigma^2}. \tag{2.5}$$

The variance component ratio and the variance ratio are estimated by substituting estimates of the variances σ^2 and τ^2 in (2.4) and (2.5). For the blood pressures example (Section 1.5.2) we have the estimates $\hat{\rho} = 6.36/396 = 0.0161$ and $\hat{\omega} = 6.36/389.7 = 0.0163$.

To assess the importance of between-cluster variation it is instructive to consider the cluster total of deviations, $(\mathbf{y}_j - \mathbf{X}_j\boldsymbol{\beta})^\top \mathbf{1}_{n_j}$. Its variance is $n_j\sigma^2 + n_j^2\tau^2$. The relative contributions of the two variance components to this quantity are $1 : n_j\omega$. When cluster sizes are large, even a small between-cluster variance matters a great deal. Alternatively, this comparison can be motivated by considering that the random term δ_j is replicated on a large number of elementary units. Here it is more appropriate to consider the population size of the cluster than its sample size. For example, a hospital has thousands of cases each year; if, hypothetically, we could replace its (positive) δ_j by zero, the accumulated benefit (e.g., lower blood pressures of its patients) would be substantial even for a very small deviation δ_j.

The random-effects model (2.1) is also known by the names compound symmetry model, equicovariance regression model, and simple two-level model. For $\tau^2 = 0$ the observations are independent and (2.1) is an ordinary regression model.

An obvious generalization of (2.1) is

$$\mathbf{y}_j = \mathbf{X}_j \boldsymbol{\beta} + \boldsymbol{\zeta}_j, \qquad (2.6)$$

where $\{\boldsymbol{\zeta}_j\}$ are independent, normally distributed random vectors, with means $\mathbf{0}$ and variance matrices $\mathbf{V}_j$. These matrices, generally of unequal size, are functions of a small number of parameters. Since the multivariate normal distribution is uniquely given by its mean vector and variance matrix, an alternative (and equivalent) specification of the general model (2.6) is as follows:

$$\mathbf{E}(\mathbf{y}) = \mathbf{X}\boldsymbol{\beta},$$
$$\text{var}(\mathbf{y}) = \{\mathbf{V}_j\} \otimes \mathbf{I}_{N_2} \quad (= \mathbf{V}), \qquad (2.7)$$

with additional specification of a parametric form of the variance matrices $\mathbf{V}_j$. The random-effects model (2.1) is a special case of (2.7) with a restricted (equicovariance) pattern of within-cluster covariance given by (2.3), involving only two parameters.

2.1.1 THE LOG-LIKELIHOOD

The log-likelihood for the vector of outcomes $\mathbf{y}$ in (2.6) is

$$l(\mathbf{y}; \boldsymbol{\beta}, \boldsymbol{\theta}) = -\frac{1}{2}\left\{ N \log(2\pi) + \log(\det \mathbf{V}) + \mathbf{e}^\top \mathbf{V}^{-1}\mathbf{e}\right\}, \qquad (2.8)$$

where $\mathbf{e} = \mathbf{y} - \mathbf{X}\boldsymbol{\beta}$ is the vector of deviations of the outcomes from their expectations and $\boldsymbol{\theta}$ is the set of all parameters involved in $\mathbf{V}$. Evaluation of the log-likelihood is straightforward when the clusters are small because $\mathbf{V}$ is block-diagonal and inversion and evaluation of the determinant of each block $\mathbf{V}_j$ raise no computational problems. For the random-effects

model (2.1) with large cluster sizes the following equations can be used
with advantage:

$$\mathbf{V}_j^{-1} = \frac{1}{\sigma^2}\mathbf{I}_{n_j} - \frac{\tau^2}{\sigma^4 g_j}\mathbf{J}_{n_j} \tag{2.9}$$

$$\det \mathbf{V}_j = \sigma^{2n_j} g_j, \tag{2.10}$$

where $g_j = (1 + n_j\tau^2/\sigma^2)$. The proof of these identities is simple. The
inversion formula is proved by checking that the product of $\mathbf{V}_j$ and (2.9)
is the identity matrix:

$$(\sigma^2\mathbf{I} + \tau^2\mathbf{J}) \left(\frac{1}{\sigma^2}\mathbf{I} - \frac{\tau^2}{\sigma^4 g_j}\mathbf{J}\right)$$

$$= \mathbf{I} + \frac{\tau^2}{\sigma^2}\left(1 - \frac{1}{g_j} - n_j\frac{\tau^2}{\sigma^2 g_j}\right)\mathbf{J} = \mathbf{I},$$

using the identity $\mathbf{J}_n\mathbf{J}_n = n\mathbf{J}_n$. To prove the determinant formula (2.10)
we find an eigenvalue decomposition of $\mathbf{V}_j$: since $\mathbf{V}_j\mathbf{1} = \sigma^2 g_j\mathbf{1}$, $\sigma^2 g_j$
is an eigenvalue of $\mathbf{V}_j$. Further, let $\mathbf{v}$ be a vector containing one entry
each equal to 1 and –1, and the rest of the elements equal to zero, $\mathbf{v} = (0,\ldots,0,-1,\ldots,1,0,\ldots,0)$. There are n_j linearly independent vectors
with such a pattern, and since for each of them $\mathbf{V}_j\mathbf{v} = \sigma^2\mathbf{v}$, σ^2 is an
$(n_j - 1)$-multiple eigenvalue of $\mathbf{V}_j$. The determinant of a matrix is equal
to the product of its eigenvalues; therefore (2.10) holds.

When there are no explanatory variables in (2.1), $\operatorname{var}(\bar{y}_j) = g_j\sigma^2/n_j$,
while $\operatorname{var}(\bar{y}_j\,|\,\delta_j) = \sigma^2/n_j$, so that g_j is the inflation factor for the un-
conditional variance of the mean $\bar{y}_j$ over the conditional variance, given
the true mean $\mu + \delta_j$.

In a modelling approach we seek a shortlist of important explanatory
variables which provide an adequate description of the outcome variable.
In ordinary regression we often regard it as a success if we find a model with
relatively few explanatory variables and low residual variance. In random-
effects models there are two variance parameters, one associated with each
level of nesting, and it may be too optimistic to expect that both variances
can be reduced by adjustment for a suitable set of explanatory variables.

Addition of a cluster-level explanatory variable leaves the elementary
variance unchanged because the conditional variance of the deviations $y_{ij} - \mathbf{x}_{ij}\boldsymbol{\beta} - \delta_j$ remains unaltered. Additional adjustment for an elementary-level
variable can have unpredictable and counterintuitive outcomes. Suppose
in a setting similar to the blood pressures example (Section 1.5.2) we have
a more extensive dataset in which between-hospital variance is very small.
Severity of the case may be an important explanatory variable. If hospitals
that tend to treat more severe cases provide higher quality treatment, then

after adjustment for severity the between-hospital variance will be much greater. Without adjustment the hospitals appear to provide treatment of similar quality, but after adjustment there appear to be differences in quality. However, the unadjusted elementary-level variance would be substantially reduced by adjustment for severity.

Section 2.2 introduces five examples which will be used for illustration of the developments in this and later sections. Sections 2.3, 2.4, and 2.5 describe three methods of maximum likelihood estimation with the random-effects ANCOVA. We recommend reading all three sections, although after reading either one the reader is well-equipped for the rest of the book. We emphasize the Fisher scoring algorithm (Section 2.3) because of its good convergence properties, easy description and computer implementation, and general applicability.

2.2   Examples

This section introduces five examples of clustered data which are then analysed in Chapters 3 and 5. Here we describe the data and their context, and outline the main inferential problems.

2.2.1   FINANCIAL RATIOS

Financial institutions use so-called 'financial ratios', such as the ratio of the liabilities by the assets of a company, to assess the financial viability, credit-worthiness, risk of non-payment of future debts, performance of shares, and the like. The value of assets is used to adjust for the size of the company. For example, in Britain there are companies with assets of $£10^9$, and others with much less than $£100\,000$.

In reexamining the use of the financial ratio of companies' liabilities by their assets the following questions arose:

- Is the ratio an appropriate means of adjustment of the liabilities for the size of the company?
- Is it appropriate to use the ratio as a criterion for the company's performance without regard to its industrial sector?

The companies are classified into industrial sectors, of which there are about 70 in Britain, but their definitions are altered through the years so as to reflect structural changes in the economy. Also, companies sometimes change the main line of their business. The classification is particularly ambiguous for the largest (multinational) companies, which may be involved in several sectors (then the sector in which they are primarily engaged is declared). In some sectors there are only a handful of companies, in others there more than 100.

We explore the applicability of the random-effects model (2.1) for this problem. First, we note that the data, values of liabilities, and values of assets of all the British companies listed on the London Stock Exchange in 1977, do not involve any sampling scheme or experimental design. By law all such companies have to publish these financial summaries annually. Next, we lack any information about which companies are 'healthy', and which are in bad shape, or are likely to be declared bankrupt in the future, so that we could explore whether the ratio differentiates between such companies. Further, measurement 'error' may be considerable because of uncertainty about how to assess the liabilities and assets in the presence of inflation, fluctuating prices, various forms of payment for products and services, differences in methods of accounting, and so on. It is obvious, though, that these errors (more appropriately termed 'uncertainty') are likely to be much larger for companies with large assets than for small companies. An effective way of removing this heterogeneity is to apply the log-transformation to both variables. This has a strong intuitive appeal. Now instead of *additive* uncertainty we have *multiplicative* uncertainty. The ratio of the original variables becomes the difference of the transformed variables, and the first question stated above becomes:

- Is the difference of logarithms of the liabilities and assets an appropriate measure of companies' financial performance?

The financial ratios may be appropriate for comparing companies within a sector, but not across sectors, because the ratios for the healthy companies may have different distributions across the sectors. The fortunes of the industrial sectors are bound to be uneven, due to the effects of general market conditions. In some sectors companies tend to have high liabilities because they require substantial capital investment; in other sectors companies require much less extensive capital investment but have higher expenditures that are not reflected in liabilities. It is meaningful to consider 'company' and 'sector' as two sources of variation, and to describe the association of liabilities and assets by the simple random-effects model,

$$y_{ij} = \alpha + x_{ij}\beta + \delta_j + \varepsilon_{ij}, \qquad (2.11)$$

where y and x are the respective logarithms of the liabilities and assets, and i and j are the respective indices of the company and sector. The regression slope β is essentially the exponent in the modelled adjustment

$$\frac{liabilities}{(assets)^\beta}, \qquad (2.12)$$

and $\exp(\alpha)$ is the *geometric average* of these ratios. Use of the geometric average can also be motivated by the need to reduce the influence of the largest companies.

A segment of the data is given in Table 2.1. The units for the log-liabilities and log-assets are $\log(£1000)$. The bottom panel of the table contains the numbers of companies in each sector. Fieldsend *et al.* (1987)

Table 2.1.   A segment of the financial ratios dataset. The records are sorted by sectors. The units are $\log(\pounds 1000)$. The bottom panel contains the numbers of companies within sectors (listed by rows). The order of the sectors is the same in both panels

Sector no.	Company no.	Log-assets	Log-liabilities
1	1	8.162	7.545
1	2	12.406	11.584
1	3	8.158	7.303
1	4	11.370	10.724
⋮	⋮	⋮	⋮
1	29	10.175	9.877
2	1	8.212	7.854
2	2	11.513	10.962
2	3	9.823	9.482
⋮	⋮	⋮	⋮

Sectors	Sector sizes								
1–10 :	29	9	17	44	10	10	19	119	56	8
11–20 :	20	39	19	10	15	9	118	30	12	7
21–30 :	16	25	24	8	13	16	23	15	11	42
31–40 :	64	19	34	25	26	53	94	10	37	17
41–50 :	32	59	21	10	7	62	65	9	21	42
51–60 :	8	15	9	18	39	44	31	47	31	19
61–70 :	97	8	21	47	4	1	1	3	1	2
71 :	58								

and references therein give a more detailed background of the dataset and its context.

2.2.2   RAT WEIGHTS

In teratogenic studies a treatment (typically, a diet) is administered to pregnant animals of a species, and the effect of the treatment is observed on their offspring. Some elements of experimental design can be implemented in such studies in a relatively straightforward way: the animals to be impregnated are assigned mates by a random process and the treatments are assigned by another random process (and these two processes are mutually independent). However, the number of offspring, and their male–female distribution, is outside the control of the experimenter.

Dempster *et al.* (1984) analysed such a study on rats. The outcome variable was the weight of the newborn rats. Larger (smaller) weight of a newborn rat is associated with beneficial (detrimental) effects of the treatment. Even if the treatments to be compared were identical the outcomes (weights of the newborns) are likely to be associated with litter size and there may be systematic sex differences. The treatments themselves may affect litter size and, conceivably, the distribution of the sexes within litters. This constitutes an important departure from the assumptions of experimental design on which standard approaches to analysis rely.

Although the experimental animals are genetically selected and bred in a clinical environment for the purpose of reducing their variability, they can in no way be regarded as identical clones. The litters exhibit substantial heterogeneity which can be described as within-litter covariance. For a comparison of treatments, and assessment of accuracy of such a comparison, this feature has to be taken into account.

Table 2.2. A segment of the rat weights data. The weights are in milligrams. The symbols 'M' and 'F' in the column 'Sex' stand for male and female respectively (adapted from Dempster *et al.* (1984) with permission of the Royal Statistical Society)

Rat-level data					Litter-level data		
Litter	Rat	Weight	Sex		Litter	Treatment	Litter size
1	1	6600	M		1	1	12
1	2	7400	M		2	1	14
1	3	7150	M		3	1	4
1	4	7240	M		4	1	14
1	5	7100	M		5	1	13
⋮	⋮	⋮	⋮		6	1	9
1	12	6570	F		⋮	⋮	⋮
2	1	6370	M				
2	2	6370	M				
2	3	6900	M				
⋮	⋮	⋮	⋮				

Table 2.2 contains a segment of the data. The rat- and litter-level data can be considered as two connected datasets. Parts of these datasets are given in separate panels in the table.

2.2.3 MONITORING PREGNANCY

In a study of concentration of the human placental lactogen (HPL), conducted in Aalborg, Denmark, 69 healthy pregnant volunteers with reliable information about gestational age were taken blood samples between the 25th week of pregnancy and childbirth. Participation of the women in the study was patchy; most women joined the study later than the 25th week of their pregnancy. Naturally, data from days 280 and later are available only from women who delivered later than the due date. Participation of the women in the study is summarized graphically in Figure 2.1. Each woman (client) is represented by a horizontal segment on which dots indicate the gestational ages, in days, when the observations (samples) were obtained. Table 2.3 contains a segment of the data in separate panels corresponding to observation- and client-level data.

Concentration of HPL tends to increase during the later stages of pregnancy, reaches its peak at around the 36th week, and then declines. The purpose of the study is to describe the growth of concentration of HPL, that is, the growth for an average healthy woman during a normal pregnancy and the variation in growth in the population of healthy women.

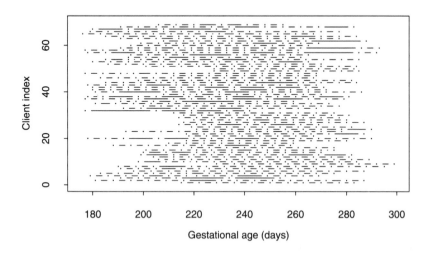

Fig. 2.1. Participation of clients in the pregnancy monitoring study. Each client is represented by a horizontal segment, with dots indicating days on which observations (samples) were collected.

Table 2.3. A segment of the pregnancy monitoring data

Visit no.	Client no.	Observation-level data	
		Gestational age (days − 252)	log(%HPL) concentration
1	1	−37	1.686
2	1	−30	1.775
3	1	−23	1.946
4	1	−16	2.001
5	1	−9	2.015
⋮	⋮	⋮	⋮
11	1	31	1.988
1	2	−71	1.099
2	2	−64	1.335
3	2	−57	1.335
⋮	⋮	⋮	⋮

Client-level data		
Client	Observ'ns	Size
1	11	1.73
2	15	1.64
3	6	1.77
4	13	1.72
⋮	⋮	⋮

Alternatively, we can consider the dataset as a set of time series, in which case a description of temporal dependence is of interest.

In this example the set of observations from a client forms a cluster. Experimental design is infeasible in this situation because the study has to rely on women who happen to be pregnant at the time of recruitment. Nevertheless, it is reasonable to assume that the clients are a representative sample of pregnant women in the geographical region and during a given period of time (e.g., over a few years). Sampling within clusters is also impossible. It would have been better if each woman had provided a blood sample in each of the weeks 25–40. However, the voluntary nature of participation has to be respected; some volunteers were past week

25 of their pregnancy at the time of recruitment, several women opted
not to provide more than a certain number of blood samples, and several
appointments were not kept by the clients for a variety of everyday reasons.
Childbirth premature by a few days intervened in a few cases.

For further background and analyses of related datasets see Lundbye-
Christensen (1991) and references therein.

2.2.4   HOUSE PRICES

Harrison and Rubinfeld (1978) reported a study of house prices in the
Boston Metropolitan Area (BMA). Their principal goal was to assess the
association of the air quality in the neighbourhood with the house price
after taking account of other influences:

- How much do people pay for clean air when they purchase a house?

From various censuses they collected the *means* of the values of owner-
occupied housing units at the time of the study, and the means of a number
of variables related to the attributes of the houses (number of rooms), of
the neighbourhood (employment opportunities, social composition), and of
the environment (concentration of pollutants), for the 906 census tracts in
BMA. Some of the means refer to the census tracts, while others refer to the
districts (towns). The BMA consists of 92 towns, which each contain 1–30
census tracts, although only 13 towns have more than ten census tracts.
Table 2.4 contains the counts of census tracts within towns. The list of
explanatory variables and their summary statistics are given in Table 2.5.
Table 2.6 contains a segment of the data with separate panels for census-
tract and town-level data.

Table 2.4.   Counts of census tracts within towns. House price data.
The 92 counts add up to 906 – the number of census tracts (adapted from
Belsley *et al.* (1980) with permission of John Wiley and Sons)

Towns	Census tracts in towns									
1–10 :	1	2	3	7	22	4	2	9	4	1
11–20 :	1	1	1	6	1	2	3	4	6	4
21–30 :	4	4	3	5	11	9	7	15	30	7
31–40 :	8	6	2	1	3	2	2	2	11	4
41–50 :	18	6	10	2	1	12	5	5	4	1
51–60 :	1	1	1	3	3	5	3	3	4	12
61–70 :	8	3	2	8	1	1	2	2	1	1
71–80 :	2	2	1	2	8	6	3	2	7	11
81–90 :	13	8	19	23	11	6	7	4	4	5
91–92 :	8	5								

Table 2.5. Definition and summaries of the outcome and explanatory variables. House price data. The census-tract level variables are marked with an asterisk * (adapted from Belsley *et al.* (1980) with permission of John Wiley and Sons)

Symbol	Mean	St. dev.	Definition
LMV^*	9.94	0.41	logarithm of the median value of owner-occupied homes (the outcome variable)
$CRIME^*$	3.62	8.60	per capita crime rate
$ZONE$	11.22	23.18	percentage of town's residential land zoned for lots greater than 25 000 square feet
$INDUS$	11.14	6.86	percentage of non-retail business acres per town
$CHAS$	0.07		set to 1 if the census tract borders the Charles River, 0 otherwise
$NOXSQ$	32.11	13.92	annual average nitric oxide concentration (square of parts per 10^8)
$ROOM^*$	39.98	9.09	average number of rooms squared
AGE^*	68.58	28.14	percentage of owner-occupied units built prior to 1940
$DIST^*$	1.19	0.54	logarithm of the weighted mean of distances of five employment centres in the Boston region
RAD^*	1.87	0.88	logarithm of index of accessibility to radial highways
TAX	408.24	168.54	full-value property-tax rate (per \$10 000 dollars)
$PTRAT$	18.46	2.16	pupil–teacher ratio by town
$BLACK^*$	0.36	0.09	$(Bk - 0.63)^2$, where Bk is the proportion of Black Americans in the census tract
$LSTAT^*$	–2.23	0.60	logarithm of the proportion of the population that is of lower status

Harrison and Rubinfeld (1978) used ordinary regression to estimate the 'effects' of the explanatory variables related to the concentration of certain pollutants. Belsley *et al.* (1980) examined the residuals of this OLS fit and concluded that they were distinctly non-normally distributed. They mentioned clustering of the census tracts within districts as a possible reason for the apparent non-normality and they reanalysed the data using robust methods.

Random-effects models provide an alternative approach to regression analysis of this dataset.

Table 2.6. A segment of the house price data. Empty lines in census-tract level data delineate towns. To conserve space some of the entries are rounded. The full dataset is published in Belsley *et al.* (1980)

Census tract	Town	Census-tract level data						
		LMV	CRIME	ROOM	AGE	DIST	BLACK	LSTAT
1	Nahant	10.09	0.006	43.23	65.2	1.41	0.397	−3.00
2	Swam't	9.98	0.027	41.23	78.9	1.60	0.397	−2.39
3	Swam't	10.45	0.027	51.62	61.1	1.60	0.393	−3.21
4	Marb'd	10.42	0.032	48.97	45.8	1.80	0.395	−3.53
5	Marb'd	10.50	0.069	51.08	54.2	1.80	0.397	−2.93
6	Marb'd	10.26	0.030	41.34	58.7	1.80	0.394	−2.96
7	Salem	10.04	0.088	36.14	66.6	1.72	0.396	−2.08
8	Salem	10.21	0.145	38.09	96.1	1.78	0.397	−1.65
9	Salem	9.71	0.211	31.71	100.0	1.80	0.387	−1.21
10	Salem	9.85	0.170	36.05	85.9	1.89	0.387	−1.77
⋮	⋮	⋮	⋮	⋮	⋮	⋮	⋮	⋮

Town	Town-level data						
	ZONE	INDUS	CHAS	NOXSQ	RAD	TAX	PTRAT
Nahant	18.0	2.31	0	28.94	0.00	296	15.3
Swampscott	0.0	7.07	0	22.00	0.69	242	17.8
Marblehead	0.0	2.18	0	20.98	1.099	222	18.7
Salem	12.5	7.87	0	27.46	1.61	311	15.2
⋮	⋮	⋮	⋮	⋮	⋮	⋮	⋮

2.2.5 VALIDITY OF AN EDUCATIONAL TEST

The Graduate Record of Examination (GRE) is a standardized test for US undergraduate students who wish to continue their studies towards a postgraduate degree. The results of the test, consisting of scores on three sections, verbal, quantitative, and analytical, form one of several criteria considered by the admissions staff of US graduate schools. The GRE scores are denoted by V, Q, and A respectively.

Educational Testing Service, the administrator of GRE, has funded a program called GRE Validity Study Service, the purpose of which is to collect data about the academic performance of first-year graduate students and relate them to the GRE scores. Evidence that the GRE scores are powerful predictors of academic performance in the graduate school would enhance the importance of GRE among both students and staff in charge of admissions at graduate schools.

In addition to the first-year grade-point average (FYA), the adopted measure of academic performance in the graduate school, and GRE scores $(Q, V,$ and $A)$, the undergraduate grade-point average (GP), the average of all the grades during the undergraduate studies, is collected. GP turns out to be the most important predictor variable; it is reasonable to expect that past exam results are a better predictor of future exam results than any standardized test. At issue, however, is the role of the GRE scores as additional predictors.

Data were obtained from graduate school departments which volunteered the requested information; the studied database comprises 748 departments with 10 322 students. Most departments have only 5–10 first-year students in any given year (departments with fewer than five students were not invited to participate). Since we want to consider four predictors (the GRE scores and the grade-point average GP), any method of estimation of the regression coefficients for a department that is based solely on the data from that department is hopelessly inefficient. Clearly, some form of parsimonious description for between-department variation is necessary. This would enable us to improve estimation for the departments by combining within- and between-department information. Another point of interest is the determination of the smallest department size for which it is meaningful to report the fitted (department-specific) regression coefficients.

The data collection involves no sampling design – departments (or universities) provide data if they wish to do so. Allocation of the students to departments is extremely non-representative: better students tend to end up in better departments because admissions staff prefer them to other students and because most students aspire to more demanding programs so as to improve their career prospects. Of course, these processes of selection (and self-selection) are very complex; any simple description is likely to contain substantial components of uncertainty and inaccuracy. The 'ranking' of schools from worst to best (from least to most competitive), as well as the ranking of students does contain an element of subjectivity, and criteria other than schools' and students' qualities enter into the decision making of both admissions staff and the applicants.

The FYA and GP scores are summaries of a student's performance, but the standards for these scores vary from institution to institution. While for FYA this variation in standards is a source of variation at the department level, variation in the standards for GP raises more serious problems,

Table 2.7. Sums of squares and cross-products for department 1. GRE validity data

Variable	1	V	Q	A	GP	FY A
1	13.00	32.97	37.05	36.57	42.76	45.29
V	32.97	87.85	96.22	96.69	109.26	116.47
Q	37.05	96.22	107.90	107.55	122.36	130.15
A	36.57	96.69	107.55	109.71	121.28	129.66
GP	42.76	109.26	122.36	121.28	144.61	150.75
FY A	45.29	116.47	130.15	129.66	150.75	160.24

the principal of which is that students' undergraduate institutions are not recorded. The GRE scores represent an assessment on a standardized scale, independent of the institution. Their main appeal is in objective scoring and in easier administration than the conventional university examinations.

In attempting to make any inference about the relationship of these educational scores we face an insurmountable problem. We wish to make an inference for one population (the applicants to graduate schools) based on a sample (not a random one) drawn from the population of *successful* applicants, with a non-random allocation of the successful applicants to graduate schools. In particular, we are unable to state whether (adjusted) differences in the *FY A* scores are due to different standards of instruction, different standards in awarding grades, differential ability to attract good students, or indeed, whether some schools tend to admit the best students from schools with lower standards for *GP*, and others the poorer students from schools with high standards.

In order to ensure confidentiality the data were made available in the form of within-department cross-products of all the variables. The matrix of sums of cross-products for a department is given in Table 2.7. For more detailed background, see Longford (1991).

2.3 Newton–Raphson and Fisher scoring algorithms

The Newton–Raphson and Fisher scoring algorithms, as general methods for maximum likelihood estimation, were sketched in Section 1.9.1. In this section we apply them to the random-effects model (2.1). The algorithms require evaluation of the scoring function $\partial l/\partial \theta$, and of either the second-order partial derivatives, $\partial l^2/(\partial \theta_1 \partial \theta_2)$, or the information function, $-\mathbf{E}\{\partial^2 l/(\partial \theta_1 \partial \theta_2)\}$, for each regression and variance parameter ($\boldsymbol{\beta}, \sigma^2$, and τ^2), or their pairs, respectively.

We consider first a general parametrization for the variance matrix $\mathbf{V}$. Suppose $\mathbf{V}$ is a function of the vector of parameters $\boldsymbol{\theta}$, and $\boldsymbol{\theta}$ is functionally independent of the regression parameters $\boldsymbol{\beta}$. In other words, the mean and the variance matrix for the observations involve disjoint sets of parameters. We refer to $\boldsymbol{\theta}$ as the vector of *covariance structure parameters*.

Using the rules for formal matrix differentiation (see Section 1.2.2) we obtain

$$\frac{\partial l}{\partial \boldsymbol{\beta}} = \mathbf{X}^{\top} \mathbf{V}^{-1} \mathbf{e},$$

$$\frac{\partial^2 l}{\partial \boldsymbol{\beta} \partial \boldsymbol{\beta}^{\top}} = -\mathbf{X}^{\top} \mathbf{V}^{-1} \mathbf{X}. \tag{2.13}$$

When the variance matrix $\mathbf{V}$ is known, the Newton–Raphson algorithm is equivalent to Fisher scoring, and is described by the updating equation

$$\hat{\boldsymbol{\beta}}_{new} = \hat{\boldsymbol{\beta}}_{old} + \left(\mathbf{X}^{\top} \mathbf{V}^{-1} \mathbf{X} \right)^{-1} \mathbf{X}^{\top} \mathbf{V}^{-1} \hat{\mathbf{e}}_{old}, \tag{2.14}$$

where $\hat{\mathbf{e}}_{old} = \mathbf{y} - \mathbf{X} \hat{\boldsymbol{\beta}}_{old}$. This is equivalent to the generalized least squares (GLS) estimator

$$\hat{\boldsymbol{\beta}} = \left(\mathbf{X}^{\top} \mathbf{V}^{-1} \mathbf{X} \right)^{-1} \mathbf{X}^{\top} \mathbf{V}^{-1} \mathbf{y}. \tag{2.15}$$

Note that $\hat{\boldsymbol{\beta}}$ is the OLS estimator when $\mathbf{V} = \sigma^2 \mathbf{I}$, and the weighted least squares estimator when $\mathbf{V}$ is diagonal.

For a general covariance structure parameter θ we have

$$\frac{\partial l}{\partial \theta} = -\frac{1}{2} \left\{ \frac{\partial \log(\det \mathbf{V})}{\partial \theta} - \mathbf{e}^{\top} \mathbf{V}^{-1} \frac{\partial \mathbf{V}}{\partial \theta} \mathbf{V}^{-1} \mathbf{e} \right\}. \tag{2.16}$$

The expectation of this scoring function is zero, $\mathbf{E}(\partial l / \partial \theta) = 0$, and so

$$\frac{\partial \log(\det \mathbf{V})}{\partial \theta} = \mathbf{E} \left(\mathbf{e}^{\top} \mathbf{V}^{-1} \frac{\partial \mathbf{V}}{\partial \theta} \mathbf{V}^{-1} \mathbf{e} \right) = \text{tr} \left(\mathbf{V}^{-1} \frac{\partial \mathbf{V}}{\partial \theta} \right). \tag{2.17}$$

For the latter identity we have used the commutative property of the trace of a matrix; for an arbitrary $N \times N$ matrix $\mathbf{W}$:

$$\mathbf{E}(\mathbf{e}^{\top} \mathbf{W} \mathbf{e}) = \mathbf{E} \left\{ \text{tr} \left(\mathbf{W} \mathbf{e} \mathbf{e}^{\top} \right) \right\} = \text{tr}(\mathbf{W} \mathbf{V}).$$

Equation (2.17) holds for any invertible matrix $\mathbf{V}$, see Graybill (1969), though its general proof is more involved. We will use (2.17) only for variance matrices. Substitution of (2.17) in (2.16) yields

$$\frac{\partial l}{\partial \theta} = -\frac{1}{2} \left\{ \mathrm{tr} \left(\mathbf{V}^{-1} \frac{\partial \mathbf{V}}{\partial \theta} \right) - \mathbf{e}^\top \mathbf{V}^{-1} \frac{\partial \mathbf{V}}{\partial \theta} \mathbf{V}^{-1} \mathbf{e} \right\}. \tag{2.18}$$

The second-order partial derivative with respect to a pair of covariance structure parameters θ_1 and θ_2 is

$$\frac{\partial^2 l}{\partial \theta_1 \partial \theta_2} = \frac{1}{2} \mathrm{tr} \left(\mathbf{V}^{-1} \frac{\partial \mathbf{V}}{\partial \theta_1} \mathbf{V}^{-1} \frac{\partial \mathbf{V}}{\partial \theta_2} \right) - \mathbf{e}^\top \mathbf{V}^{-1} \frac{\partial \mathbf{V}}{\partial \theta_1} \mathbf{V}^{-1} \frac{\partial \mathbf{V}}{\partial \theta_2} \mathbf{V}^{-1} \mathbf{e}. \tag{2.19}$$

Since $\mathbf{E}(\mathbf{e}^\top \mathbf{W} \mathbf{V}^{-1} \mathbf{e}) = \mathrm{tr}(\mathbf{W})$, the element of the information matrix corresponding to the parameters θ_1 and θ_2 is

$$-\mathbf{E} \left(\frac{\partial^2 l}{\partial \theta_1 \partial \theta_2} \right) = \frac{1}{2} \mathrm{tr} \left(\mathbf{V}^{-1} \frac{\partial \mathbf{V}}{\partial \theta_1} \mathbf{V}^{-1} \frac{\partial \mathbf{V}}{\partial \theta_2} \right). \tag{2.20}$$

Further,

$$-\mathbf{E} \left(\frac{\partial l}{\partial \beta \partial \theta} \right) = \mathbf{X} \frac{\partial \mathbf{V}^{-1}}{\partial \theta} \mathbf{E} \mathbf{e} = 0. \tag{2.21}$$

By comparing (2.19) and (2.20) we see that the Fisher scoring algorithm is somewhat simpler than the Newton–Raphson one. Also, owing to (2.21) each iteration of the Fisher scoring algorithm has two separate parts: reestimating β, (2.15), and updating the vector of covariance structure parameters θ.

Application of equations (2.18) and (2.20) for the two-level model (2.1) is relatively straightforward since

$$\frac{\partial \mathbf{V}_j}{\partial \sigma^2} = \mathbf{I}_{n_j}, \qquad \frac{\partial \mathbf{V}_j}{\partial \tau^2} = \mathbf{J}_{n_j}, \tag{2.22}$$

and, for example, $\mathrm{tr}(\mathbf{V}^{-1} \partial \mathbf{V} / \partial \tau^2) = \sum_j \mathbf{1}_{n_j}^\top \mathbf{V}_j^{-1} \mathbf{1}_{n_j}$. However, rather than proceeding with the substitutions, we consider a reparametrization that further simplifies the equations for the Fisher scoring algorithm.

The variance ratio, $\omega = \tau^2 / \sigma^2$, was defined by equation (2.5). The matrix $\mathbf{W}_j = \sigma^{-2} \mathbf{V}_j$ as a function of ω does not depend on σ^2:

$$\mathbf{W}_j = \mathbf{I}_{n_j} + \omega \mathbf{J}_{n_j}. \tag{2.23}$$

The scoring function for σ^2 is equal to

$$\frac{\partial l}{\partial \sigma^2} = -\frac{1}{2\sigma^2} \left(N - \frac{1}{\sigma^2} \mathbf{e}^\top \mathbf{W}^{-1} \mathbf{e} \right) \tag{2.24}$$

($\mathbf{W} = \{\mathbf{W}_j\} \otimes \mathbf{I}_{N_2}$), and, given $\mathbf{W}$ and β, it has a unique root

$$\hat{\sigma}^2 = \frac{\mathbf{e}^\top \mathbf{W}^{-1} \mathbf{e}}{N}. \tag{2.25}$$

The scoring function for the variance ratio ω is

$$\frac{\partial l}{\partial \omega} = \sigma^2 \frac{\partial l}{\partial \tau^2} = -\frac{1}{2} \text{tr} \left(\mathbf{W}^{-1} \frac{\partial \mathbf{W}}{\partial \omega} \right) + \frac{1}{2\sigma^2} \mathbf{e}^\top \mathbf{W}^{-1} \frac{\partial \mathbf{W}}{\partial \omega} \mathbf{W}^{-1} \mathbf{e},$$

and, since $\partial \mathbf{W}_j / \partial \omega = \mathbf{J}_{n_j} = \mathbf{1}_{n_j}^\top \mathbf{1}_{n_j}$, we have

$$\frac{\partial l}{\partial \omega} = -\frac{1}{2} \sum_j \mathbf{1}_{n_j}^\top \mathbf{W}_j^{-1} \mathbf{1}_{n_j} + \frac{1}{2\sigma^2} \sum_j \left(\mathbf{e}_j^\top \mathbf{W}_j^{-1} \mathbf{1}_{n_j} \right)^2 \qquad (2.26)$$

and

$$-\mathbf{E} \left\{ \frac{\partial^2 l}{(\partial \omega)^2} \right\} = \frac{1}{2} \text{tr} \left(\mathbf{W}^{-1} \frac{\partial \mathbf{W}}{\partial \omega} \mathbf{W}^{-1} \frac{\partial \mathbf{W}}{\partial \omega} \right) = \frac{1}{2} \sum_j \left(\mathbf{1}_{n_j}^\top \mathbf{W}_j^{-1} \mathbf{1}_{n_j} \right)^2.$$

$$(2.27)$$

The inverse and the determinant of $\mathbf{W}_j$ are obtained directly from (2.9) and (2.10). The equations for the Fisher scoring algorithm require the evaluation of various quadratic forms in $\mathbf{W}_j^{-1}$ or $\mathbf{V}_j^{-1}$, but evaluation or storage of the matrices $\mathbf{V}_j^{-1}$ or $\mathbf{W}_j^{-1}$ is not necessary: for arbitrary $n_j \times 1$ vectors $\mathbf{u}, \mathbf{u}_1$, and $\mathbf{u}_2$ we have

$$\mathbf{u}^\top \mathbf{W}_j^{-1} \mathbf{1}_{n_j} = \mathbf{u}^\top \mathbf{1}_{n_j} (1 - n_j \omega g_j^{-1}) = g_j^{-1} \mathbf{u}^\top \mathbf{1}_{n_j} \qquad (2.28)$$

and

$$\mathbf{u}_1^\top \mathbf{W}_j^{-1} \mathbf{u}_2 = \mathbf{u}_1^\top \mathbf{u}_2 - \omega g_j^{-1} \mathbf{u}_1^\top \mathbf{1}_{n_j} \mathbf{u}_2^\top \mathbf{1}_{n_j}, \qquad (2.29)$$

where $g_j = 1 + n_j \omega$. Note how (2.28) simplifies the evaluation of (2.26) and (2.27).

Iterations of the Fisher scoring algorithm require a starting solution. The OLS fit given by (1.13) and (1.14) provides suitable starting values for $\hat{\boldsymbol{\beta}}$ and $\hat{\sigma}^2$. For ω (or τ^2) any non-negative starting value is suitable (such as $\hat{\omega}_0 = 0$ when there is no prior information about ω). The Fisher scoring algorithm has very good convergence properties; usually three to nine iterations are sufficient for the algorithm to converge. A reasonable non-iterative estimator of τ^2 is given by

$$\hat{\tau}_0^2 = \frac{1}{N_2} \sum_j \frac{(\mathbf{e}^\top \mathbf{1}_{n_j})^2}{n_j^2}. \qquad (2.30)$$

It can be used as the starting solution for $\hat{\tau}^2$. Its expectation, given the regression parameters $\boldsymbol{\beta}$, is equal to $\tau^2 + \sigma^2 \sum_j n_j^{-1}/N_2$.

2.3.1 SOME TECHNICAL DETAILS

The following concerns commonly arise when using iterative fitting procedures, such as the Fisher scoring algorithm,

- existence and uniqueness of the solution,
- convergence of the algorithm,
- estimation of the standard errors associated with the estimated parameters.

We deal with these concerns one-by-one.

The information matrices for $\boldsymbol{\beta}$ and (σ^2, ω) are non-negative definite, and are singular only in degenerate cases: when the design matrix $\mathbf{X}$ is of incomplete rank (this would usually be detected while fitting the corresponding OLS), and when each cluster contains only a single elementary observation. Details are postponed till Section 2.9. If $\mathbf{X}$ is of full rank and $\sigma^2 > 0$, the observed information matrix for the regression parameters $\boldsymbol{\beta}$, $\mathbf{X}^\top \mathbf{V}^{-1} \mathbf{X}$, is positive definite, and so the log-likelihood is a convex function in $\boldsymbol{\beta}$. Therefore, it has at most one maximum, and since

$$\lim l(\mathbf{y}; \boldsymbol{\beta}, \boldsymbol{\theta}) = -\infty$$

as any component of $\boldsymbol{\beta}$ approaches infinity, the log-likelihood has a unique maximum. Equation (2.25) implies a unique MLE for σ^2 (given the rest of the parameters).

To explore the existence and uniqueness of the MLE of ω we consider first the extended parameter space for ω given by the condition

$$\omega > \frac{-1}{\max n_j}.$$

Let j^* be the cluster with the largest sample size. Then, as $\omega \to -1/n_{j^*}$, $g_{j^*} = (1 + n_{j^*}\omega) \to 0$. The scoring function (2.26),

$$\frac{\partial l}{\partial \omega} = -\frac{1}{2} \left\{ \sum_j \frac{n_j}{g_j} - \sigma^{-2} \sum_j \frac{(\mathbf{e}_j^\top \mathbf{1}_{n_j})^2}{g_j^2} \right\}, \qquad (2.31)$$

depends on ω only through negative powers of g_j. As $g_{j^*} \to 0+$ the term $(\mathbf{e}_j^\top \mathbf{1})^2 / g_j^2$ becomes dominant, and the scoring function (2.31) becomes positive. Similarly, as $\omega \to +\infty$ both summands in (2.31) converge to zero, but $\sum_j n_j/g_j$ is slower to converge to zero, and so (2.31) is negative for some large values of ω. The scoring function (2.31) changes its sign and is continuous, therefore it has a root. Of course, it may have a single negative root. Then the maximum likelihood estimate of ω is 0. The matrix of second-order partial derivatives is positive definite 'on average', giving us some comfort, though not a proof, that the log-likelihood has a unique maximum in ω.

There are two kinds of criteria for convergence. The iterations can be terminated when the consecutive solutions differ by less than a prescribed tolerance; since the solutions are represented by a vector, it is convenient to consider a norm for the difference of two consecutive solutions (vectors), such as the Euclidean norm, the sum, or the maximum, of the absolute differences of the components, or similar. It is useful to make an adjustment for the number of estimated components. Alternatively, the iterations can be stopped when the successive values of the log-likelihood differ by less than a prescribed small quantity. The two kinds of criteria can be combined by insisting on one of each kind. When the maximum likelihood solution is in the interior of the parameter space, proximity of the scoring vector to $\mathbf{0}$ can also be used as another criterion for convergence.

If the parameter vector is in the interior of the parameter space (that is, not on its boundary), then the asymptotic (large sample) variance matrix of the MLE is equal to the inverse of the information matrix. Of course, in a typical setting we do not know whether the parameter vector is in the interior or on the boundary of the parameter space. The only information we have about its location comes from the estimate itself. Note that when $\mathbf{V}$ is known the inverse of the information matrix for the regression parameters, $\left(\mathbf{X}^{\top}\mathbf{V}^{-1}\mathbf{X}\right)^{-1}$, is equal to var$(\hat{\boldsymbol{\beta}}_{ML})$ even for small samples. The standard errors are the square roots of the sampling variances.

In practice the information matrix is itself estimated by substituting the maximum likelihood estimates for the parameters in the information matrix. This practice is problematic for small datasets for two reasons: the asymptotic properties may not apply to them, and there is substantial uncertainty about the parameters involved in the information function; the information function may vary substantially as a function of the parameters.

Several other qualifications have to be attached to this universally accepted method of obtaining standard errors for MLEs. First, it is not clear what 'asymptotic' means in the context of clustered data. It turns out that we have to require that the number of clusters, rather than the number of elementary observations, grows above all bounds. In addition, the average size of the clusters has to remain well above 1. Next, the asymptotic standard errors obtained from the information matrix are valid only when $\omega > 0$. This problem can be sidestepped by formally allowing for negative values of ω; these are meaningful so long as we do not insist on the model (2.1), but merely wish to fit a constant within-cluster covariance $\sigma^2\omega$.

The information matrix is a function of unknown parameters and is estimated by substituting the maximum likelihood estimates for them. Thus, whether some of the parameters are estimated or known has no impact on the *estimated* information about the estimated parameters. This point is discussed further in Section 2.6. Asymptotic theory is discussed in more detail in Chapter 9.

Another approach is based on estimating the square root of the variance ratio, $\xi = \sqrt{\omega}$, or the square root of the variance τ^2. The scoring and information functions are derived from (2.26) and (2.27) using the chain rule for differentiation of composite functions. Note that $\hat{\xi} = 0$ is always a root of the scoring function, and so care has to be exercised when setting the starting solution for ξ.

2.4   Generalized least squares

The generalized least squares (GLS) estimator (2.15) is also the MLE for $\boldsymbol{\beta}$ when the outcomes $\mathbf{y}$ are normally distributed. However, (2.15) has some desirable properties even when the outcomes are not normally distributed. For example, when $\mathbf{V}$ is known (2.15) is an unbiased estimator. For estimating the variance parameters we may consider GLS for the within-cluster cross-products $\{e_{ij}e_{i'j}\}$; their expectations, given the regression parameters $\boldsymbol{\beta}$, are linear functions of the variances τ^2 and σ^2:

$$\mathbf{E}\left(\mathbf{e}_j\mathbf{e}_j^\top\right) \;=\; \mathbf{V}_j; \tag{2.32}$$

the components of $\mathbf{V}_j$ are equal to $\sigma^2 + \tau^2$ (diagonal) and τ^2 (off-diagonal). We can align all the cross-products $\{e_{ij}e_{i'j}\}$ in a single vector $\mathbf{e}^*$ (of length $\sum_j n_j^2$), and estimate the variances τ^2 and σ^2 by the corresponding GLS estimator. The variance matrix for $\mathbf{e}^*$ is equal to

$$\mathbf{V}^* \;=\; \{\mathbf{V}_j \otimes \mathbf{V}_j\} \otimes \mathbf{I}_{N_2}. \tag{2.33}$$

Note that the diagonal elements of $\mathbf{V}^*$ are equal to $(\tau^2 + \sigma^2)^2$ or $2(\tau^2 + \sigma^2)^2$. The values of the off-diagonal elements depend on the configuration of the observations that contribute to the two cross-products. The pattern of this matrix can be exploited to avoid numerical inversion of large matrices; for example, we have

$$(\mathbf{V}_j \otimes \mathbf{V}_j)^{-1} \;=\; \mathbf{V}_j^{-1} \otimes \mathbf{V}_j^{-1}.$$

For the iterations of GLS the off-diagonal entries of $\mathbf{V}^*$ can all be replaced by zeros. The standard errors for the estimated parameters are obtained from the information matrices $\mathbf{X}^\top\mathbf{V}^{-1}\mathbf{X}$ and $\mathbf{X}^{*\top}\mathbf{V}^{-1}\mathbf{X}^*$, where $\mathbf{X}^*$ is the design matrix associated with $\mathbf{e}^*$, $\mathbf{E}(\mathbf{e}^*) = \sigma^2\mathbf{X}^*(1,\omega)^\top$.

The GLS method is closely related to the method of moments. Since

$$\mathbf{E}(\mathbf{X}^\top\mathbf{V}^{-1}\mathbf{y}) \;=\; \mathbf{X}^\top\mathbf{V}^{-1}\mathbf{X}\boldsymbol{\beta}, \tag{2.34}$$

the GLS estimator is also a method of moments estimator for suitably transformed outcomes $\mathbf{y}$. For estimation of the variances we may consider two quadratic statistics, such as the sum of squares of residuals, $\mathbf{e}^\top\mathbf{e}$, and the sum of squares of the within-cluster totals of residuals, $\sum_j \left(\mathbf{1}_{n_j}^\top\mathbf{e}_j\right)^2$. Their expectations, assuming known $\boldsymbol{\beta}$, are:

$$\mathbf{E}(\mathbf{e}^\top \mathbf{e}) = N(\sigma^2 + \tau^2),$$

(2.35)

$$\mathbf{E}\left\{\sum_j \left(\mathbf{1}_{n_j}^\top \mathbf{e}_j\right)^2\right\} = N\sigma^2 + \sum_j n_j^2 \tau^2.$$

Matching the observed moments with those of (2.35) involves the solution of a simple system of two linear equations in the variances τ^2 and σ^2. The GLS estimators (2.34) and (2.35) coincide with the MLE because the multivariate distribution given by the density (2.8) belongs to the exponential family.

2.5 EM algorithm

In statistical practice we frequently encounter situations where the selected method of analysis would be much simpler if the data at our disposal were supplemented by further information. Familiar examples are various situations involving unbalanced data; if observations were added to an unbalanced dataset in such a way that a balanced dataset were obtained, a straightforward analysis could be applied. In the random-effects model we are in a similar position; if the random variables δ_j were observed the model (2.1) could be fitted by OLS.

More generally, suppose the dataset at our disposal, $\mathbf{y}$, is a subset of a (hypothetical) dataset $(\mathbf{y}, \boldsymbol{\delta})$. We refer to $\mathbf{y}$ as the *incomplete data*, to $(\mathbf{y}, \boldsymbol{\delta})$ as the *complete data*, and to $\boldsymbol{\delta}$ as the *missing data*. The respective likelihoods for the incomplete and the complete data, l and l_c, are related through the identity

$$l(\boldsymbol{\theta}; \mathbf{y}) = \int \cdots \int l_c(\boldsymbol{\theta}; \mathbf{y}, \boldsymbol{\delta}) dF(\boldsymbol{\delta}),$$

(2.36)

where F denotes the distribution function of $\boldsymbol{\delta}$. A naïve solution would be to estimate the missing data, and then maximize the complete likelihood with the estimated missing data substituted for the unobserved missing data, that is, maximize $l_c(\boldsymbol{\theta}; \mathbf{y}, \hat{\boldsymbol{\delta}})$. Such an approach may lead to substantial biases. For illustration, consider estimating the between-cluster variance σ_B^2 in the blood pressure example, Section 1.5.2. The variance of the (observed) within-cluster means, $\bar{y}_j$, is an overestimate of σ_B^2; in fact $\text{var}(\bar{y}_j) = \sigma_B^2 + \sigma_W^2/n$. This simple example highlights the problems of aggregation in the presence of substantial elementary-level variance.

The missing data are represented in the complete likelihood by certain functions of their sufficient statistics. Thus, it suffices to estimate these sufficient statistics and use them in the complete data analysis. In general, the EM algorithm is an iterative procedure with each iteration consisting of two parts: the E-step (expectation), in which the conditional expectations of the functions of the sufficient statistics for the missing data are calculated, and the M-step (maximization), in which the complete likelihood, with the functions of the missing data replaced by their conditional expectations, is maximized. The conditioning in the E-step is on the incomplete (available) data and the current estimates of the parameters. The estimates are updated in the subsequent M-step.

The EM algorithm tends to converge rather slowly, especially when the missing data contain a lot of information about the estimated parameters. But the undeniable virtue of the EM algorithm is that problems which cannot be solved by easy-to-implement methods, can often be conceived as an analysis of a larger dataset, for which a method of estimation is available and is easy to implement. The specification of the missing dataset is up to the analyst; in principle, there may be several essentially different choices.

In the context of the random-effects model (2.1) the likelihood for the incomplete data is given by the equation

$$\int \cdots \int l_c(\boldsymbol{\beta}, \sigma^2, \tau^2; \mathbf{y}, \boldsymbol{\delta}) d\boldsymbol{\delta}$$

$$= (2\pi\sigma^2)^{-N/2} (2\pi\tau^2)^{-N_2/2} \prod_j \int_{-\infty}^{+\infty} \exp\left(-\frac{\boldsymbol{\varepsilon}_j^\top \boldsymbol{\varepsilon}_j}{2\sigma^2} - \frac{\delta_j^2}{2\tau^2}\right) d\delta_j,$$

(2.37)

where $\boldsymbol{\varepsilon}_j = \mathbf{e} - \delta_j$, suggesting an application of the EM algorithm along the lines described above. But the integral in (2.37) has the form (2.8) and can be evaluated analytically, as described in Section 2.3. For completeness, we give details of the EM algorithm for the random-effects model.

2.5.1   THE E-STEP

The complete log-likelihood is a quadratic function of the missing data $\{\delta_j\}$, and so we require the conditional expectations of δ_j and δ_j^2, given the outcomes $\mathbf{y}$. Equation (1.10) for the conditional distribution under normality implies that

$$(\delta_j \mid \mathbf{y}_j) \sim \mathcal{N}(\tau^2 \mathbf{1}_{n_j}^\top \mathbf{V}_j^{-1} \mathbf{e}_j, \ \tau^2 - \tau^4 \mathbf{1}_{n_j}^\top \mathbf{V}_j^{-1} \mathbf{1}_{n_j}),$$

(2.38)

and by applying (2.28) it simplifies to

$$(\delta_j \mid \mathbf{y}_j) \sim \mathcal{N}\left(\frac{\tau^2 \mathbf{1}_{n_j}^\top \mathbf{e}_j}{\sigma^2 g_j}, \ \frac{\tau^2}{g_j}\right).$$

(2.39)

2.5.2 THE M-STEP

The M-step is essentially an ordinary regression with outcome vector $\mathbf{y}$, in which the explanatory variables are augmented by the cluster-level variable $\boldsymbol{\delta}^*$ given by the unobserved values δ_j:

$$\mathbf{y}_j = \mathbf{X}_j\boldsymbol{\beta} + \delta_j^*\tau + \boldsymbol{\varepsilon}_j, \qquad (2.40)$$

where $\delta_j^* = \tau^{-1}\delta_j\mathbf{1}_{n_j}$. We denote the corresponding regression design matrix by $\mathbf{X}^*$, $\mathbf{X}^* = (\mathbf{X}, \boldsymbol{\delta}^*)$, where $\boldsymbol{\delta}^* = \{\delta_j^*\} \otimes \mathbf{1}_{N_2}$. The regression parameters in (2.40) are $\boldsymbol{\beta}^* = \begin{pmatrix} \boldsymbol{\beta} \\ \tau \end{pmatrix}$. Thus the matrix of cross-products for the complete data is

$$\mathbf{X}^{*\top}\mathbf{X}^* = \begin{pmatrix} \mathbf{X}^\top\mathbf{X} & \sum_j \delta_j\mathbf{1}_{n_j}^\top\mathbf{X}_j \\ \sum_j \delta_j\mathbf{X}_j^\top\mathbf{1}_{n_j} & \sum_j n_j\delta_j^2 \end{pmatrix}, \qquad (2.41)$$

and

$$\mathbf{X}^*\mathbf{y} = \left(\mathbf{X}^\top\mathbf{y}, \sum_j \delta_j\mathbf{1}_{n_j}^\top\mathbf{y}_j\right)^\top. \qquad (2.42)$$

The conditional expectations are

$$\mathbf{E}(\delta_j^* \mid \mathbf{y}_j) = \frac{\tau\mathbf{e}_j^\top\mathbf{1}_{n_j}}{\sigma^2 g_j}$$

and

$$\mathbf{E}(\delta_j^{*2} \mid \mathbf{y}_j) = \frac{1}{g_j} + \frac{\tau^2(\mathbf{e}_j^\top\mathbf{1}_{n_j})^2}{\sigma^4 g_j^2}.$$

They are evaluated at the current estimates of $\boldsymbol{\beta}^*$ and σ^2, and substituted in (2.41) and (2.42); they yield the new estimate of $\boldsymbol{\beta}^*$:

$$\hat{\boldsymbol{\beta}}_{new}^* = \left(\mathbf{X}^{*\top}\mathbf{X}^*\right)^{-1}\mathbf{X}^{*\top}\mathbf{y}.$$

The variance σ^2 is estimated as the residual variance of the model in (2.40).

2.6 Restricted maximum likelihood

A justified criticism of the maximum likelihood estimation in ordinary regression is that the estimator of the residual variance,

$$\hat{\sigma}^2 = \frac{(\mathbf{y} - \mathbf{X}\hat{\boldsymbol{\beta}})^\top(\mathbf{y} - \mathbf{X}\hat{\boldsymbol{\beta}})}{N} = \frac{\mathbf{y}^\top\left\{\mathbf{I}_N - \mathbf{X}(\mathbf{X}^\top\mathbf{X})^{-1}\mathbf{X}^\top\right\}\mathbf{y}}{N}, \qquad (2.43)$$

is biased:

$$\mathbf{E}(\hat{\sigma}^2) \;=\; \frac{N-p}{N}\sigma^2.$$

The familiar unbiased estimator is obtained by replacing the denominator N in (2.43) with $N-p$, that is, by taking account of the *degrees of freedom* due to regression. If the estimator $\hat{\beta}$ in (2.43) were replaced by the 'true' parameter vector β, $\hat{\sigma}^2$ would be conditionally unbiased. In other words, the estimator $\hat{\sigma}^2$ fails to take account of the uncertainty about the regression parameters.

This issue remains relevant in the random-effects models: in maximum likelihood estimation no distinction is made between known and unknown (estimated) regression parameters. When several regression parameters are estimated it is reasonable to expect that the MLEs of the variances and their standard errors are downward biased. Correction for this bias is desirable, especially when it involves little or no loss of efficiency.

A general solution to this problem, due to Patterson and Thompson (1971), is to base the estimation of the variances on a complete set of *error contrasts* for the outcomes $\mathbf{y}$. An error contrast is defined as any linear combination of the outcomes $\mathbf{y}$, $\mathbf{u}^\top\mathbf{y}$, which has zero expectation, $\mathbf{E}(\mathbf{u}^\top\mathbf{y}) = 0$. The error contrasts form an $(N-p)$-dimensional linear space. A set of $N-p$ linearly independent error contrasts is called a complete set of error contrasts.

It is easy to show that a set of $N-p$ independent rows of the matrix

$$\mathbf{P}_X \;=\; \mathbf{I}_N - \mathbf{X}\left(\mathbf{X}^\top\mathbf{X}\right)^{-1}\mathbf{X}^\top$$

form a complete set of error contrasts: first,

$$\mathbf{E}(\mathbf{P}_X\mathbf{y}) \;=\; \left\{\mathbf{X} - \mathbf{X}\left(\mathbf{X}^\top\mathbf{X}\right)^{-1}\mathbf{X}^\top\mathbf{X}\right\}\beta \;=\; 0,$$

so that the rows of $\mathbf{P}_X$ are error contrasts. Next, $\mathbf{P}_X^2 = \mathbf{P}_X$, and therefore the eigenvalues of $\mathbf{P}_X$ are either 0 or 1; the rank of $\mathbf{P}_X$ is equal to the multiplicity of the eigenvalue 1, and this is equal to

$$\mathrm{tr}(\mathbf{P}_X) \;=\; \mathrm{tr}\left\{\mathbf{I}_N - \mathbf{X}\left(\mathbf{X}^\top\mathbf{X}\right)^{-1}\mathbf{X}^\top\right\}$$

$$=\; N - \mathrm{tr}(\mathbf{I}_p) \;=\; N - p.$$

Let $\mathbf{y}^*$ be a complete set of error contrasts, and l^* the corresponding log-likelihood. The complete sets of error contrasts form an equivalence class; one complete set of error contrasts can be obtained from another by a non-singular linear transformation, $\mathbf{y}_2^* = \mathbf{A}\mathbf{y}_1^*$. It is easy to see that the log-likelihoods for $\mathbf{y}_1^*$ and $\mathbf{y}_2^*$ differ by the constant $-\frac{1}{2}\log(\det \mathbf{A})$. Thus the log-likelihoods corresponding to all the sets of error contrasts have the same maxima and the same information functions.

The joint distribution of the error contrasts is normal with mean $\mathbf{0}$ and variance matrix $\mathbf{BV}^{-1}\mathbf{B}^{\top}$, where $\mathbf{B}$ is the non-singular transformation matrix $(\mathbf{y}^* = \mathbf{By})$.

Maximization of the log-likelihood for $\mathbf{y}$ is referred to as the *full* maximum likelihood (ML), and maximization of a set of error contrasts $\mathbf{y}^*$ as the *restricted* maximum likelihood (REML) estimation. We also use the acronyms ML and REML for the corresponding log-likelihood, MLEs, and so on. The adjectives 'full' and 'restricted' are not to be interpreted as the former being superior to the latter in any way.

It is reasonable to ask how much information about the covariance structure parameters is lost by, essentially, reducing the analysed dataset by p observations. Of interest is the case when p/N is large; otherwise ML and REML estimators almost coincide. For independent outcomes the OLS estimator of the residual variance *is* based on a complete set of error contrasts; in this case no loss of information is incurred. The MLEs of the variances in the random-effects model depend on the vector of outcomes $\mathbf{y}$ only through the vector of residuals $\mathbf{e}$:

$$\mathbf{e} \;=\; \mathbf{y} - \mathbf{X}\hat{\beta} \;=\; \mathbf{P}_X\mathbf{y},$$

which is a (singular) linear transformation of a complete set of error contrasts. This implies that no information about the covariance structure parameters is discarded in exchange for unbiasedness.

The distribution of the error contrasts does not depend on the regression parameters β, and therefore REML involves maximization only over the covariance structure parameters. This is made more complex, though, because the variance matrix of the error contrasts no longer has a tractable form. Patterson and Thompson (1971) selected a particular set of error contrasts which yield an analytic expression for the log-likelihood. Harville (1974) derived a more convenient formula for the log-likelihood of the error contrasts:

$$l^* = C - \frac{1}{2}\log(\det \mathbf{V}) - \frac{1}{2}\log\left\{\det\left(\mathbf{X}^{\top}\mathbf{V}^{-1}\mathbf{X}\right)\right\} - \frac{1}{2}\mathbf{e}^{\top}\mathbf{V}^{-1}\mathbf{e}, \quad (2.44)$$

that is, the log-likelihood for the error contrasts differs from the log-likelihood for the original observations $\mathbf{y}$ by a constant (C, dependent only on the choice of the set of contrasts) and the additive term

$$-\frac{1}{2}\log\left\{\det\left(\mathbf{X}^{\top}\mathbf{V}^{-1}\mathbf{X}\right)\right\}.$$

The Newton–Raphson and Fisher scoring algorithms can be adapted for REML estimation. The first-order partial derivatives with respect to a covariance structure parameter θ have to be adjusted by the term

$$-\frac{1}{2}\frac{\partial\left(\mathbf{X}^{\top}\mathbf{V}^{-1}\mathbf{X}\right)}{\partial\theta} = \frac{1}{2}\text{tr}\left\{\left(\mathbf{X}^{\top}\mathbf{V}^{-1}\mathbf{X}\right)^{-1}\mathbf{X}^{\top}\mathbf{V}^{-1}\frac{\partial\mathbf{V}}{\partial\theta}\mathbf{V}^{-1}\mathbf{X}\right\}. \quad (2.45)$$

Using the (σ^2, ω) parameterization the REML estimator of the elementary-level variance σ^2 can be obtained directly from the first-order partial derivative of (2.44):

$$\frac{\partial l^*}{\partial\sigma^2} = \frac{\partial l}{\partial\sigma^2} + \frac{p}{2\sigma^2},$$

and so (2.24) implies that

$$\hat{\sigma}^2 = \frac{\mathbf{e}^{\top}\mathbf{W}^{-1}\mathbf{e}}{N-p}. \quad (2.46)$$

This is an unbiased estimator of σ^2. To prove it let

$$\mathbf{P}_V = \mathbf{I}_N - \mathbf{X}\left(\mathbf{X}^{\top}\mathbf{V}^{-1}\mathbf{X}\right)^{-1}\mathbf{X}^{\top}\mathbf{V}^{-1}.$$

The matrix $\mathbf{P}_V$ is idempotent ($\mathbf{P}_V^2 = \mathbf{P}_V$) and $\mathbf{P}_V\mathbf{V}\mathbf{P}_V^{\top} = \mathbf{P}_V\mathbf{V}$. Since $\text{tr}(\mathbf{P}_V) = N - \text{tr}(\mathbf{I}_p) = N - p$ we have

$$\mathbf{E}(\mathbf{e}^{\top}\mathbf{W}^{-1}\mathbf{e}) = \sigma^2\text{tr}(\mathbf{P}_V^{\top}\mathbf{V}^{-1}\mathbf{P}_V\mathbf{V})$$

$$= \sigma^2\text{tr}(\mathbf{P}_V) = (N-p)\sigma^2, \quad (2.47)$$

which implies unbiasedness of $\hat{\sigma}^2$ in (2.46).

2.7   Balanced design

In the least general sense, balanced design refers to equal numbers of elementary observations in each cluster, $n_j \equiv N/N_2$; we refer to this as the balanced *nesting* design. More generally, we may consider identical distributions (or sets of values) of some or all the explanatory variables within the clusters (balanced *regression* design). An example of balanced regression design is the complete dataset from a longitudinal study with rectangular design, i.e., when each subject is observed at each of a fixed set of time-points.

An important property of the regression balanced design is that the OLS and ML estimators coincide. Let $\mathbf{e} = \mathbf{y} - \mathbf{X}\hat{\beta}_{OLS}$ denote the OLS residuals. Then $\mathbf{X}^\top \mathbf{e} = \mathbf{0}$, and the scoring vector for the regression parameters, evaluated at $\hat{\boldsymbol{\beta}}_{OLS}$, is

$$\mathbf{X}^\top \mathbf{V}^{-1} \mathbf{e} = -\frac{\sigma^2}{g} \sum_j \mathbf{X}_j^\top \mathbf{1}_n \mathbf{1}_n^\top \mathbf{e}_j, \tag{2.48}$$

where $g = 1 + n\omega$ is the common value of all g_j, $j = 1, \ldots, N_2$, and $\mathbf{e}_j$ is the segment of $\mathbf{e}$ corresponding to cluster j. When the design is balanced with respect to all the explanatory variables, $\mathbf{X}_j^\top \mathbf{1}_n$ is constant for all j, and so (2.48) is equal to $\mathbf{0}$. Therefore, the OLS solution is also the ML (and the REML) solution.

Similarly, for cluster-level variables in a nesting balanced design we have $\mathbf{X}_j^\top \mathbf{1}_n = n\mathbf{x}_j$, and so the scoring vector for $\boldsymbol{\beta}$ is

$$\mathbf{X}^\top \mathbf{V}^{-1} \mathbf{e} = \frac{n}{g} \sum_j \mathbf{x}_j \mathbf{1}_n^\top \mathbf{e}_j = \frac{n}{g} \mathbf{X}^\top \mathbf{e} = \mathbf{0}.$$

Estimation of the variance components in the regression balanced design is also simplified. The scoring equation for ω, (2.31), implies that

$$N\hat{g}\hat{\sigma}^2 = \sum_j e_j^2 \tag{2.49}$$

($e_j = \mathbf{e}_j^\top \mathbf{1}_{n_j}$ and $\hat{g} = 1 + n\hat{\omega}$), and the equation for σ^2, (2.25), implies that

$$N\hat{g}\hat{\sigma}^2 = \hat{g}\mathbf{e}^\top \mathbf{e} - \hat{\omega} \sum_j e_j^2. \tag{2.50}$$

By equating the right-hand sides of (2.49) and (2.50) we obtain

$$\hat{\omega} = \frac{\sum_j e_j^2 - \mathbf{e}^\top \mathbf{e}}{n\mathbf{e}^\top \mathbf{e} - \sum_j e_j^2} \tag{2.51}$$

and

$$\hat{\sigma}^2 = \frac{\sum_j e_j^2}{N\hat{g}}. \tag{2.52}$$

Note that the numerator of (2.52) is the n-multiple of the within-cluster sum of squares of the residuals. It is non-negative, and is equal to zero almost surely if and only if $\hat{\sigma}^2 = 0$. The estimators (2.51) and (2.52), with $\bar{n} = N/N_2$ instead of n, can be used as a starting solution for the variances in an iterative estimation procedure, even in an unbalanced case.

2.8   The price of ignoring clustering

Between-cluster variation is often a nuisance feature in the data. It is reasonable to ask, then, whether we could get away without fitting the random-effects model for a two-level dataset, and still obtain estimates of the regression parameters with almost full efficiency. The answer depends on the actual value of the unknown between-cluster variance. Often only extreme values of the between-cluster variance can be excluded *a priori* as infeasible, and so the answer is at best incomplete. Nevertheless, being aware of this dependence is useful.

If the model (2.1) holds, then both OLS and ML estimators of the regression parameters are unbiased:

$$E\left\{\left(\mathbf{X}^\top\mathbf{X}\right)^{-1}\mathbf{X}^\top\mathbf{y}\right\} = \beta,$$

$$E\left\{\left(\mathbf{X}^\top\mathbf{V}^{-1}\mathbf{X}\right)^{-1}\mathbf{X}^\top\mathbf{V}^{-1}\mathbf{y}\right\} = \beta.$$

(2.53)

The latter estimator, in which $\mathbf{W} = \sigma^{-2}\mathbf{V}$ can be used instead of $\mathbf{V}$, assumes that the variance ratio ω is known. The variances of these estimators are

$$\mathrm{var}(\hat{\beta}_{OLS}) = \left(\mathbf{X}^\top\mathbf{X}\right)^{-1}\mathbf{X}^\top\mathbf{V}\mathbf{X}\left(\mathbf{X}^\top\mathbf{X}\right)^{-1}$$

$$= \sigma^2\left(\mathbf{X}^\top\mathbf{X}\right)^{-1}\left\{\mathbf{I} + \omega\sum_j\mathbf{U}_j\left(\mathbf{X}^\top\mathbf{X}\right)^{-1}\right\}$$

(2.54)

and

$$\mathrm{var}(\hat{\beta}_{ML}) = \left(\mathbf{X}^\top\mathbf{V}^{-1}\mathbf{X}\right)^{-1} = \sigma^2\left(\mathbf{X}^\top\mathbf{X} - \omega\sum_j g_j^{-1}\mathbf{U}_j\right)^{-1},$$

(2.55)

where $\mathbf{U}_j = \mathbf{X}_j^\top\mathbf{J}_{n_j}\mathbf{X}_j = n_j^2\bar{\mathbf{x}}_j\bar{\mathbf{x}}_j^\top$. We say that a (vector) estimator $\hat{\beta}_A$ is more efficient than another estimator $\hat{\beta}_B$ of the same parameter vector β if $\mathbf{c}^\top\hat{\beta}_A$ is a more efficient estimator of $\mathbf{c}^\top\beta$ than $\mathbf{c}^\top\hat{\beta}_B$ for any non-zero vector $\mathbf{c}$ of the same length as β. For unbiased estimators, efficiency is essentially a comparison of the sampling variances of the estimators. Let $\mathbf{U}_A$ and $\mathbf{U}_B$ be the respective sampling variance matrices of the unbiased estimators of β, $\hat{\beta}_A$, and $\hat{\beta}_B$. Then $\mathbf{c}^\top\hat{\beta}_A$ is more efficient than $\mathbf{c}^\top\hat{\beta}_B$ for all vectors $\mathbf{c}$ if and only if

$$\mathbf{c}^\top(\mathbf{U}_B - \mathbf{U}_A)\mathbf{c} > 0,$$

which is equivalent to the condition that $\mathbf{U}_A - \mathbf{U}_B$ be positive definite.

We now explore the expression $\mathbf{B} = \text{var}(\hat{\boldsymbol{\beta}}_{OLS})\left\{\text{var}(\hat{\boldsymbol{\beta}}_{ML})\right\}^{-1} - \mathbf{I}$; it is positive definite if and only if $\hat{\boldsymbol{\beta}}_{ML}$ is a more efficient estimator than its OLS counterpart $\hat{\boldsymbol{\beta}}_{OLS}$. By substituting (2.54) and (2.55) we obtain

$$
\begin{aligned}
\mathbf{B} &= \omega \left(\mathbf{X}^\top \mathbf{X}\right)^{-1} \sum_j \mathbf{U}_j - \omega \left(\mathbf{X}^\top \mathbf{X}\right)^{-1} \sum_j g_j^{-1} \mathbf{U}_j \\
&\quad - \omega^2 \left(\mathbf{X}^\top \mathbf{X}\right)^{-1} \sum_j \mathbf{U}_j \left(\mathbf{X}^\top \mathbf{X}\right)^{-1} \sum_j g_j^{-1} \mathbf{U}_j \\
&= \omega^2 \left(\mathbf{X}^\top \mathbf{X}\right)^{-1} \left\{ \sum_j n_j g_j^{-1} \mathbf{U}_j - \sum_j \mathbf{U}_j \left(\mathbf{X}^\top \mathbf{X}\right)^{-1} \sum_j g_j^{-1} \mathbf{U}_j \right\}.
\end{aligned}
$$

(2.56)

Discussion of this expression for the general case is difficult, but for balanced data $(n_j \equiv n)$ we have

$$
\mathbf{B} = n^2 \omega^2 g^{-1} \left(\mathbf{X}^\top \mathbf{X}\right)^{-1} \mathbf{T}_W \left(\mathbf{X}^\top \mathbf{X}\right)^{-1} \mathbf{T}_B, \qquad (2.57)
$$

where $\mathbf{T}_B = n^{-1} \sum_j \mathbf{U}_j$ and $\mathbf{T}_W = \mathbf{X}^\top \mathbf{X} - \mathbf{T}_B$ are the respective matrices of between- and within-cluster sums of squares and cross-products, and $g = 1 + n\omega$ is the common value of the factors g_j. Since all the matrix factors in (2.57) are non-negative definite, so is $\mathbf{B}$.

The identity (2.57) also implies that $\mathbf{B} = \mathbf{0}$ when $\omega = 0$, and is singular when the design matrix $\mathbf{X}$ contains a cluster-level variable. If the design matrix contains only cluster-level variables, $\mathbf{B} = \mathbf{0}$. Then the estimators $\hat{\boldsymbol{\beta}}_{OLS}$ and $\hat{\boldsymbol{\beta}}_{ML}$ not only have equal variances but they coincide, as shown in Section 2.7. In general, for an unbalanced design $\hat{\boldsymbol{\beta}}_{OLS}$ and $\hat{\boldsymbol{\beta}}_{ML}$ are not identical, although the difference is likely to be substantial only in extremely unbalanced designs.

Unless equal to zero, each element of the difference matrix $\mathbf{B}$ is an increasing function of ω; higher between-cluster variation is associated with greater loss of efficiency of $\hat{\boldsymbol{\beta}}_{OLS}$. To illustrate the dependence of $\mathbf{B}$ in (2.57) on the common within-cluster sample size n we consider the following situation. Suppose the design matrix for each cluster j is replicated h times, so that the new design matrices are $hn \times p$ matrices $\mathbf{X}_j^* = \left(\mathbf{X}_j^\top, ..., \mathbf{X}_j^\top\right)^\top = \mathbf{X}_j \otimes \mathbf{1}_h$. Then the corresponding matrices $\mathbf{B}_n$ and $\mathbf{B}_{hn}$ differ by the multiplicative scalar

$$
\frac{h^2(1 + n\omega)}{1 + hn\omega},
$$

which lies between h and h^2. Thus efficiency of $\hat{\boldsymbol{\beta}}_{OLS}$ is most problematic for data with large clusters, unless $\omega = 0$.

Similarly, we can discuss the bias of the estimator of the standard error of $\hat{\boldsymbol{\beta}}_{OLS}$. To simplify the algebra, we compare the information matrix with the inverse of the true variance. The former is $\sigma_{OLS}^{-2}\mathbf{X}^\top\mathbf{X}$, while the latter is given by (2.54). Here we have to distinguish between the elementary-level variance σ^2 and the 'OLS' residual variance σ_{OLS}^2, which is estimated by OLS.

When the regression parameters $\boldsymbol{\beta}$ are known, the OLS estimator of the residual variance is $\hat{\sigma}_{OLS}^2 = N^{-1}\mathbf{e}^\top\mathbf{e}$, and its expectation is

$$\mathbf{E}(\hat{\sigma}_{OLS}^2) \;=\; \frac{1}{N}\mathbf{1}_N^\top\mathbf{V}\mathbf{1}_N \;=\; \sigma^2 + \frac{1}{N}\sum_j n_j^2\tau^2 \;=\; \frac{\sigma^2}{N}\sum_j n_jg_j. \qquad (2.58)$$

The variance estimated by the OLS method depends on the clustering design. It is equal to $\sigma^2 + \tau^2$ when each cluster sample consists of a single elementary observation, and to σ^2 when the sample contains a single cluster. We denote by σ_{OLS}^2 the expected variance (2.58). The 'nominal' variance matrix of $\hat{\boldsymbol{\beta}}_{OLS}$ (assuming that the observations are independent) is $\mathbf{H}_{OLS} = \sigma_{OLS}^2\left(\mathbf{X}^\top\mathbf{X}\right)^{-1}$, and the variance matrix for $\hat{\boldsymbol{\beta}}_{ML}$, denoted by $\mathbf{H}_{ML}$, is given by (2.55). We have

$$\mathbf{H}_{OLS}\mathbf{H}_{ML}^{-1} \;=\; \frac{1}{N}\sum_j n_jg_j\left\{\mathbf{I}_p - \omega\left(\mathbf{X}^\top\mathbf{X}\right)^{-1}\sum_j g_j^{-1}\mathbf{U}_j\right\}^{-1}. \qquad (2.59)$$

This is a non-negative definite matrix since $\omega < g_j$ and $\mathbf{I} - \left(\mathbf{X}^\top\mathbf{X}\right)\sum_j\mathbf{U}_j$ is non-negative definite. The bias of the estimator of the sampling variance of $\hat{\boldsymbol{\beta}}_{OLS}$ depends on two factors: the decomposition of the total sum of squares and cross-products into their within- and between-cluster components, and the variance ratio ω.

2.9   Cluster size and information

When setting up the clustering design for an experiment or an observational study it is important to ensure that the resulting sample contains adequate information about the between-cluster variance. A standard approach to this issue is based on the analysis of the expected information matrix $\mathbf{H}$ for the estimated parameters as a function of the nesting design. For the random-effects model this information matrix is block-diagonal, with a block each corresponding to the regression and variation parameters. Although the information matrix depends on the variances σ^2 and τ^2, it is independent of the regression parameters $\boldsymbol{\beta}$. Also, the submatrix of

$\mathbf{H}$ corresponding to (σ^2, τ^2) does not depend on the regression design $\mathbf{X}$. This greatly simplifies discussion of the relationship of the design and the precision of estimation.

In Section 2.8 we established that the OLS estimator of the regression parameters is less efficient than its ML counterpart. Hence, the predicted sampling variances of the OLS estimators, the diagonal elements of $\sigma^2_{OLS} \left(\mathbf{X}^\top \mathbf{X} \right)^{-1}$, are upper bounds for the predicted sampling variances of the ML estimators. Prediction based directly on the information matrix, $\mathbf{X}^\top \mathbf{V}^{-1} \mathbf{X}$, requires knowledge of the between-cluster variance τ^2 and of the within-cluster summaries $\mathbf{X}_j^\top \mathbf{1}$. Such summaries are rarely available, though.

For balanced data, substantial simplification takes place. Let $\mathbf{T}_W$ and $\mathbf{T}_B$ be the decomposition of the matrix of total sums of squares and cross-products of the explanatory variables, $\mathbf{X}^\top \mathbf{X} = \mathbf{T}_W + \mathbf{T}_B$, where

$$\mathbf{T}_W = \sum_j (\mathbf{X}_j - \mathbf{1}_{n_j} \bar{\mathbf{x}}_j)^\top (\mathbf{X}_j - \mathbf{1}_{n_j} \bar{\mathbf{x}}_j),$$

$$\mathbf{T}_B = \sum_j n_j \bar{\mathbf{x}}_j \bar{\mathbf{x}}_j^\top$$

($\bar{\mathbf{x}}_j$ is the row-vector of means of the variables in cluster j). Then

$$\mathbf{X}^\top \mathbf{V}^{-1} \mathbf{X} = \frac{1}{\sigma^2} \left(\mathbf{X}^\top \mathbf{X} - \frac{n\omega}{g} \mathbf{T}_B \right) = \frac{1}{\sigma^2} \left(\mathbf{T}_W + \frac{1}{g} \mathbf{T}_B \right), \qquad (2.60)$$

and so the information matrix depends on the data only through the within- and between-cluster summaries $\mathbf{T}_W$ and $\mathbf{T}_B$. When n $(n\omega)$ is small, $\mathbf{X}^\top \mathbf{V}^{-1} \mathbf{X} \approx \sigma^{-2} \mathbf{X}^\top \mathbf{X}$. Taking clustering into account has little effect on the information, and therefore also on the standard errors of the regression parameters. If an unbalanced dataset is to be collected, approximations to the sampling variances can be obtained by evaluating (2.60) for a balanced dataset with the average planned cluster size. Decisions about the cluster size n can be based on the requirement that (certain) diagonal elements of (2.60) be greater than preset values, although it would be more appropriate to impose similar conditions on the inverse of (2.60), the form of which is less tractable.

The information matrix for the variance parameters (σ^2, τ^2) is

$$\frac{1}{2\sigma^4} \begin{pmatrix} N - N_2 + \sum_j g_j^{-2} & \sum_j n_j g_j^{-2} \\ \sum_j n_j g_j^{-2} & \sum_j n_j^2 g_j^{-2} \end{pmatrix}. \qquad (2.61)$$

Its elements are obtained by substituting the identities (2.22) and (2.28) in (2.20). For example, the element (σ^2, σ^2) of (2.61) is

$$
\frac{1}{2}\sum_j \text{tr}(\mathbf{V}_j^{-2}) = \frac{1}{2\sigma^4} \sum_j \text{tr}\left(\mathbf{I}_{n_j} - 2\frac{\omega}{g_j}\mathbf{1}_{n_j}\mathbf{1}_{n_j}^\top + \frac{\omega^2}{g_j^2}\mathbf{1}_{n_j}\mathbf{1}_{n_j}^\top\mathbf{1}_{n_j}\mathbf{1}_{n_j}^\top\right)
$$

$$
= \frac{1}{2\sigma^4}\left\{N - N_2 + \sum_j\left(1 - 2\frac{n\omega}{g_j} + \frac{n^2\omega^2}{g_j^2}\right)\right\}
$$

$$
= \frac{1}{2\sigma^4}\left(N - N_2 + \sum_j g_j^{-2}\right).
$$

The information matrix (2.61) is positive definite unless each cluster contains a single unit, $n_j \equiv 1$. Discussion of the information matrix (2.61) and its inverse for the general nesting design is not feasible, and so we focus on the balanced case. The determinant of (2.61) in the nesting balanced case is

$$
\frac{1}{4\sigma^8}(N - N_2)Nng^{-2},
$$

and the asymptotic sampling variance matrix for (σ^2, τ^2) is

$$
\frac{2\sigma^4}{(N - N_2)n}\left(\begin{array}{cc} n & -1 \\ -1 & \frac{n-1}{n}g^2 + \frac{1}{n} \end{array}\right).
$$

The asymptotic sampling variance of $\hat{\sigma}^2_{ML}$ is $2\sigma^4/(N - N_2)$. In ordinary regression the sampling variance of $\hat{\sigma}^2_{OLS}$ is $2\sigma^4_{OLS}/(N - 1)$. Thus estimating τ^2 involves inflation of the sampling variance of $\hat{\sigma}^2$ equivalent to the loss of about one observation per cluster. Such a loss of efficiency is of importance only in datasets with small cluster sizes where, incidentally, taking account of between-cluster variation is least important since OLS is then quite efficient. Note that $\text{var}(\hat{\sigma}^2_{ML})$ does not depend on ω.

The elementary-level variance σ^2 is often regarded as a nuisance parameter, and the focus of inference is on the between-cluster variance τ^2 (or ω). The asymptotic variance of $\hat{\tau}^2$ is

$$
\text{var}(\hat{\tau}^2) = \frac{2\sigma^4}{N}\left\{\frac{1}{n-1} + 2\omega + n\omega^2\right\}. \tag{2.62}
$$

When collection of data for each observation is associated with a fixed amount of financial and other expenditures, it is useful to consider the

nesting design which minimizes the variance (2.62) subject to the constraint
of fixed sample size N. The optimal cluster size n can easily be determined
by minimizing the function $f(n) = N\mathrm{var}(\hat{\tau}^2)/2\sigma^4$, which depends on N_2
only through n. By setting the derivative of $f(n)$ to zero,

$$- (n-1)^{-2} + \omega^2 = 0, \qquad (2.63)$$

we obtain the unique positive solution $n^* = 1 + \omega^{-1}$. Thus, when ω is
small, large clusters are preferable. Also, the form of the derivative in
(2.63) implies that, for small ω, $f(n)$ is very flat around n^*. Little efficiency
is lost by choosing a sample size other than the nearest integer to n^* but
not too distant from it. On the other hand, for very large ω (e.g., $\omega > 2$),
any observation after the second one in a cluster is of little value for estimating τ^2. Of course, in practice several possibly conflicting requirements
on sampling variances have to be reconciled, but sampling variation for
the regression parameters usually depends much less on the nesting design
(given N) than does the sampling variance of $\hat{\tau}^2$.

A more realistic scheme assumes that the cost of data collection is a
linear function of the number of clusters and the number of elementary
units, i.e., accessing each cluster involves expenditure in excess of the expenditure associated with each elementary unit. Then the optimal nesting design is given by the minimum of $\mathrm{var}(\hat{\tau})$ subject to the constraint
$N + cN_2 = C$, where c is the relative cost of accessing a cluster, and C the
relative available funds, expressed on a scale for which accessing an individual is associated with unit cost. This constrained maximization task does
not have a closed-form solution, but the minimum of $\mathrm{var}(\hat{\tau})$ can be found
by plotting its values for appropriate combinations of cluster and sample
sizes.

When σ^2 is known,

$$\mathrm{var}(\hat{\tau}^2) = \frac{2\sigma^4 g^2}{Nn}, \qquad (2.64)$$

and its minimum, subject to fixed N, is attained for $n^* = \omega^{-1}$. For small ω,
such as $\omega < 0.1$, the difference of variances (2.64) for n^* and $n^* + 1$ is very
small, and so whether σ^2 is known or not is of little importance for setting a
suitable nesting design. For $\omega > 1$ the optimality criterion, assuming known
σ^2, leads to single-observation clusters. This is not a counterintuitive result; the variance τ^2 can be estimated as the variance of the independent
observations (one in each cluster) in excess of the elementary-level variance
σ^2.

Equations (2.62) and (2.64) offer a straightforward assessment of the
effect of knowing σ^2 on the efficiency of estimation of τ^2. The difference of
these sampling variances is $2\sigma^4/\{Nn(n-1)\}$; knowing σ^2 is of diminishing
utility when σ^2 is small and sample and cluster sizes are large.

2.10 Residuals. Model checking

In ordinary regression, formal and informal procedures for assessing the appropriateness of the fitted model rely on the vector of residuals $\hat{\mathbf{e}} = \mathbf{y} - \mathbf{X}\hat{\boldsymbol{\beta}}$. A simple (and naïve) description of such a procedure is as follows. Having fitted an ordinary regression model by OLS we regard the (fitted) residuals $\hat{\mathbf{e}}$ for the true (model) deviations $\boldsymbol{\varepsilon} = \mathbf{y} - \mathbf{X}\boldsymbol{\beta}$. According to the adopted model, this vector should be a random sample from the univariate normal distribution, and it should not be related to any of the explanatory variables. In models for clustered observations, each elementary observation and each cluster is associated with univariate random terms, and these cannot be recovered even when the regression parameters are known exactly.

If all the parameters of the model (2.1) were known, our information about each random variable δ_j could be summarized by its conditional distribution given the vector of outcomes $\mathbf{y}$,

$$(\delta_j \mid \mathbf{y}; \boldsymbol{\beta}, \sigma^2, \omega) \sim \mathcal{N}\left(\omega g_j^{-1} \mathbf{e}_j^\top \mathbf{1}_{n_j}, \; \sigma^2 \omega g_j^{-1}\right). \tag{2.65}$$

We regard the conditional expectation of δ_j as the *cluster-level residual* and denote it by $\hat{\delta}_j$. The conditional distribution for the vector of elementary-level random terms $\boldsymbol{\varepsilon}_j$ is

$$(\boldsymbol{\varepsilon}_j \mid \mathbf{y}; \boldsymbol{\beta}, \; \sigma^2, \tau^2) \sim \mathcal{N}\left(\sigma^2 \mathbf{V}_j^{-1} \mathbf{e}_j, \; \sigma^4 \mathbf{V}_j^{-2}\right). \tag{2.66}$$

The conditional expectation of $\boldsymbol{\varepsilon}_j$, denoted by $\hat{\boldsymbol{\varepsilon}}_j$, can be used as the vector of *elementary-level residuals*. Note that

$$\hat{\boldsymbol{\delta}}_j^\top \mathbf{1}_{n_j} + \hat{\boldsymbol{\varepsilon}}_j \; = \; \mathbf{e}_j, \tag{2.67}$$

which simplifies calculation of the elementary-level residuals.

In practice, the conditional expectations $\{\hat{\delta}_j\}$ and $\{\hat{\varepsilon}_j\}$ are themselves estimated by substituting the maximum likelihood estimates in place of the parameters $(\boldsymbol{\beta}, \sigma^2, \omega)$ in (2.65). This is justified when the standard errors of the estimated parameters are small, otherwise the conditional variances are downward biased because they ignore this additional uncertainty.

It is instructive to regard the model (2.1) as two related regression models, one for each level of clustering. Since we have the corresponding sets of residuals, we also have to consider two kinds of possible outliers: clusters and elementary units.

Detection of outliers in the random-effects model creates no additional difficulties; for each set of residuals the usual methods of inspection of the residuals are applicable. Note, however, that the decomposition of each residual e_{ij} into its within- and between-cluster components involves an additional element of uncertainty.

2.10.1 SHRINKAGE ESTIMATORS

In a slightly different perspective we may be interested in the residuals $\hat{\delta}_j$ as *estimators* of the deviation of the adjusted cluster-means from the population mean, or, more generally, as estimators of the departure of each cluster from the pooled regression $\mathbf{x}\boldsymbol{\beta}$. The conditional expectation of δ_j is the estimator of the realized value of δ_j with the minimal mean squared error. The conditional variance $g_j^{-1}\tau^2$ can then be used as a measure of uncertainty of this estimator. Note that when $\tau^2 > 0$ the conditional variance is smaller than τ^2, even for clusters with only a single observation.

In the standard definition, estimation is associated with an unknown *parameter*. This is not in conflict with apparently estimating *random* variables; the conditional expectations $\{\hat{\delta}_j\}$ are estimates of the realized values of the random terms $\{\delta_j\}$.

The conditional expectation $\hat{\delta}_j$ is closely related to shrinkage estimators. Let $\hat{\boldsymbol{\beta}} = \begin{pmatrix} \hat{\beta}^{(1)} \\ \hat{\boldsymbol{\beta}}^{(x)} \end{pmatrix}$ be the partitioning of the MLE of $\boldsymbol{\beta}$ into its intercept and subvector of the slopes. Then $\hat{\beta}_j^{(1)} = \hat{\beta}^{(1)} + \hat{\delta}_j$ is a natural estimator of the intercept $\beta_j^{(1)}$ for cluster j. Since the intercept term $\mathbf{1}$ is part of the regression design, it is a linear combination of the columns of $\mathbf{X}\mathbf{V}^{-1}$ for any $N \times N$ non-singular matrix $\mathbf{V}$. The equation

$$\mathbf{X}^\top \mathbf{V}^{-1}(\mathbf{y} - \mathbf{X}\hat{\boldsymbol{\beta}}) = \mathbf{0}$$

then implies

$$\hat{\beta}^{(1)} = \bar{y} - \bar{\mathbf{x}}\hat{\boldsymbol{\beta}}^{(x)}$$

($\bar{y}$ and $\bar{\mathbf{x}}$ are the sample means for y and $\mathbf{x}$ respectively). Hence

$$\hat{\beta}^{(1)} + \hat{\delta}_j = \bar{y} - \bar{\mathbf{x}}\hat{\boldsymbol{\beta}}^{(x)} + \frac{\hat{g}_j - 1}{\hat{g}_j}(\bar{y}_j - \bar{\mathbf{x}}_j\hat{\boldsymbol{\beta}})$$

$$= \frac{\hat{\beta}^{(1)}}{\hat{g}_j} + \frac{\hat{g}_j - 1}{\hat{g}_j}\left\{\bar{y}_j - \bar{y} - (\bar{\mathbf{x}}_j - \bar{\mathbf{x}})\hat{\boldsymbol{\beta}}^{(x)}\right\} \qquad (2.68)$$

($\bar{y}_j$ and $\bar{\mathbf{x}}_j$ are the within-cluster sample means of y and $\mathbf{x}$, respectively, and $\hat{g}_j = 1 + n_j\hat{\omega}$). This is a shrinkage estimator that combines the within-cluster estimator of the intercept, $\bar{y}_j - \bar{y} - (\bar{\mathbf{x}}_j - \bar{\mathbf{x}})\hat{\boldsymbol{\beta}}^{(x)}$, with the the MLE $\hat{\beta}^{(1)}$. The weight given to the 'pooled' estimator $\hat{\beta}^{(1)}$ decreases with the product of the cluster size and the (estimated) variance ratio.

2.11   Measures of quality of the model fit

Reduction of the residual variance to as small a quantity as possible is a goal underlying many an ordinary regression analysis. It is perhaps bad practice to make it an overriding priority, but we associate explanatory variables that have contributed to the reduction of the residual variance with 'explanatory power'. In the random-effects models we have two variances, and if both of them were reduced to zero the adopted model would describe ('explain') the differences within and between clusters perfectly.

In ordinary regression, addition of an explanatory variable reduces, or leaves unchanged, the residual sum of squares on which the estimator of the residual variance is based. The 'proportion of variation explained' is frequently used as a measure of the reduction of the variance of the outcomes by conditioning on the explanatory variables. Its definition is

$$R^2 = 1 - \frac{(N-1)\hat{\sigma}_p^2}{(N-p)\hat{\sigma}_0^2}, \qquad (2.69)$$

where $\hat{\sigma}_0^2$ is the *raw* estimated variance of y, and $\hat{\sigma}_p^2$ is the estimated residual variance in the adopted model (with p regression parameters).

A natural extension of this definition for the random-effects model is the pair of proportions, referring to each of the levels:

$$R_1^2 = 1 - \frac{\hat{\sigma}_0^2}{\hat{\sigma}_p^2}, \qquad R_2^2 = 1 - \frac{\hat{\tau}_0^2}{\hat{\tau}_p^2}, \qquad (2.70)$$

where the subscript 0 refers to the estimators for the *empty model*

$$y_{ij} = \mu + \delta_j + \varepsilon_{ij}, \qquad (2.71)$$

the subscript p to the adopted model (with p regression parameters), and all the estimators are maximum likelihood. If REML estimators are available, the appropriate adjustment by the respective factors $(N-p)/(N-1)$ and $(N_2 - p)/(N_2 - 1)$ is necessary.

The proportion R^2 for the ordinary regression is always in the interval $[0, 1]$. The proportions (2.70) do not have the analogous property, and so some care is necessary when interpreting them as proportions of the variation explained at each level.

2.12   Bibliographical notes

Harville (1977) contains a comprehensive review of earlier developments of random coefficient models, emphasizing direct, full and restricted maximum likelihood methods (Fisher scoring and Newton–Raphson). The definition of the EM algorithm by Dempster *et al.* (1977) diverted many researchers from the then established methods for maximum likelihood estimation. Strenio *et al.* (1983) applied the EM algorithm for a random-effects analysis relevant in animal experimentation.

Rosner (1984) applied variance component methods for problems in ophthalmology. There is also an extensive literature on the analysis of familial data: Donner and Koval (1980), Muñoz, *et al.* (1986), or Paul (1990) are suitable entry points into the literature on this subject. Jones and Moon (1991) applied multilevel analysis in health care research. Sterne *et al.* (1988) discussed an application of variance components in dentistry.

Goldstein (1986a) is the principal reference on the iteratively reweighted least squares method. Goldstein (1989) describes an adaptation of this method for REML. These methods cater for models with complex patterns of variation dealt with in Chapters 4 and 6.

Random-effects and related methods have been applied fruitfully in small area estimation; see Fay and Herriot (1979), Malec and Sedransk (1985) and Battese *et al.* (1988) for examples.

3
Examples. Random-effects models

In this chapter random-effects methods are illustrated for the five datasets introduced in Section 2.2. The analyses are incomplete and they highlight the need for analysis with more general patterns of between-cluster variation. After the theoretical groundwork layed in Chapter 4, the same datasets are reanalysed in Chapter 5.

Throughout the chapter, statistical significance at the 5 per cent level is used.

3.1 Financial ratios

The background to this example and some of the research questions were given in Section 2.2.1. The data contain records of monetary values (in £1000 UK) of the assets and liabilities of 2004 British companies whose shares were quoted on the London Stock Exchange in 1977. The data are available to academic researchers through the ESRC Data Archive, University of Essex, England. The companies are classified into 71 industrial sectors. Three sectors are represented by one company each, and another 18 sectors are represented by ten or fewer companies. For brevity, we refer to the number of companies in a sector as the *sector size*. Table 3.1 contains the stem-and-leaf plot of the sector sizes. The distribution of the sector sizes is severely skewed to the right. The assets and liabilities are plotted in Figure 3.1; in the left-hand panel the original scale and in the right-hand panel the logarithmic scale are used. The logarithms of these variables (log-assets and log-liabilities) appear to be linearly related, with the variance of observations a decreasing function of the log-assets.

Since there are several sectors which have substantial sector sizes, it is worth while fitting ordinary regressions for some of them and comparing the estimated regression parameters informally. The four largest sectors are 8 (clothing), 17 (department stores), 61 (laundries and cleaners), and 37 (conglomerates), with respective sector sizes 119, 118, 97, and 94. The plots of log-assets and log-liabilities, with the fitted regressions, plots of the residuals against the log-assets, histograms of the residuals, and quantile

Table 3.1. Stem-and-leaf plot of the sector sizes (numbers of companies in the sectors) in financial ratios data. The stems are in multiples of 10. The median sector size is 19, the mean is 28

0	:	1 1 1 2 3 4 7 7 8 8 8 8 9 9 9 9
1	:	0 0 0 0 0 1 2 3 5 5 5 6 6 7 7 8 9 9 9 9
2	:	0 1 1 1 3 4 5 5 6 9
3	:	0 1 1 2 4 7 9 9
4	:	2 2 4 4 7 7
5	:	3 6 8 9
6	:	2 4 5
7	:	
8	:	
9	:	4 7
10	:	
11	:	8 9

plots of the residuals for sectors 8 and 17 are displayed in Figures 3.2 and 3.3. Scatterplots of sectors 37 and 61 are given in the top panels of Figure 3.4 (the scatterplots of the other sectors in the diagram are referred to later in this section). The ordinary regression summaries for the four largest sectors are given in Table 3.2. Of principal interest is the proximity of the regression slopes to 1, because, then, the log-ratios

$$\log \left(\frac{liabilities}{assets} \right)$$

are deviations from the average log-ratio and are independent of the assets. The distribution of the log-ratios then provides a simple description for the companies in the sector. For example, a company with a log-ratio in the right tail of this distribution would be interpreted as having excessive liabilities, and, consequently, financial institutions may give it a low credit rating.

The simple, within-sector regressions appear to fit well, although there is an obvious outlier in each of the four largest sectors except sector 37. The estimated slopes on log-assets are very close to unity, and are not significantly different from it (the corresponding t-ratios are smaller than 1.0). Also, the differences between the estimated residual variances are insubstantial.

Based on these four fitted regressions the random-effects model

Table 3.2. Regression model fits for the largest sectors in financial ratios data

Sector no.	Sector size	Fitted within-sector regressions			
		Intercept	Slope	(St. error of slope)	Residual variance
8	119	−0.482	1.016	(0.024)	0.098
17	118	−0.549	0.997	(0.015)	0.068
61	97	−0.372	0.991	(0.019)	0.076
37	94	−0.230	0.984	(0.020)	0.094

$$y_{ij} \;=\; a + x_{ij}b + \delta_j + \varepsilon_{ij}, \tag{3.1}$$

$\delta_j \sim \mathcal{N}(0, \tau^2)$, i.i.d., appears to be reasonable, and we would expect the sector-level variance to be of the order 10^{-2}.

To curb the optimism generated by these observations consider sector 36 (bricks and roofing tiles), which has 53 companies. Its plot of log-assets and log-liabilities is drawn in Figure 3.4. There is a clear pattern of variance

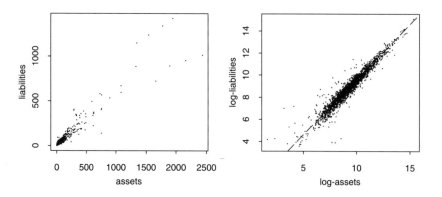

Fig. 3.1. Plots of the current liabilities and assets. Financial ratios data. The left-hand panel contains the plot of liabilities and assets on the original scale (the units are £1 million) and the right-hand panel contains the plot of the logarithms (the units are $\log(£1000)$). The largest company in the data, with assets of £4.57×10⁹ and liabilities £3.51×10⁹, is excluded from the plot in the left-hand panel so as to improve the resolution of the plot. The broken line in the right-hand panel, $y = x - 0.4$, is a reasonable fit to the data with unit slope.

heterogeneity; the variance of an observation is a decreasing function of log-assets. The fitted ordinary regression is

$$0.536 + 0.908x \quad (\hat{\sigma}^2 = 0.378). \tag{3.2}$$

The estimated regression slope is much smaller than unity (its standard error is 0.040), possibly caused by a set of influential observations at the lower end of the scale of log-assets. The estimate of the variance is much larger than the estimated variances for the four largest sectors; compare with Table 3.2.

For the moment we ignore these problems and proceed with fitting the random-effects model (3.1). We return to the issues of model inappropriateness later. The pooled ordinary regression fit is

$$-0.102 + 0.963x \quad (\hat{\sigma}^2 = 0.200). \tag{3.3}$$

We choose the starting solution for the within- and between-sector variances as 0.1 each, that is, the starting solution for the variance ratio is 1.0. The subsequent iterations of the Fisher scoring algorithm change the regression slope estimate only slightly (the changes in the intercept are

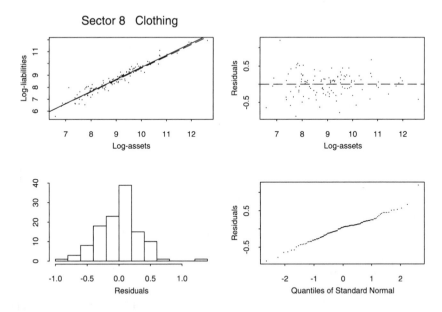

Fig. 3.2. Regression diagnostics for sector 8 (clothing), 119 companies. Financial ratios data. The panels contain the plot of log-sales and log-assets (with the OLS fit drawn by a solid line, and $x - 0.4$ drawn by a dashed line), plot of the OLS residuals and the log-assets, histogram of the residuals, and the quantile plot of the residuals.

Table 3.3. Iterations of the Fisher scoring algorithm for financial ratios data. Iteration 0 stands for the starting solution

	Fisher scoring iterations			
Iteration	Regression		Within-sector	Variance
	intercept	slope	variance	ratio
0	−0.101	0.963	0.100	1.000
1	−0.229	0.974	0.168	0.327
2	−0.220	0.973	0.170	0.181
3	−0.210	0.972	0.172	0.179
4	−0.211	0.972	0.172	0.176
5	−0.211	0.972	0.172	0.176
St. error		(0.0062)		(0.039)

of lesser interest). The estimate of the between-sector variance is becoming much smaller than the initial guess. Four iterations of the algorithm

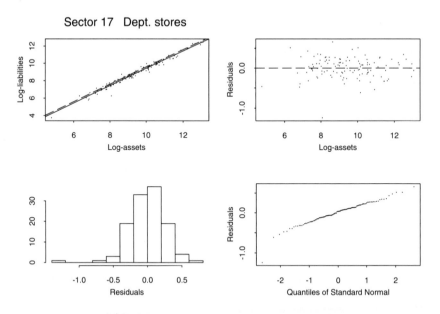

Fig. 3.3. Regression diagnostics for sector 17 (department stores), 118 companies. Financial ratios data. The layout is the same as in Figure 3.2.

are sufficient to achieve convergence; the progression of the intermediate
estimates is displayed in Table 3.3.

The estimate of the between-sector variance is $0.172 \times 0.176 = 0.0304$;
the standard deviation of the within-sector intercepts is $\sqrt{0.0304} = 0.174$.
The standard errors associated with these estimates are 0.039 (for the variance ratio), 0.0068 (for the between-sector variance), and 0.0195 (for the
between-sector standard deviation). They are obtained from the fitted
information matrix and by application of the chain rule.

The t-ratios for the between-sector variance and for the standard deviation differ a great deal. But there is strong evidence that between-sector
variation is an important feature of the data. Also, the likelihood ratio
test statistic for the comparison of the pooled ordinary regression with
the random-effects model fit is $2457.8 - 2274.3 = 183.5$, supporting this
conclusion.

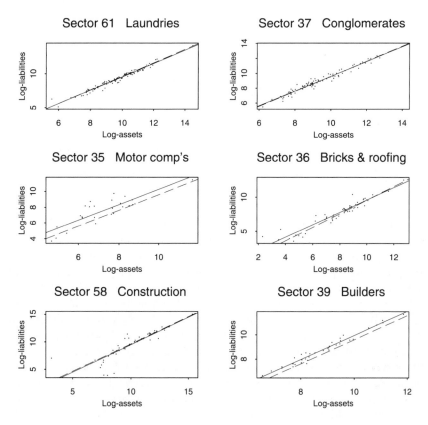

Fig. 3.4. Scaterplots for some of the sectors. Financial ratio data. The OLS
fits are drawn by solid lines, and $x - 0.4$ by a dashed line.

Table 3.4.   Order and ranks of the sector-level residuals. Financial ratios data

	Order									
Ranks	**Sector**									
1–10	52	14	21	6	34	53	18	7	13	10
11–20	50	49	26	47	16	24	28	11	70	17
21–30	42	43	9	22	12	56	29	1	48	45
31–40	46	58	32	2	66	55	30	23	27	51
41–50	20	4	54	38	68	19	41	61	67	57
51–60	15	5	40	69	3	44	65	37	8	64
61–70	62	25	36	63	31	71	33	60	59	39
71	35									
Sector	**Ranks**									
1–10	28	34	55	42	52	4	8	59	23	10
11–20	18	25	9	2	51	15	20	7	46	41
21–30	3	24	38	16	62	13	39	17	27	37
31–40	65	33	67	5	71	63	58	44	70	53
41–50	47	21	22	56	30	31	14	29	12	11
51–60	40	1	6	43	36	26	50	32	69	68
61–70	48	61	64	60	57	35	49	45	54	19
71	66									

The standard error for the regression slope is 0.0062. It is only 2.5–4 times smaller than the standard errors for the slopes of the four largest sectors (fitted by OLS), see Table 3.2. Unlike the sector-wise regression analyses, the hypothesis of the regression slope being equal to unity would now be rejected; the corresponding t-ratio is equal to $(1 - 0.972)/0.0062 = 4.46$. As an alternative way of testing this hypothesis, the random-effects model with the slope on log-assets constrained to unity can be fitted and the deviance of this fit compared with the model fit for estimated slope.

In order to assess the appropriateness of the adopted model we explore the conditional expectations of the random terms (the residuals). The sector-level residuals are also of considerable interest because they may indicate sectors with exceptional characteristics. Apart from plotting the sector-level residuals it is also informative to inspect the *order* and the *ranks* of the residuals. The order of the sector-level residuals is defined as the permutation of the integers 1, 2, ..., $N_2 = 71$, the kth component of which is the sector with the kth lowest residual. The ranks of the sector-level residuals are also a permutation of the indices 1, 2, ..., N_2. Its kth component is the order of sector k. The ranks and the order are mutually inverse permutations. Table 3.4 displays the order and ranks of the sector-level residuals. For example, sector 52 (newspapers and periodicals) has the lowest, and sector 35 (motor components) the highest residual; sector 1 (pumps and valves) has the 28th lowest residual, sector 2 (machine and other tools) the 34th lowest, and so on.

Figure 3.5 contains the quantile plots of the company- and sector-level residuals. Both quantile plots indicate substantial departure from normality. In particular, sector 35 has by far the highest residual, equal to 0.606; the next highest residual, 0.357, is that of sector 39 (builders merchants). Sector 35, comprising 26 companies, has a single highly influential observation with assets much higher than the rest of the companies; see Figure 3.4.

The company residuals also contain several outlying values, see the left-hand panel of Figure 3.5. Since the number of companies is much larger it is more informative to inspect only the tails of the vector of orders, say, the ten companies with the highest and lowest residuals, and associate them with their sectors, as in Table 3.5. Among the ten highest and ten lowest company residuals, eight belong to sector 58, contracting and construction (three positive and five negative residuals, out of 47 companies in the sector), and three each to sectors 35, motor components (two positive and one negative, out of 26), and 36, bricks and roofing tiles (three positive residuals out of 53). The three lowest residuals belong to companies in sector 58. Whereas sector 35 has an extreme (positive) sector-level residual, the residual for sector 36 is 0.207, the ninth largest, and the residual for sector 58 is -0.034, the 33rd smallest.

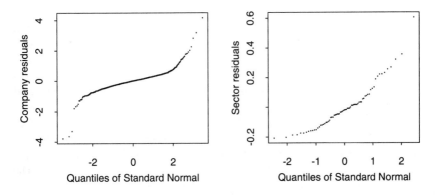

Fig. 3.5. Quantile plots for the company- and sector-level residuals. Financial ratios data. Companies with extreme residuals can be identified from Tables 3.4 and 3.5.

Table 3.5. The extreme company-level residuals in financial ratios data. The sample mean of log-liabilities is 9.21, the standard deviation is 1.59

Order	Company	(Sector)	Residual	Log-liabilites
The largest negative residuals				
1	1704	(58)	−3.80	8.54
2	1678	(58)	−3.63	7.99
3	1689	(58)	−3.30	7.69
4	1898	(64)	−1.79	9.29
5	1274	(43)	−1.66	7.95
6	1706	(58)	−1.61	7.38
7	587	(19)	−1.50	7.68
8	452	(17)	−1.28	8.07
9	1705	(58)	−1.28	7.48
10	959	(35)	−1.22	4.66
The largest positive residuals				
1995	1690	(58)	1.64	9.47
1996	1004	(36)	1.76	8.90
1997	963	(36)	1.79	3.64
1998	942	(35)	1.83	7.75
1999	949	(35)	1.90	6.59
2000	993	(36)	2.01	2.30
2001	1707	(58)	2.25	9.69
2002	1569	(55)	2.82	1.61
2003	1968	(71)	3.18	4.09
2004	1671	(58)	4.17	3.18

The plot of log-assets and log-liabilities for sector 58 is displayed in Figure 3.4. The smallest companies of the sector depart from the pattern of the rest of the companies. There is an outlier with substantially higher liabilities (with unrealistically low assets, though), and three companies with very low liabilities relative to their assets. These four companies also have extreme residuals in the random-effects model.

The plot of the company-level residuals against log-assets (Figure 3.6) reveals that most of the large positive residuals occur among the smaller companies. This confirms our observation in Figure 3.1. In general, the larger companies appear to vary less than the smaller ones. Clearly, the assumption of homogeneity of the within-sector variances (adjusted for assets) is problematic.

There are two avenues that could be pursued. We could exclude some of the exceptional sectors (e.g., sectors 35 and 58), and refit the random-effects model. Also, we could consider different accounting rules for large and for small companies (the financial institutions probably do so anyway) and consider separate random-effects models for these subsets of data. An important difficulty in substantive interpretation is that we want to make an inference about 'healthy' companies (and 'healthy' sectors) based on data about *all* the companies and sectors. In principle, information can be obtained about which companies are likely to go bankrupt in future, or are likely to have severe financial problems, although that would require substantial resources and expertise.

Excluding smaller companies leads to better fitting models; clearly, the performance of the smallest companies is much less predictable than that of the larger ones. The sectors, with the smallest companies excluded, have much smaller within- and between-sector variances. Also the fitted regression slope is much closer to unity. Of course, the choice of the upper limit of assets (liabilities) for the companies to be excluded from the analysis is subject to an arbitrary decision. We omit the details for now, but reanalyse the data in Chapter 5.

The conditional variance of the sector-level residuals is $\omega\sigma^2/(1 + n_j\omega)$, where σ^2 is the within-sector variance (estimated as 0.172) and ω the variance ratio (0.176); see (2.65). It is a decreasing function of the sector size n_j. The fitted conditional standard deviation for sector size $n_j = 100$ is

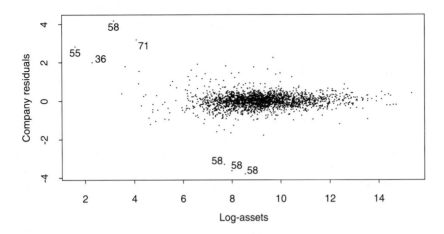

Fig. 3.6. Scatterplot of company-level residuals and log-assets. The sectors of some of the companies with outlying residuals are indicated by their order numbers.

about 0.040, which appears to be only a marginal improvement over the standard deviation from the ordinary regression for the largest sectors in the data. The improvement is substantial for the smaller sectors, though. For example, for a sector with ten companies the fitted conditional standard deviation is $\sqrt{0.172 \times 0.176/(1 + 1.76)} = 0.105$. The fitted standard error of the mean for a sector with ten companies, each with the same value of assets, is $\sqrt{0.0172} = 0.131$. Of course, if the ten companies have unequal assets this standard error may be even larger.

In summary, we have strong evidence of between-sector variation in log-liabilities. The regression slope of log-liabilities on log-assets is close to unity, but differs from it significantly. Based on the analysis thus far we would conclude (ignoring model criticism and inappropriateness of the sample) that a uniform criterion for all the industrial sectors is not appropriate (because of substantial between-sector variation), and that in the ratio of liabilities and assets the latter should be replaced by the power of assets with the estimated exponent 0.972.

In this example we used full ML because we are interested in both regression and variation parameters. In any case, differences between ML and REML are trivial because the number of regression parameters, two, is much smaller than the number of observations and clusters). Differences between ML and REML are discussed in other examples.

3.2   Rat weights

The elementary units in this experiment are newborn rats clustered within litters. The litters (dams) were divided into three treatment groups according to the diet administered to the dam. The diets, denoted A, B, and C, contained, respectively, none, a low dose, and a high dose of an experimental compound; ten dams were each given diets A and B, seven dams each had diet C. Another three dams assigned to diet C produced no offspring that survived until the measurement. Since large litters tend to contain smaller offspring we consider the size of the litter as a litter-level explanatory variable. There was one litter each of size 2, 3, and 4, the rest of the dams had litters of 8–19. Table 3.6 gives the litter sizes with the distribution of the sexes within litters. The sex of the rat is the only elementary-level explanatory variable. While assignment of dams to treatments (diets) was random (subject to the constraint of ten dams in each group), the litter size and sex of the offspring are, naturally, outside the experimenter's control. While sex of the offspring may be allocated 'by nature' at random, the litter size is likely to be affected by the treatment. In fact, the three dams in treatment group C that did not produce any surviving offspring are strong evidence that this is the case (one dam did not conceive, another one had a single still-born offspring, and the third one cannibalized all her newborns).

Table 3.6.   Litter sizes in rat weights data

Diet	Litter sizes (males + females)				
A	12 (8+ 4)	14 (9+ 5)	4 (2+ 2)	14 (8+ 6)	13 (8+ 5)
	9 (6+ 3)	18 (9+ 9)	17 (8+ 9)	17 (13+ 4)	13 (6+ 7)
B	16 (10+ 6)	2 (0+ 2)	12 (8+ 4)	15 (9+ 6)	13 (7+ 6)
	13 (3+10)	14 (7+ 7)	15 (3+12)	10 (4+ 6)	16 (10+ 6)
C	14 (11+ 3)	10 (5+ 5)	3 (1+ 2)	12 (3+ 9)	8 (5+ 3)
	9 (2+ 7)	9 (6+ 3)			

The outcome variable is the weight of the rat at birth. The lightest rat weighs 3680 milligrams (mg), 800 mg less than the next lightest, and then there are several rats with weights around 5000 mg. Among the heaviest rats there are no outlying weights. Among the ten heaviest weights, litter no. 6 is represented by four rats; other litters are represented by no more than two rats. Since the weights have a wide range of values it is meaningful to consider a transformation, such as the logarithm. We present the analysis for both the original and log-transformed values. Figure 3.7 contains the histograms of the weights and their logarithms.

The ordinary regression of the weights is problematic because the litters are bound to vary a great deal. The OLS estimates are displayed in Table 3.7, and those for ML and REML in Table 3.8.

We see that there is little to choose between the ML and the REML estimates of the regression parameters and their standard errors, but these

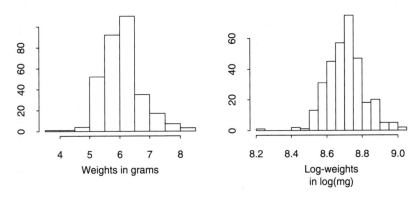

Fig. 3.7.   Histogram of the rat weights. The left-hand panel contains the histogram with the original data and the right-hand panel the histogram with the log-transformed data.

Table 3.7. Ordinary regression fit for rat weights data

Parameter	Estimate	(St. error)
Intercept	8228.0	
Female–male	−275.3	(55.9)
Diet contrast B–A	−425.1	(62.4)
Diet contrast C–A	−907.7	(85.2)
Litter size	−125.1	(10.1)
Residual variance	251 000	
Deviance		3429.33

Table 3.8. Maximum likelihood and restricted maximum likelihood estimates. Rat weights data

Parameter	ML		REML	
	Estimate	(St. error)	Estimate	(St. error)
Intercept	8307.8		8307.6	
Female–male	−357.9	(47.3)	−357.7	(47.7)
Diet contrast B–A	−428.1	(139.3)	−428.1	(139.6)
Diet contrast C–A	−861.0	(169.0)	−861.2	(169.4)
Litter size	−128.8	(17.6)	−128.8	(17.7)
Rat variance	162 100		162 800	
Variance ratio	0.504	(0.164)	0.598	(0.185)
Deviance		3344.41		

differ substantially from their OLS counterparts. The diets are associated with differences in weight of about 430 mg and 860 mg (A being the best and C the worst diet). Larger litters are associated with lower weight; the fitted difference between the average weight of a rat in a litter of, say, size 10 and the average in a litter of 15 is $5 \times 129 = 645$ mg.

The estimates of the litter-level variance differ quite substantially. The REML estimate of the variance ratio is 20 per cent larger than the ML estimate, and the standard error for the former is also larger by about 12 per cent, reflecting the relatively large number of regression parameters compared with the number of level-2 units (litters).

Table 3.9. Maximum likelihood fit for rat weights data, after log-transformation

Parameter	ML Estimate	(St. error)	REML Estimate	(St. error)
Intercept	9.0526		9.0529	
Female–male	–0.0591	(0.0080)	–0.0594	(0.0090)
Diet contrast B–A	–0.0641	(0.0234)	–0.0695	(0.0252)
Diet contrast B–A	–0.1339	(0.0284)	–0.1336	(0.0305)
Litter size	–0.0199	(0.0030)	–0.0199	(0.0032)
Rat variance	0.00458		0.00460	
Variance ratio	0.502	(0.163)	0.597	(0.186)
Deviance	–769.60			
OLS deviance	–679.81			

Figure 3.8 contains the normal quantile plots for the two sets of residuals from the ML fit. The quantile plots appear not to contradict the assumptions of normality, except for one elementary-level residual which corresponds to the lightest rat in the sample. The residual for its litter, –54 mg, is not exceptional; it is the 17th largest (out of 27 litters).

To a large extent, the analysis of log-weights confirms our findings from the analysis of the original data. A summary of the results is given in Table 3.9. Instead of the differences between the sexes and among the

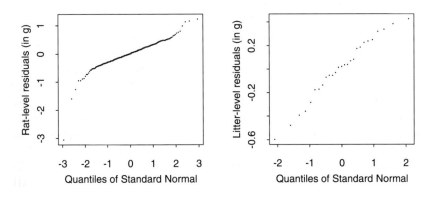

Fig. 3.8. Quantile plots of the residuals. Rat weights' data. The left-hand panel contains the plot for the rat-level residuals, the right-hand panel the plot for the litter-level residuals.

treatments, the logarithms of the corresponding ratios are estimated. Thus the fitted ratio for the sexes is $\exp(-0.0591) = 0.9426$, that is, females are on average 5.74 per cent lighter than males. The average weight of the rats is 6050 mg, and so 5.74 per cent correspond to an average difference of about 347 mg. This is very close to 360 mg, the corresponding estimate based on the original data (Table 3.8). Similarly, the fitted ratios for the treatment contrasts reproduce the fitted differences closely. The weight differences between litter sizes are about 2 per cent per rat, matching the fitted difference of 125 mg.

The within-litter variance in the original analysis is 162 100 (standard deviation 402.5), while the corresponding figure in the analysis of log-weights is 0.004 58 (standard deviation 0.0676). The two standard deviations are in close agreement since the ratio of the standard error and the mean weight, $402.5/6050 = 0.0665$, is close to the standard deviation in the analysis of log-weights. Similarly, the variance ratios and their standard errors in both analyses are very close to their counterparts in the analysis of the original data. The data are not extensive enough to assess with any confidence which of the models fits better.

Consistency of sex differences across the treatments and litter sizes can be explored by including the relevant interaction terms in the regression model. We reanalyse the data in Section 5.2 with the focus on consistency of sex differences within the litters.

3.3 Monitoring pregnancy

This dataset, described in Section 2.2.3, has an irregular longitudinal design. The outcome variable, the logarithm of the concentration of human placental lactogen (HPL), is known to attain its maximum around the 252nd day of pregnancy. The subject of the analysis is the relationship of HPL concentration to time (gestational age). It is likely to be nonlinear, and therefore we consider a cubic regression on time up to day 252 and a different cubic regression thereafter. Continuity, or even smoothness (differentiability), of the resulting regression on time can be ensured by a suitable parametrization. Figure 3.9 contains the plot of the concentration against time. The solid line in the plot is a model fit discussed below. We define the dichotomous variable d, which indicates whether the measurement took place before (0) or after (1) day 252 since conception, and its interactions (products) with the powers of time. Thus we have eight explanatory variables at the elementary level. There is one variable at the client level, *size*, a proxy measure of the size of the placenta, based on certain physiological/anatomical measures of the client. The terms representing the polynomial are

$$t_1 = t - 252, \quad t_2 = \frac{(t - 252)^2}{100}, \quad t_3 = \frac{(t - 252)^3}{1000},$$

Table 3.10.   Ordinary regression for pregnancy monitoring data. The estimate of the residual variance is 0.062 and the deviance of the fitted model is 44.26

	Until day 252		Adjustment after day 252	
Parameter	Estimate	(St. error)	Estimate	(St. error)
Intercept	2.987	(0.121)	−0.0024	(0.0678)
Linear term	0.0061	(0.0046)	0.0013	(0.0127)
Quadratic term	−0.0093	(0.0150)	−0.0466	(0.0690)
Cubic term	−0.0007	(0.0014)	0.0083	(0.0109)
Size	−0.589	(0.071)		

where t is the gestation age in days, and the regression considered is

$$\beta_0 + \gamma_0 d + (\beta_1 + \gamma_1 d)t_1 + (\beta_2 + \gamma_2 d)\,t_2 + (\beta_3 + \gamma_3 d)\,t_3 + b\,size.$$

The OLS fit is displayed in Table 3.10. The corresponding residual variance and deviance are 0.062 and 44.26 respectively. The ML estimates for the random-effects model are given in Table 3.11. We see that the regression parameter estimates have changed only marginally, but their standard errors are substantially different. The estimate of the variance ratio is 6.40 ($\hat{\sigma}^2 = 0.0085$ and $\hat{\tau}^2 = 0.0544$). A large proportion of the variation is at the

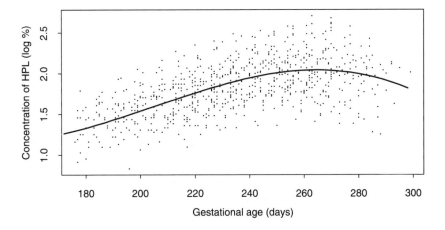

Fig. 3.9.   Plot of concentration of HPL against time for all the clients. Pregnancy monitoring data. The curve drawn is fitted by maximum likelihood, based on the sample mean of *size* equal to 1.633.

Table 3.11. Maximum likelihood estimates for pregnancy monitoring data. The deviance of the fitted model is –1197.7

	Until day 252		Adjustment after day 252	
Parameter	Estimate	(St. error)	Estimate	(St. error)
Intercept	3.024	(0.371)	–0.0030	(0.0256)
Linear term	0.0070	(0.0017)	–0.0030	(0.0048)
Quadratic term	–0.0071	(0.0057)	–0.0056	(0.0261)
Cubic term	–0.0004	(0.0005)	0.0008	(0.0041)
Size	–0.602	(0.227)		
Within-client var.	0.0085			
Variance ratio	6.40	(1.11)		

client level; there are consistent differences among the pregnant women in their concentration of HPL. The deviance of the model fit is -1197.7, much lower than the corresponding figure for the OLS fit, 44.26. The quantile plots of the residuals at both observation and client levels, displayed in Figure 3.10, do not contradict the assumptions of normality.

Continuity of the growth curve at day 252 can be ensured by setting the parameter γ_0 associated with the indicator variable d to zero. In fact, the parameter estimate for this term has a low t-ratio ($-0.0298/0.0256 = -1.16$). The corresponding model fit is given in the leftmost columns of Table 3.12. The regression fit for this model is drawn in Figure 3.9, based

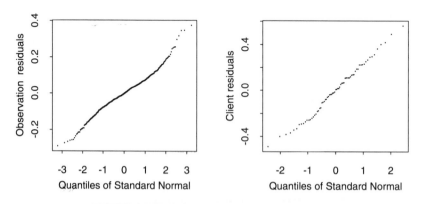

Fig. 3.10. Diagnostic plots for pregnancy monitoring data. The left-hand panel contains the normal quantile plot of the observation-level residuals and the right-hand panel the plot for the client-level residuals.

Table 3.12. Maximum likelihood estimates for pregnancy monitoring data. Reduced models

Parameter	Estimate	(St. error)	Estimate	(St. error)	Estimate	(St. error)
			Until day 252			
Intercept	3.016	(0.371)	3.002	(0.371)	3.001	(0.371)
Linear term	0.0060	(0.0015)	0.0043	(0.0009)	0.0041	(0.0004)
Quadratic term	−0.0098	(0.0052)	−0.0149	(0.0037)	−0.0157	(0.0021)
Cubic term	−0.0006	(0.0005)	−0.0011	(0.0004)	−0.0011	(0.0003)
			Adjustment after day 252			
Intercept	0		0		0	
Linear term	−0.0058	(0.0041)	0		0	
Quadratic term	−0.0098	(0.0052)	−0.0032	(0.0129)	0	
Cubic term	−0.0032	(0.0036)	0.0010	(0.0020)	0.0006	(0.0011)
Size	−0.602	(0.227)	−0.603	(0.226)	−0.603	(0.226)
Level-1 variance	0.0085		0.0085		0.0085	
Variance ratio	6.38	(1.11)	6.37	(1.10)	6.37	(1.10)
Deviance	−1196.35		−1194.60		−1194.32	

on the mean value of *size*, equal to 1.633. The deviance for this reduced
model is –1196.35. Deletion of the variable in question is well justified. We
can also delete its interaction with time (i.e., set $\gamma_1 = 0$), so that the fitted
curve has a continuous derivative at day 252 (see the central part of Table
3.12). The t-ratio for the corresponding estimate is $-0.0058/0.0041 = -1.4$.
The deviance after this model reduction is –1194.60. For further reduction
we consider the quadratic interaction term. Its deletion from the regression
model increases the deviance to only –1194.32 (see the rightmost part of
Table 3.12). The cubic interaction term can also be deleted, and all the
estimates are changed only slightly. The variance ratio estimate remains
at 6.37 (the client-level variance is 0.0544).

A remarkable feature of the data is the high within-client covariance of
the observations. After adjustment for gestational age and *size* a pair of
observations on the same client has the fitted correlation $6.37/7.37 = 0.86$.

Importance of the time as an explanatory (descriptive) variable can be
assessed by comparing the fitted variances with the variances in the model
that uses no explanatory variables:

$$y_{ij} = \mu + \delta_j + \varepsilon_{ij}. \tag{3.4}$$

The maximum likelihood estimates of the within-client variance and the
variance ratio are 0.050 and 1.161 respectively. Thus the 'raw' within-
client correlation of the HLP concentrations is only $1.161/2.161 = 0.54$.
The 'raw' within-client variance has been reduced by the adjustment for
time to 0.0085, so that the within-client proportion of variation explained,
R_1^2, is 0.83. This is due to substantial variation of the concentration of
HPL over time for each client. On the other hand, the client-level variance has been reduced only marginally, from 0.058 to 0.050 ($R_2^2 = 0.14$).
A large proportion of the within-client variation has been explained by
the polynomial regression on time. But we were much less successful in
describing the between-client variation. Note that a component of raw
between-client variation is due to uneven distribution of the inspection
times t across clients. Therefore, the proportion $R_2^2 = 0.14$ is a somewhat
inflated assessment of the descriptive power of the client-level variable *size*.
A more appropriate comparison of the descriptive power of *size* is obtained
by comparison of the estimated variances in the models fitted in Table 3.12,
with the regression on *size* constrained to zero.

Thus, adjustment for time is essential. After appropriate adjustment
future observations on a client are predictable with high precision based
on just one or a small number of observations. This could enable detection
of departures from the normal progress of pregnancy. On the other hand,
confident prediction of a new client's level of HPL concentration is not
possible because of the large variation among pregnant women in general,
even after adjustment for gestational age.

According to the model assumptions and the values of the estimated variances, each client has a curve parallel with the average fitted curve given by the regression estimates, with very small standard deviation ($\sqrt{0.0085} \doteq 0.09$), whereas the distances of the curves from the average curve have a standard deviation about $\sqrt{6.36} \doteq 2.5$ times larger. The assumption that the curves are parallel is assessed in the reanalysis of this dataset in Section 5.3.

3.4 House prices

The data originate from several sources, such as the 1970 US census, Boston Metropolitan Area Planning Commission, and others. The clustering structure is that of 902 census tracts in 96 towns in the Boston Metropolitan Area (BMA). The dataset is a census, with no data missing. The variables are defined either for census tracts or for towns, see Table 2.5, but several variables are aggregates (averages) of variables defined for housing units, such as the average number of rooms in an owner-occupied housing unit and the median value of housing units. The outcome variable, *LMV* (logarithm of the median house price), and *ROOM*, *AGE*, *CRIME*, *DIST*, *BLACK*, and *LSTAT* are defined for census tracts; the rest of the variables are defined for clusters. Altogether there are 14 explanatory variables.

The purpose of the analysis is to assess the association of house prices with the air quality of the neighbourhood, after adjustment for all the other relevant factors. In a strict experimental-design setup we would consider the price of each indispensible 'component' of a home: number of rooms (floor space), various attributes of the neighbourhood, employment opportunities in the area, schools, and so on. Then it would be meaningful to ask what is the (unit) price for clean air, or, alternatively, the discount for pollution.

The explanatory variables describe the typical (average) housing unit (*ROOM* and *AGE*), the (socio-) economic composition, educational facilities, employment opportunities in the neighbourhood (*CRIME*, *ZONE*, *CHAS*, *INDUS*, *TAX*, *PTRAT*, *BLACK*, and *LSTAT*), accessibility by roads (*DIST* and *RAD*), and air pollution (*NOXSQ*).

The ordinary regression estimates are given in the left-hand part and the random-effects model fit is summarized in the right-hand part of Table 3.13. We see that the OLS and random-effects model estimates differ substantially for several parameters. For example, the estimates for *CHAS* have opposite signs, the t-ratio for *AGE* is much smaller than unity for OLS, but negative and smaller than −2 for the random-effects model. Comparison of the deviances implies that the random-effects model provides a much better fit (difference of deviances 172 for one degree of freedom). The census-tract and town-level variances are about equal; the town-level variation is substantial even though several town-level variables have been

Table 3.13. Ordinary regression (OLS) and random-effects model (ML) fits. House prices data. The ML fit required seven iterations

		OLS		ML	
	Variable	Estimate	(St. error)	Estimate	(St. error)
1	*G. MEAN*	9.755		9.677	
3	*CRIME*	−0.0119	(0.0012)	−0.0072	(0.0010)
4	*ZONE*	1.71E–4	(4.95E–4)	8.11E–5	(6.60E–4)
5	*INDX*	2.81E–4	(2.34E–3)	2.25E–4	(4.34E–3)
6	*CHAS*	0.0913	(0.0327)	−0.0118	(0.0285)
7	*NOXSQ*	−6.38E–3	(1.12E–3)	−5.88E–3	(1.223E–3)
8	*ROOM*	6.30E–3	(1.29E–3)	9.15E–3	(1.16E–3)
9	*AGE*	9.80E–5	(5.19E–4)	−9.49E–4	(4.59E–4)
10	*DIST*	−0.194	(0.0327)	−0.132	(0.0450)
11	*RAD*	0.0962	(0.0189)	0.097	(0.0284)
12	*TAX*	−4.27E–4	(1.21E–4)	−3.78E–4	(1.89E–4)
13	*PTRAT*	−0.0308	(4.92E–3)	−0.0296	(9.72E–3)
14	*BLACK*	0.362	(0.102)	0.576	(0.0995)
15	*LSTAT*	−0.371	(0.0246)	−0.284	(0.0235)
Census tract					
Variance		0.0324		0.0171	
St. deviation		0.180		0.131	
Town					
Variance				0.0177	(0.0039)
St. deviation				0.133	(0.0127)
	Deviance	−300.15		−472.02	

fitted. For comparison, the estimates for the model using no explanatory variables are:

$$\hat{\mu} = 10.056, \quad \hat{\sigma}^2 = 0.0517, \quad \hat{\tau}^2 = 0.0963,$$

and the random-effects model fit with *CRIME* and *ROOM*, the two most important census-tract level variables, is

$$\widehat{LMV} = 9.2088 - 0.0085 \ CRIME + 0.0208 \ ROOM$$
$$(0.0013) \qquad\qquad (0.0011)$$

($\hat{\sigma}^2 = 0.0296$ and $\hat{\tau}^2 = 0.0400$). Most of the unadjusted variation is among the towns and much less among the census tracts. The explanatory

variables reduce the town-level variance much more than the census-tract
level variance.

The regression parameter estimates associated with *NOXSQ* are –0.0064
(standard error 0.0011) for the OLS and –0.0059 (0.0012) for the random-
effects model fit; the corresponding t-ratios are equal to about 5. The sign
of the estimates is in agreement with our expectation; higher concentration
of nitric oxide is associated with lower house prices. Most towns have values
of *NOXSQ*, the squared annual average concentration, in the range of 10–
65. Reduction by ten units, a substantial amount, is associated with the
increase in the values of the housing units by about 6 per cent.

Belsley *et al.* (1980) analysed this dataset using robust regression
methods assuming independence of the observations. After examining the
residuals for census tracts they pointed out that outliers tended to be
concentrated in a small number of towns. It is of interest whether outliers remain an issue when a covariance structure is accounted for by the
random-effects model. Figure 3.11 contains the histograms and the normal
quantile plots for the OLS fit (top panels), and for the random-effects model
fit (bottom panels). We see that the departure from normality is moder-
ated somewhat, although instead of a larger number of outliers among the
OLS residuals there are two distinct ones at either extreme (a census tract
in town 76, Back Bay, with the lowest, and a census tract in town 80, East
Boston, with the highest residual), but the rest of the census-tract level

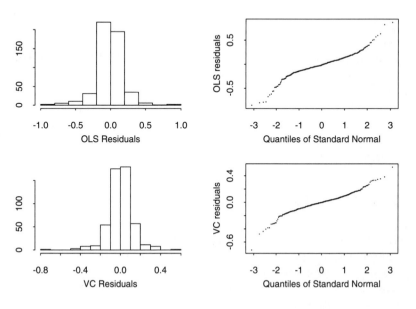

Fig. 3.11. Residuals from the ordinary regression and the maximum likelihood
fit. House prices data.

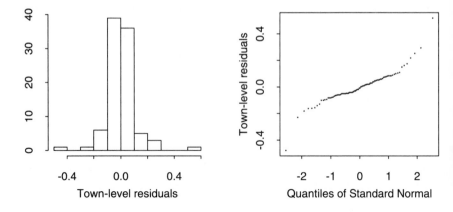

Fig. 3.12. Diagnostics for town-level residuals. House prices data. Two outlying towns are identified by their order numbers.

residuals are somewhat better aligned along a straight line in the normal quantile plot. Figure 3.12 contains diagnostic plots for the towns. These are somewhat misleading because the residuals for the towns with one or two census tracts are greatly shrunk towards zero. However, two towns, one with very high and the other one with very low adjusted house prices are easy to identify in the normal quantile plot. The former, 77 (Beacon Hill), is a town with only three census tracts; the latter, 81 (South Boston), has 13 census tracts.

We assume that the data have been checked thoroughly and that they do not contain any gross errors. Thus the census tracts and towns with exceptional residuals should be regarded as an integral part of the data description. We could proceed by discarding these units and refitting the random-effects model, presumably obtaining a fit with much better looking diagnostics, and accept this result as a description of the data, supplemented with the small number of towns and census tracts that do not conform to this formula.

3.5   GRE validity study

The purpose of the study is to describe the association of graduate school performance (FYA) on the GRE test scores (V, Q, and A) and the undergraduate grade-point average, GP. The administrator of the GRE tests wishes to establish the 'predictive validity' of the test scores, that is, to assess the importance of the scores as predictors of graduate school performance. Graduate school departments would like to assess their selection procedures, and possibly alter the relative emphasis given to the various admissions criteria.

The assumption of within-cluster (within-department) correlation, or of between-department variation, is particularly natural in this example. The students in a department have the same instructors, they share educational facilities and research equipment, and are subjected to the same methods of instruction; they are exposed to the same academic environment. The grades awarded are on a scale common to all students in the department, but the scales may vary across departments (even within the same university).

Interpretation of the between-department variation is difficult because it is an amalgam of several sources of variation: students' selection of the department, admissions staff's selection of students (and an interplay of these), scales of examination results, quality of instruction, and influence of the academic environment. On the one hand, the GRE test scores have a universal scale, on the other hand their universal validity for students of all subjects is somewhat contentious. We will demonstrate that the GRE scores are much weaker predictors of the first-year average grade score (FYA) than is the undergraduate grade-point average GP; it appears that past exam results are the best predictors of future exam results. However, GRE scores are important as predictors additional to GP.

Another problematic issue is non-normality of the exam scores. Students' composite exam scores (GP and FYA) are usually averages of small numbers of exam results, each with a discrete number of possible outcomes. A large proportion of students have perfect scores on FYA (4.0), GP (4.0), or even on both, and scores of 3.0 and various fractions with small denominators are also attained by disproportionately many students. A number of approaches could be adopted to alleviate this problem. For example, the recorded FYA and GP scores could be regarded as right-censored at 4.0. For the purposes of illustration we ignore these issues, but point out that definition of suitable alternative measures for outcomes of higher education is a well-identified problem in educational research.

Naturally, we would expect all the regression coefficients, perhaps with the exception of the intercept, to be positive; higher scores and better exam performance are associated with better academic performance in the graduate school.

The range of the GRE test scores is 200–800. To simplify interpretation of the regression parameters we divide these scores by 200, so that each GRE score is in the same range as GP and FYA (1–4). In order to maintain confidentiality of the data, only the matrices of within-department sums of squares and cross-products for the six variables (including the intercept) are available. The matrix of cross-products for the first department in the dataset is displayed in Table 2.7. For orientation, this department has 13 students, and its mean FYA score is $45.29/13 = 3.48$. Note that from such a matrix all the cross-products that involve a within-department mean can be calculated. For example, the cross-product total for Q and $\overline{V}$ is

$$c_j(Q, \overline{V}) \;=\; c_j(Q, 1)\frac{c_j(V, 1)}{c_j(1, 1)}\,,$$

where c_j is the sum of cross-products for the two arguments in department j.

The sample means of the scores are

$$\overline{V} = 2.685, \qquad \overline{Q} = 2.839, \qquad \overline{A} = 2.845, \qquad \overline{U} = 3.271, \qquad \overline{FYA} = 3.534.$$

The effective range of the grade-point average scores GP and FYA is much narrower than three points (1–4); the ranges of the department means for GP and FYA are 2.57–3.88 and 2.85–3.95 respectively. The range of the GRE score means is 1.90–3.90.

The ordinary regression fit,

$$\widehat{FYA} \;=\; 2.633 \;+\; \underset{(0.0081)}{0.090\,V} \;-\; \underset{(0.0081)}{0.033\,Q} \;+\; \underset{(0.0092)}{0.022\,A} \;+\; \underset{(0.0080)}{0.211\,GP}$$

$$(3.5)$$

($\hat{\sigma}^2 = 0.1259$), contains a negative parameter estimate. This appears to be counterintuitive; higher Q scores are associated with lower FYA scores. Would FYA scores in a department therefore be higher if admissions staff preferred students with lower Q scores while admitting students with the same distribution of scores on the other predictors? Highly unlikely. Equation (3.5) describes (among other phenomena) the complex selection processes (how an applicant is offered a place and accepts a position). If these processes are changed, so is the prediction equation.

The standard errors for the parameter estimates associated with the GRE exams are smaller than 10^{-2}; the parameter estimate for Q is (nominally) significantly smaller than zero.

Making inferences by separate department-wise regressions is hopelessly inefficient because most of the departments have small sample sizes relative to the within-department variance and the number of explanatory variables. The average first-year enrolment is about 14 students per department (the median is 10). For illustration, we consider ordinary regression for the largest department in the dataset. This department provided data from 106 students. The OLS fit for this department is

$$\widehat{FYA} \;=\; 1.944 \;+\; \underset{(0.071)}{0.121\,V} \;-\; \underset{(0.083)}{0.010\,Q} \;+\; \underset{(0.061)}{0.007\,A} \;+\; \underset{(0.087)}{0.379\,GP}$$

($\hat{\sigma}^2 = 0.0633$). The standard errors are so large that comparisons with other departments (or with the 'national' regression given by (3.5)) are

meaningless because only extreme differences would be detected as statistically significant. Addition of any quadratic terms, or interactions of the
predictors, improves the model fit only slightly but inflates the standard
errors substantially.

The OLS solution (3.5) should be close to the ML fit for the random-
effects model because both estimators are consistent and the analysed
dataset is quite large. But the ML fit is

$$\widehat{FYA} = 2.346 + \underset{(0.0076)}{0.093\,V} + \underset{(0.0082)}{0.057\,Q} + \underset{(0.0081)}{0.030\,A} + \underset{(0.0078)}{0.212\,GP}$$
$$(3.6)$$

($\hat{\sigma}^2 = 0.0881$, $\hat{\omega} = 0.371$); now all the regression parameter estimates are
positive (and significantly greater than zero). The fitted variance ratio is
0.371 and its standard error is 0.020. Much smaller between-department
variance would also be detected as significant in such a large dataset. The
departments vary substantially; students with the same background scores
tend to attain FYA scores varying both within and between departments.

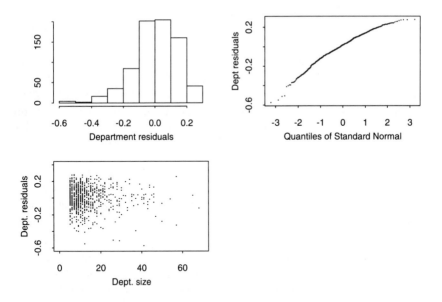

Fig. 3.13. Diagnostics for department-level residuals. GRE validity data.
Histogram and normal quantile plot of the residuals, and a plot of the residuals
against department size. Department 140, with 106 students, is excluded from
the latter plot so as to improve its resolution. The residual for department 140
is equal to −0.094.

The department-level residuals contain a small number of obvious out-liers. The top panels of Figure 3.13 contain the histogram and the normal quantile plot of the department-level residuals. The bottom panel of the diagram contains the plot of the residuals (conditional means) against the department sizes.

Even though the random-effects model fit given by (3.6) accounts for an important feature of the data, its interpretation as evidence of predictive validity is problematic because of the numerous caveats discussed above. It is important to note that the nominal standard errors obtained from the inverse of the fitted information matrix represent only one component of uncertainty in our inference. Departure from the assumptions of random sampling, random allocation, and normality erode our ability to make an inference with the confidence usually associated with large samples.

Educational researchers consider the *context* of a cluster (classroom, school, or department) as an important predictor of educational outcomes. The context is described by cluster-level variables, and their choice in modelling should reflect substantive as well as statistical criteria. In the absence of any cluster-level variables in the dataset, the within-department means for one or several variables could be used as variables describing the context. This raises several issues. First, if we consider the students' scores as outcomes of a random process, then such a constructed cluster-level variable is associated with variation akin to measurement error. Since the error variance is smaller for larger clusters this 'measurement error' is heteroscedastic. If the context of a department were assumed constant across a few years then the mean of the student-level variables across these years could be used as a contextual variable with smaller error variance. This would resolve the 'measurement error' issue, but not the validity issue, that is, appropriateness of representation of the context by one or several department-means of student-level variables.

For illustration we consider the within-department means of V as a contextual variable. The use of such a variable can be motivated by describing the cluster-level random term δ_j by a regression model using the contextual variable:

$$\delta_j = a + b\overline{V}_j + \gamma_j, \qquad (3.7)$$

where a and b are unknown parameters and $\{\gamma_j\}$ is a centred normal random sample. Substitution of (3.7) in the random-effects model is equivalent to supplementing its regression part with $\overline{V}_j$. The fit for this model is

$$\widehat{FYA} = 2.923$$

$$+ \ \underset{(0.0089)}{0.143V} \ - \ \underset{(0.0082)}{0.019Q} \ + \ \underset{(0.0092)}{0.022A} \ + \ \underset{(0.0086)}{0.226GP} \ - \ \underset{(0.0143)}{0.193\,\overline{V}}$$

$$(3.8)$$

($\hat{\sigma}^2 = 0.0881$, $\hat{\omega} = 0.289$). The large negative parameter estimate for the context $\overline{V}$ has a simple interpretation. In departments with lower background scores FYA scores tend to be inflated; a typical student with a given background would do well and obtain higher grades in a department with poorer composition than in a very competitive department.

Note that the estimate of the elementary-level variance has not been altered by the inclusion of $\overline{V}$ in the regression part, but the department-level variance has been reduced substantially. Cluster-level variables describe (reduce) only between-cluster variation.

Plots of the department-level residuals against the within-department means $\overline{V}_j$ are given in Figure 3.14. In the left-hand panel the residuals from (3.6), and in the right-hand panel the residuals from (3.8), are used. It appears that our attempt to remove possible dependence of the residuals on the contextual variable $\overline{V}$ has been a failure. The residuals from the contextual model (3.8) display a stronger association with $\overline{V}$ than do the residuals from (3.6).

In addition to the problems discussed above we have to consider which value of $\overline{V}_j$ to use in prediction: the recorded (past) value, or to supplement the prediction procedure with uncertainty about the future value of $\overline{V}_j$.

3.5.1 ESTIMATION BASED ON THE WITHIN-DEPARTMENT MEANS

Analysis of large datasets, such as the validity study dataset, is often deemed burdensome, time-consuming, and expensive, especially when up-to-date computing facilities are not available. Computing the within-department means, or other within-department statistics, and carrying out an analysis of these statistics as if they were the original observations, appears to be an attractive alternative. In addition to the substantial

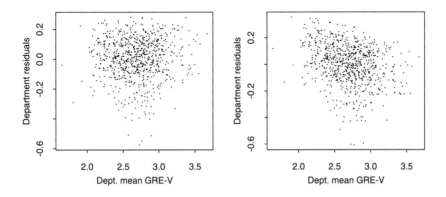

Fig. 3.14. Department-level residuals. GRE validity data. The left-hand panel is based on residuals from the model without the contextual variable $\overline{V}$ and the right-hand panel is based on residuals from the model with the $\overline{V}$.

reduction of the sample and data sizes, the problem of clustering of the observations appears to have disappeared. In this section we carry out such an analysis of the GRE dataset to demonstrate the fallacy of this approach.

The ordinary regression of $\overline{FYA}$ on the within-department means yields the equation

$$\widehat{\overline{FYA}} = 3.162 + \underset{(0.031)}{0.136\,\overline{V}} - \underset{(0.030)}{0.150\,\overline{Q}} - \underset{(0.047)}{0.024\,\overline{A}} + \underset{(0.034)}{0.158\,\overline{GP}},$$

$$(3.9)$$

with $\hat{\sigma}^2 = 0.0309$. Since the aggregate observations (means) are averages of unequal numbers of observations, the weighted least squares, with weights equal to department sizes, may be more appropriate. The corresponding model fit is

$$\widehat{\overline{FYA}} = 2.980 + 0.084\,\overline{V} - 0.116\,\overline{Q} - 0.074\,\overline{A} + 0.266\,\overline{GP}, \qquad (3.10)$$

with similar residual variance and standard errors as in (3.9). The weights appear to make a lot of difference. In both model fits the regression estimate for $\overline{Q}$ is negative and very large in absolute value.

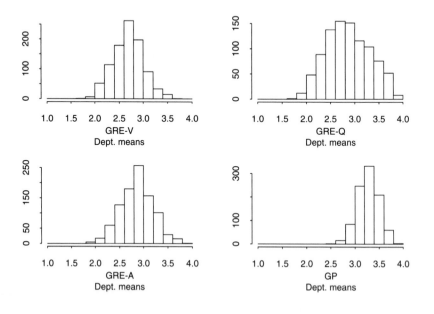

Fig. 3.15. Histograms of the department-means. GRE validity data.

The analyses of department means yield results that are substantially different from those based on student-level data. The regression for the means refers to different relationships than the regression for the student-level data, and so the two cannot be interchanged.

Whereas the student-level data cover almost the entire range of 1–4 for each variable, the range of the means is much narrower, as can be observed in Figure 3.15. The variances of the four means $\overline{V}$, $\overline{Q}$, $\overline{A}$, and $\overline{GP}$ are 0.098, 0.205, 0.115, and 0.050 respectively. Their correlation matrix is

$$
\begin{pmatrix}
1.0 & 0.579 & 0.709 & 0.505 \\
0.579 & 1.0 & 0.879 & 0.370 \\
0.709 & 0.879 & 1.0 & 0.447 \\
0.505 & 0.369 & 0.447 & 1.0
\end{pmatrix}.
$$

The corresponding correlation matrix for the student-level variables is

$$
\begin{pmatrix}
1.0 & 0.475 & 0.559 & 0.253 \\
0.475 & 1.0 & 0.707 & 0.214 \\
0.559 & 0.707 & 1.0 & 0.238 \\
0.253 & 0.214 & 0.238 & 1.0
\end{pmatrix},
$$

and the variances are 0.284, 0.378, 0.330, 0.182. The student-level variables exhibit lower correlations and higher variances than their department-level means. We return to the analysis of this dataset in Section 5.5.

4
Random regression coefficients

4.1 Models

In Chapter 2 we studied the analysis of covariance with random effects as a method for the description of between-cluster differences. In essence, the random-effects model (2.1) assumes that each cluster has its own regression,

$$\mathbf{E}(\mathbf{y}_j \,|\, \boldsymbol{\beta}_j) \;=\; \mathbf{X}_j \boldsymbol{\beta}_j + \boldsymbol{\varepsilon}_j\,, \tag{4.1}$$

and that the conditional (within-cluster) regression coefficients $\boldsymbol{\beta}_j$ have common 2nd–pth components $(\beta^{(2)}, \ldots, \beta^{(p)})$, but that their respective first components vary according to the normal law,

$$\beta_j^{(1)} \;\sim\; \mathcal{N}(\beta^{(1)}, \tau^2). \tag{4.2}$$

This implies a natural ordering for the clusters based on the (unobserved) realizations of the random terms $\beta_j^{(1)}$; in practice we consider the conditional expectations $\hat{\beta}_j^{(1)}$ in place of $\beta_j^{(1)}$. An ordering based on $\beta_j^{(1)}$ may be desired in some applications, but its existence may be a rather restrictive feature in others. Consider the validity study example in Section 3.5. Suppose each graduate-school department has a large number of students and in most departments several students fall into each of a set of narrow bands of GRE scores. If we decide to rank the departments by the means of the outcomes for these narrow bands of GRE scores, we may find that there are departments which rank high on performance of students with high GRE scores, but rank much lower on performance of students with low GRE scores. Such departments would appear to exaggerate the initial differences among the students. In other departments these differences may be reduced.

We consider first the 'fixed-effects' version of this problem. Suppose we wish to compare a small number of specific (identified) departments. The classical ANCOVA can accomodate differing regression slopes by the inclusion of certain interaction terms in the model, such as

$$\mathbf{y}_j \;=\; \mathbf{X}_j \boldsymbol{\beta} + b_j^{(1)} + \mathbf{x}_j^{(2)} b_j^{(2)} + \boldsymbol{\varepsilon}_j\,, \tag{4.3}$$

where the term $x_j^{(2)} b_j^{(2)}$ represents the interaction of the variable $x^{(2)}$ with the departments (cluster-by-variable interaction).

When the data contain a large number of clusters (departments) it is advantageous to adopt the 'random-effects' version of this model and regard each department as a random draw from an infinite population of departments. In this situation the pattern of variation allowed by the random-effects model given by (4.1) and (4.2) is insufficient because the regression slopes with respect to $x^{(2)}$ are likely to vary among departments. This chapter deals with an extension of the random-effects model which parallels the ANCOVA models with interactions. In the classical ANCOVA model the interactions induce differing within-group regressions. The 'random-effects' version of this model provides a description for the variation of the within-group regression coefficients.

Let the outcomes $\mathbf{y}_j$ have the conditional distribution, given the vectors of coefficients $\boldsymbol{\beta}_j$,

$$(\mathbf{y}_j \mid \boldsymbol{\beta}_j) \sim \mathcal{N}(\mathbf{X}_j \boldsymbol{\beta}_j, \, \sigma^2 \mathbf{I}_{n_j}), \tag{4.4}$$

or, equivalently,

$$\mathbf{y}_j = \mathbf{X}_j \boldsymbol{\beta}_j + \boldsymbol{\varepsilon}_j, \tag{4.5}$$

where $\boldsymbol{\varepsilon}_j \sim \mathcal{N}(\mathbf{0}_{n_j}, \sigma^2 \mathbf{I}_{n_j})$. The vectors $\boldsymbol{\varepsilon}_j$, $j = 1, \dots, N_2$, are mutually independent. Further, we assume that $\{\boldsymbol{\beta}_j\}$ is a random sample from a multivariate normal distribution,

$$\boldsymbol{\beta}_j \sim \mathcal{N}_p(\boldsymbol{\beta}, \boldsymbol{\Sigma}^*),$$

independent of $\{\boldsymbol{\varepsilon}_j\}$. This is equivalent to the model

$$\mathbf{y}_j = \mathbf{X}_j \boldsymbol{\beta} + \mathbf{X}_j \boldsymbol{\gamma}_j + \boldsymbol{\varepsilon}_j, \tag{4.6}$$

where $\boldsymbol{\gamma}_j = \boldsymbol{\beta}_j - \boldsymbol{\beta}$ is the vector of deviations of the regression coefficients $\boldsymbol{\beta}_j$ from their expectation, so that $\boldsymbol{\gamma}_j \sim \mathcal{N}(\mathbf{0}, \boldsymbol{\Sigma}^*)$.

The model in (4.6) is fully specified by the elementary-level variance σ^2 and the distribution of the *cluster-level* random terms $\{\boldsymbol{\beta}_j\}$, that is, owing to normality, by their vector of expectations $\boldsymbol{\beta}$ and the variance matrix $\boldsymbol{\Sigma}^*$. The parameters in $\boldsymbol{\beta}$ describe the linear regression of the outcomes on the explanatory variables, $\mathbf{E}(\mathbf{y}) = \mathbf{X}\boldsymbol{\beta}$, and so we refer to them as *regression parameters*. The linear regression formula $\mathbf{X}\boldsymbol{\beta}$, or the list of regressors in $\mathbf{X}$, is referred to as the *regression part* of the model. Selection of the explanatory variables $\mathbf{x}$, that is, the structure of the mean of the outcomes, $\mathbf{X}\boldsymbol{\beta}$, entails essentially the same difficulties as in ordinary regression.

The ANCOVA version of the model in (4.6), assuming that $\boldsymbol{\gamma}_j$ are parameters, contains interactions of the clusters with all the explanatory variables. This is often a vastly overparametrized model; usually it is practical to consider interactions with only a small number of variables. In the random coefficient formulation this corresponds to constraining some of the variances in $\boldsymbol{\Sigma}^*$ to zero. For example, the absence of interaction of the second variable in (4.3), $b_j^{(2)} \equiv b^{(2)}$, corresponds to zero variance of a component of $\boldsymbol{\delta}$, $\Sigma_{22}^* = 0$. Note that a zero variance implies zero covariances in the same row and column of $\boldsymbol{\Sigma}^*$.

In ANCOVA we cannot define interactions of the cluster with a cluster-level variable; such interactions are confounded with the cluster-parameters $b_j^{(1)}$. In the random coefficient model (4.6) it is therefore reasonable to set the variance associated with each cluster-level variable to zero.

Thus, it is more appropriate to formulate the model in (4.6) as

$$\mathbf{y}_j = \mathbf{X}_j \boldsymbol{\beta} + \mathbf{Z}_j \boldsymbol{\delta}_j + \boldsymbol{\varepsilon}_j , \qquad (4.7)$$

where the matrices $\mathbf{Z}_j$ are formed by selection of certain columns (variables) of $\mathbf{X}_j$. We refer to the matrix $\mathbf{X}$ ($\mathbf{X}_j$) as the *regression design matrix* (for cluster j) and to $\mathbf{Z} = \mathbf{Z}_j \otimes \mathbf{1}_{N_2}$ (or $\mathbf{Z}_j$) as the *variation design matrix* (for cluster j). Also, we say that an explanatory variable (a column of the regression design matrix $\mathbf{X}$) is *associated with cluster-level variation* (or, briefly, with variation) if the variable is included in the variation design $\mathbf{Z}$. The random-effects model given by (4.1) and (4.2) corresponds to $\mathbf{Z} = \mathbf{1}_N$.

In classical ANCOVA we consider cluster-by-variable interactions only for variables for which the main effect (regression parameter) is unknown and would be estimated. This convention can also be described as a hierarchy of the model parameters, and can be expressed for the random coefficient models as follows:

- The columns of the variation design matrix $\mathbf{Z}$ are selected from the columns of the regression design matrix $\mathbf{X}$.
- If the matrix $\mathbf{Z}$ is non-empty then it contains the intercept as its first column, $\mathbf{Z}^{(1)} = \mathbf{1}$.

These conventions are extended further when the regression design contains variable-by-variable interactions. The two constituent variables for such an interaction have to be included in the design, and any higher-order interaction in the design has to be accompanied by all its constituent subinteractions. This rule does not treat interactions with the clusters in any way different from other (variable-by-variable) interactions. In random coefficient models it is usually meaningful to adhere to the analogous conventions. However, the theoretical development of the methods and of algorithms for model fitting presented in this chapter does not depend on them.

The number of regression parameters (columns of $\mathbf{X}$) is p. We denote the number of columns of the variation design matrix $\mathbf{Z}$ by r. As discussed, usually $r \leq p$. Note, however, that the number of parameters for between-cluster variation, $r(r+1)/2$, may be greater than p.

4.2 Invariance and linear transformations

Statistical models are simplified descriptions of real situations in terms of random variables (distributions). Desirable (realistic or 'good') models possess several of the properties of the modelled phenomena and yet are simple enough to permit discussion of their consequences. Properties of phenomena observed universally, or in a wide variety of contexts, are the most important features of the models. Invariance is one such property.

4.2.1 INVARIANCE IN ORDINARY REGRESSION

Ordinary regression models have the following property. Suppose a random vector $\mathbf{y}$ satisfies the ordinary regression model

$$\mathbf{y} = \mathbf{X}\boldsymbol{\beta} + \boldsymbol{\varepsilon}, \tag{4.8}$$

with $\text{var}(\boldsymbol{\varepsilon}) = \sigma^2\mathbf{I}$. Let $\mathbf{X}^*$ be the design matrix formed by a non-singular linear transformation of $\mathbf{X}$, that is, $\mathbf{X}^* = \mathbf{X}\mathbf{A}$, where $\mathbf{A}$ is a non-singular $p \times p$ matrix. We assume that the first column of $\mathbf{A}$ is equal to $(1, 0, \ldots, 0)^\top$, so that the first column of $\mathbf{X}^*$ is the intercept $\mathbf{1}$. Then the linearly rescaled vector of outcomes $\mathbf{y}^* = a_y\mathbf{y} + b_y$ satisfies the ordinary regression model

$$\mathbf{y}^* = \mathbf{X}^*\boldsymbol{\beta}^* + \boldsymbol{\varepsilon}^* \tag{4.9}$$

with $\boldsymbol{\beta}^* = a_y\mathbf{A}^{-1}\boldsymbol{\beta} + (b_y, 0, \ldots, 0)^\top$ and $\boldsymbol{\varepsilon}^* = a_y\boldsymbol{\varepsilon}$ (so that the variance of ε^* is $\sigma^{*2} = a_y^2\sigma^2$). This can be proved directly by substitution; (4.8) implies that

$$a_y\mathbf{y} + b_y = a_y\mathbf{X}^*\mathbf{A}^{-1}\boldsymbol{\beta} + b_y + a_y\boldsymbol{\varepsilon}.$$

This property of ordinary regression is called invariance with respect to linear transformations. Various generalizations of ordinary regression also possess this property. An important practical consequence is that appropriateness of a description by ordinary regression is unaffected by a linear transformation of the variables. Straightforward equations apply for conversion from one unit (set of units) to another. The OLS estimators of the regression parameters and the residual variance, as well as the MLEs in random coefficient models (Section 4.4), also have this property.

For example, when time is one of the explanatory variables the choice of its units (seconds, minutes, or hours), as well as the choice of the time-point that corresponds to the origin (zero), are of little significance, other than their effect on the nominal sizes of the regression parameters. Furthermore, when polynomial regression is considered, the variables corresponding to the polynomials can be chosen arbitrarily. For instance, the model

$$y = (1,\ x,\ x^2,\ x^3\,)\boldsymbol{\beta} + \varepsilon$$

can be rewritten as, say,

$$y = \left\{1,\ x - x_0,\ (x - x_0)^2,\ (x - x_0)^3\right\} \beta^* + \varepsilon,$$

where x_0 is an arbitrary real number and

$$\beta^* = \begin{pmatrix} 1 & x_0 & x_0^2 & x_0^3 \\ 0 & 1 & 2x_0 & 3x_0^2 \\ 0 & 0 & 1 & 3x_0 \\ 0 & 0 & 0 & 1 \end{pmatrix} \beta,$$

so that the regression parameter corresponding to the highest (cubic) term is unaltered, $\beta_4^* = \beta_4$, but all the other regression parameters are transformed.

Of course, ordinary regression models are not invariant with respect to non-linear transformations. In fact, there are several methods for searching for a suitable non-linear transformation of the data that renders the assumptions of the ordinary regression model acceptable (e.g., Box–Cox transformations).

4.2.2 RANDOM COEFFICIENTS AND INVARIANCE

We now explore invariance with respect to linear transformations in the random coefficient model (4.5). For notational simplicity we use the formulation (4.6) rather than (4.7). If we replace the design matrix $\mathbf{X}$ by $\mathbf{X}^* = \mathbf{X}\mathbf{A}$, and therefore $\mathbf{X}_j$ by $\mathbf{X}_j^* = \mathbf{X}_j\mathbf{A}$ for all j, then

$$\mathbf{y}_j = \mathbf{X}_j^*\boldsymbol{\beta}^* + \mathbf{X}_j^*\boldsymbol{\gamma}_j^* + \boldsymbol{\varepsilon}_j, \tag{4.10}$$

where $\boldsymbol{\beta}^* = \mathbf{A}^{-1}\boldsymbol{\beta}$ and $\boldsymbol{\gamma}_j^* = \mathbf{A}^{-1}\boldsymbol{\gamma}_j$. The consequences of a linear transformation of y are obvious and, for simplicity, we do not consider it here. Thus, a linear transformation of the explanatory variables results in the same changes of the regression parameters $\boldsymbol{\beta}$ as in ordinary regression. This is not surprising since in both cases a linear transformation amounts to a reexpression of the expectation $\mathbf{X}\boldsymbol{\beta}$. However, for the cluster-level variance matrix the situation is more complex. If $\boldsymbol{\Sigma} = \mathrm{var}(\boldsymbol{\gamma}_j)$ then $\boldsymbol{\Sigma}^* = \mathrm{var}(\mathbf{A}^{-1}\boldsymbol{\gamma}_j) = \mathbf{A}^{-1}\boldsymbol{\Sigma}(\mathbf{A}^{-1})^\top$; a given pattern (e.g., of zeros) in the matrix $\boldsymbol{\Sigma}$ may not be reproduced in $\boldsymbol{\Sigma}^*$.

By way of illustration we discuss the following two cases:

Example A. Suppose the transformation matrix $\mathbf{A}$ has the form

$$\mathbf{A} = \begin{pmatrix} \mathbf{A}_1 & \mathbf{A}_{12} \\ \mathbf{0} & \mathbf{A}_2 \end{pmatrix}, \tag{4.11}$$

and the compatible partitioning of the variance matrix $\boldsymbol{\Sigma}$ is

$$\boldsymbol{\Sigma} = \begin{pmatrix} \boldsymbol{\Sigma}_{11} & \boldsymbol{\Sigma}_{12} \\ \boldsymbol{\Sigma}_{21} & \boldsymbol{\Sigma}_{22} \end{pmatrix}.$$

Then

$$\mathbf{A}^{-1} = \begin{pmatrix} \mathbf{A}_1^{-1} & -\mathbf{A}_1^{-1}\mathbf{A}_{12}\mathbf{A}_2^{-1} \\ \mathbf{0} & \mathbf{A}_2^{-1} \end{pmatrix}$$

and

$$\boldsymbol{\Sigma}^* = \mathbf{A}^{-1}\boldsymbol{\Sigma}\left(\mathbf{A}^{-1}\right)^{\top} = \begin{pmatrix} \boldsymbol{\Sigma}_{11}^* & \boldsymbol{\Sigma}_{12}^* \\ \boldsymbol{\Sigma}_{21}^* & \mathbf{A}_2^{-1}\boldsymbol{\Sigma}_{22}\left(\mathbf{A}_2^{-1}\right)^{\top} \end{pmatrix},$$

for certain matrices $\boldsymbol{\Sigma}_{11}^*$ and $\boldsymbol{\Sigma}_{12}^*$, and $\boldsymbol{\Sigma}_{21}^* = \left(\boldsymbol{\Sigma}_{12}^*\right)^{\top}$. Therefore, $\boldsymbol{\Sigma}_{22} = \mathbf{0}$ implies that $\boldsymbol{\Sigma}_{22}^* = \mathbf{A}_2^{-1}\boldsymbol{\Sigma}_{22}(\mathbf{A}_2^{-1})^{\top} = \mathbf{0}$. In other words, let $\mathbf{x} = (\mathbf{x}_1, \mathbf{x}_2)$ and suppose the subset of explanatory variables $\mathbf{x}_2$ is not associated with variation. Then after any non-singular linear transformation $\mathbf{x}_1^* = \mathbf{x}_1\mathbf{A}_1 + \mathbf{x}_2\mathbf{A}_{12}$, $\mathbf{x}_2^* = \mathbf{x}_2\mathbf{A}_2$, the variables $\mathbf{x}_2^*$ remain not associated with variation.

Example B. Suppose the transformation matrix $\mathbf{A}$ has the form

$$\begin{pmatrix} 1 & \mathbf{a}_{12} \\ \mathbf{0} & \mathbf{A}_2 \end{pmatrix}, \tag{4.12}$$

and $\mathbf{A}_2$ is diagonal (and non-singular). The corresponding transformation of the vector of variables $\mathbf{x}$ linearly rescales its components $k = 2, \ldots, p$:

$$x^{(k)*} = \frac{x^{(k)} - A_{k1}}{A_{kk}}.$$

If $\boldsymbol{\Sigma}_{22}$ is diagonal, then so is $\boldsymbol{\Sigma}_{22}^* = \mathbf{A}_2^{-1}\boldsymbol{\Sigma}_{22}(\mathbf{A}_2^{-1})^{\top}$. If in addition $\mathbf{A}_2 = \mathbf{I}$, that is, the linear transformations are changes of origin, then $\boldsymbol{\Sigma}_{22}^* = \boldsymbol{\Sigma}_{22}$. Thus, the random coefficient model in (4.5), in which the only non-zero covariances are in the first row and column,

$$\begin{pmatrix} \Sigma_{11} & \Sigma_{12} & \Sigma_{13} & \cdots & \Sigma_{1r} \\ \Sigma_{12} & \Sigma_{22} & 0 & \cdots & 0 \\ \Sigma_{13} & 0 & \Sigma_{33} & \ddots & \vdots \\ \vdots & \vdots & \ddots & \ddots & 0 \\ \Sigma_{1r} & 0 & \cdots & 0 & \Sigma_{rr} \end{pmatrix}, \tag{4.13}$$

is invariant with respect to linear rescaling. Such a reparametrization is of importance because it is the *minimal* one that satisfies linear invariance for unrelated variables associated with variation.

A set of variables $\mathbf{x}$ is said to be *linearly compatible,* if it is meaningful to define the linear transformation (rotation) $\mathbf{x}\,\mathbf{A}$ for an arbitrary non-singular matrix $\mathbf{A}$. For example, the dummy variables for a categorical variable are usually linearly compatible, because their rotation corresponds to a reparametrization (though Section 4.3.1 contains an example to the contrary). For instance, the GRE scores in Section 3.5 are linearly compatible because their rotation corresponds to a different representation of the academic skills measured by the test. But the GRE and *GP* scores are not compatible; linear combinations of the exam scores and the GRE scores are not meaningful. Linear compatibility is important for selection of the parametric form of the variance matrix $\boldsymbol{\Sigma}$.

In general, we wish to select from a list of potential explanatory variables those that make a nontrivial contribution to the description of the expectation $\mathbf{X}\boldsymbol{\beta}$ (the regression part of the model). This is a task very much akin to its counterpart in ordinary regression. From the regression part of the model we want to further select a subset of explanatory variables that are associated with variation, that is, are included in the variation design of the model. These variables form the *variation part* of the model.

Information about variation is usually much more scarce than is information about regression; this is a recurrent theme of the examples discussed in Chapter 5. Therefore, parsimony of the description of between-cluster variation is paramount for efficient modelling with random coefficients. This issue is parallel to the choice of interactions in ANCOVA models. Model parsimony for the variation part is achieved by judicious (economical) choice of the variation part. In addition, for a given design matrix $\mathbf{Z}$ ($N \times r$) we can consider a parametrization more parsimonious than that given by the r variances and $r(r-1)/2$ covariances. Example B implies that if the origins of the variables in the variation part are chosen arbitrarily (e.g., by a generally accepted convention) then all the covariances in the first row and column (the intercept-by-slope covariances) have to be free parameters. Further, for any pair of linearly compatible variables in the variation part their covariance in $\boldsymbol{\Sigma}$ has to be a free parameter. All the other covariances can be constrained to zero without affecting the invariance of the model.

4.3 Patterns of variation

The variance matrix for the vector of observations $\mathbf{y}$ is block-diagonal, $\mathbf{V} = \{\mathbf{V}_j\} \otimes \mathbf{I}_{N_2}$, with the blocks

$$\mathbf{V}_j = \sigma^2 \mathbf{I}_{n_j} + \mathbf{Z}_j \boldsymbol{\Sigma} \mathbf{Z}_j^{\top} \qquad (4.14)$$

corresponding to the clusters. The variance of an observation,

$$(\mathbf{V}_j)_{ii} = \mathrm{var}(\mathbf{y}_{ij}) = \sigma^2 + \mathbf{z}_{ij} \boldsymbol{\Sigma} \mathbf{z}_{ij}^{\top}, \qquad (4.15)$$

is a quadratic function of the variables in the variation part. Thus, random (regression) coefficients imply variance heterogeneity.

We use the term *pattern of variation* for any verbal, graphical, or algebraic description of features of the joint distribution of the random coefficients. We do not have any rigorous definition in mind. Since the expectation of the random coefficients is always subsumed in the regression part of the model, $\mathbf{X}\boldsymbol{\beta}$, the variance matrix $\boldsymbol{\Sigma}$ contains all the information about the pattern of variation.

Properties of the variance function (4.15) are closely related to the covariance structure of $\boldsymbol{\Sigma}$. We partition the matrix $\boldsymbol{\Sigma}$ as

$$\boldsymbol{\Sigma} \;=\; \begin{pmatrix} \Sigma_{11} & \boldsymbol{\Sigma}_{12} \\ \boldsymbol{\Sigma}_{21} & \boldsymbol{\Sigma}_{22} \end{pmatrix}$$

and $\mathbf{z} = (1, \mathbf{z}_2)$, so that the variance of an observation is equal to

$$\mathrm{var}(y) \;=\; \sigma^2 + \Sigma_{11} + 2\mathbf{z}_2\boldsymbol{\Sigma}_{21} + \mathbf{z}_2\boldsymbol{\Sigma}_{22}\mathbf{z}_2^{\mathsf{T}}. \tag{4.16}$$

The minimum of this quadratic function of $\mathbf{z}_2$ can be found as the root of the first partial derivative,

$$\frac{\partial \mathrm{var}(y)}{\partial \mathbf{z}_2} \;=\; 2\boldsymbol{\Sigma}_{21} + 2\mathbf{z}_2\boldsymbol{\Sigma}_{22}. \tag{4.17}$$

The random-effects model corresponds to $\boldsymbol{\Sigma}_{22} = \mathbf{0}$, in which case the variance in (4.16) is constant. If $\boldsymbol{\Sigma}_{22}$ is a non-singular matrix (note that $\boldsymbol{\Sigma}$ may still be singular), then (4.17) has a unique root,

$$\mathbf{z}_{MV} \;=\; -\boldsymbol{\Sigma}_{12}\boldsymbol{\Sigma}_{22}^{-1}, \tag{4.18}$$

and the minimum of the variance in (4.16) is

$$\mathrm{var}(y\,;\,\mathbf{z}_{MV}) \;=\; \sigma^2 + \Sigma_{11} - \boldsymbol{\Sigma}_{12}\boldsymbol{\Sigma}_{22}^{-1}\boldsymbol{\Sigma}_{21}. \tag{4.19}$$

When all the slope-by-slope covariances are equal to zero, that is, $\boldsymbol{\Sigma}_{22}$ is diagonal (but still positive definite), then $(\mathbf{z}_{MV})_k = -\Sigma_{1k}/\Sigma_{kk}$, and $\mathrm{var}(y\,;\,\mathbf{z}_{MV}) = \sigma^2 + \Sigma_{11} - \sum_k \Sigma_{1k}^2/\Sigma_{kk}$. For a singular non-zero variance matrix $\boldsymbol{\Sigma}_{22}$ the partial derivative in (4.17) has a continuum of roots.

Singularity of $\boldsymbol{\Sigma}$ corresponds to linear dependence of the components of the random vector $\boldsymbol{\delta}_j$ in (4.7). In many applications this is a rather anomalous situation, because usually the purpose of random coefficient models is to allow for a less restricted pattern of variation.

Note, however, that singularity of $\boldsymbol{\Sigma}$ depends on the way we declare the design matrix $\mathbf{Z}$. If $\mathbf{Z}$ contains variables which correspond to zero variances then $\boldsymbol{\Sigma}$ is necessarily singular. But such a variable (a column of $\mathbf{Z}$) could be deleted from the variation design without affecting the model. Similarly, if a correlation $\Sigma_{kh}/(\Sigma_{kk}\Sigma_{hh})^{\frac{1}{2}}$ in the matrix $\boldsymbol{\Sigma}$ is constrained to $+1$ or -1 (although there is unlikely to be a good reason for it), then $\boldsymbol{\Sigma}$ is singular.

We adopt the convention that for any random coefficient model the minimal variation design matrix is declared. That is, if a variance in $\boldsymbol{\Sigma}$ is set to zero then the corresponding variable is regarded as not included in the variation part. To simplify matters, without restricting applicability, we consider the covariances in $\boldsymbol{\Sigma}$ only as either free parameters or as set equal to zero.

Note that the minimal variance of (4.16) may not be realizable; the vector $\mathbf{z}_{MV}$ may lie outside the region of realistic (realized) values of $\mathbf{z}_{ij}$. A case in point, discussed in more detail in the next section, is a dichotomous variable.

To illustrate the relationship of $\mathbf{z}_{MV}$ to the pattern of variation we consider a 2×2 variance matrix $\boldsymbol{\Sigma}$. If $\Sigma_{22} = 0$ then $\mathrm{var}(y) = \sigma^2 + \Sigma_{11}$ is

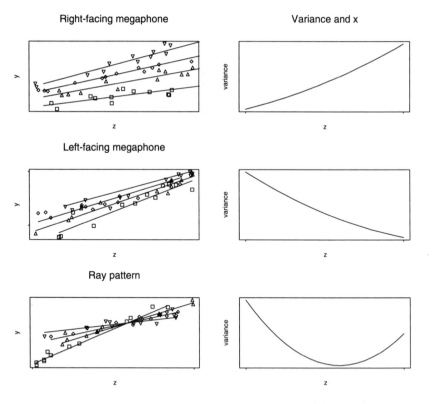

Fig. 4.1. Patterns of variation. Right-facing megaphone, left-facing megaphone, and ray pattern. Simulated examples. The left-hand panels contain the plots of the variables z and y and the right-hand panels contain the plots of the variance of y as a function of z. Each plotting symbol represents a cluster.

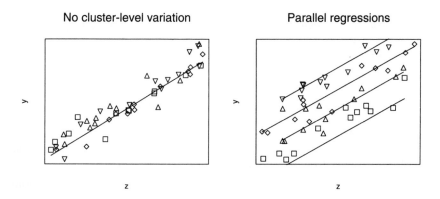

Fig. 4.2.   Patterns of variation. No between-cluster variation (left-hand panel) and parallel regressions (right-hand panel). Simulated examples. The points on the horizontal axis (z) are generated as independent draws from $U(0,1)$; the corresponding points on the vertical axis are generated by the regression $y = z + \delta_j + \varepsilon$, where $\varepsilon \sim N(0, 0.04)$; the values of δ_j are equal to 0 for all four clusters on the left, and equal to -0.33, -0.12, 0.05, and 0.20 on the right. The different plotting symbols represent four clusters.

constant. Otherwise $\operatorname{var}(y; z_2)$ has a unique minimum of $\sigma^2 + \Sigma_{11} - \Sigma_{12}^2 / \Sigma_{22}$ attained at $z_{MV} = -\Sigma_{12}/\Sigma_{22}$. The minimum is equal to σ^2 if and only if the intercept and slope are perfectly correlated, $\Sigma_{21}^2 = \Sigma_{11}\Sigma_{22}$.

Suppose the variable $z = z^{(2)}$ (the second column of $\mathbf{Z}$) has values in the interval (z_L, z_H). If $z_{MV} \leq z_L$ then $\operatorname{var}(y; z)$ increases in z, if $z_{MV} \geq z_L$ then $\operatorname{var}(y; z)$ decreases in z, and for $z_L < z_{MV} < z_H$ the variance decreases in (z_L, z_{MV}) and increases in (z_{MV}, z_H). When $\boldsymbol{\Sigma}$ is singular these three cases correspond to the right- and left-facing megaphones and

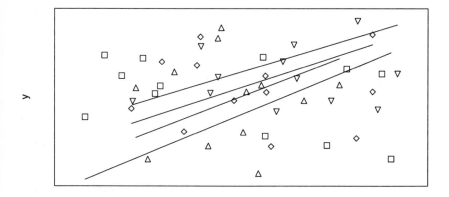

Fig. 4.3.   Patterns of variation. Near megaphone pattern. Simulated example.

the ray pattern, as depicted in Figure 4.1. The right-hand panels contain the plots of the variance as a function of z. For contrast, the case of no between-cluster variation ($\boldsymbol{\Sigma} = \mathbf{0}$) and the pattern for random effects ($\Sigma_{22} = 0$) are depicted in Figure 4.2. Note that, except for the ray pattern, in all these cases there is a natural ordering of the clusters because the within-cluster regression lines do not intersect within the range of the data.

When $\boldsymbol{\Sigma}$ is non-singular it is still possible that the regression lines do not intersect, especially when the correlation ρ_{12} of the intercept and slope is close to $+1$ or -1, or the range (z_L, z_H) is narrow. In these cases z_{MV} is outside the interval (z_L, z_H). Figure 4.3 depicts an example.

4.3.1 CATEGORICAL VARIABLES AND VARIATION

A *continuous* (quantitative) variable in the variation part corresponds to varying within-cluster regression slopes on that variable. A *categorical* variable in the variation part corresponds to varying within-cluster differences among the categories. In this section we discuss patterns of between-cluster variation induced by a categorical variable.

For simplicity we consider first a dichotomous variable in the variation part, say, male/female. Suppose the within-cluster male–female differences have the distribution $\mathcal{N}(d, \tau^2)$. The proportion of clusters with negative male–female differences is $\Phi(-d/\tau)$. Even if the average difference d is positive, but τ is large relative to d, this proportion is sizeable; it approaches 50 per cent as τ diverges to $+\infty$.

Note that τ^2 does not contain all the information about the pattern of variation of the within-cluster differences. Suppose τ^2 is positive and the within-cluster means for men and women have different variances. If we interchange the categories, the variance of their differences remains equal to τ^2 but the pattern of variation is altered.

Not even the distributions of within-cluster means for men and women contain complete information about the pattern of variation. Each of the four panels in Figure 4.4 plot the within-cluster means of an outcome variable for men (left-hand side of the panel) and women (right-hand side). Points corresponding to the same cluster are joined by dotted lines. Men and women have the same distributions of within-cluster means in all four panels, but in one case the ranks of the means for men and women are identical (synchronous pattern), in another case the ranks are reversed (criss-cross pattern); in fact, any ranking of the means for men and women can be realized (in this illustration, it can be accomplished by a suitable permutation of one set of the means). Note that the synchronous pattern does not correspond to zero variance of within-cluster male–female differences; for that the dotted lines in the panel have to be parallel, and then, necessarily, the means for men and women have equal variances. Also, the criss-cross pattern corresponds to a singular cluster-level variance matrix only if the dotted lines have a common intersection.

All four patterns of variation depicted in Figure 4.4 involve differing unconditonal variances for men and women. Another means of allowing category-specific variances is by having two elementary-level variances, say, σ_M^2 and σ_F^2, one for each category. Now the two categories have different (unconditional) variances even if they have constant within-cluster differences. Thus, for clustered observations there are two alternative ways of modelling category-specific variance:

- random within-cluster differences among the categories;
- category-specific elementary-level variances.

These two ways are not identical. The former implies within-cluster covariances, whereas the latter does not. In practice, it is difficult to distinguish between the corresponding patterns of variation unless the data contain a large number of clusters in which both categories are represented by many observations.

Categorical variables often contain the category 'not known' or 'none of the above' (*dnk*). When using such a category as a predictor, the fact that the observations in this category may have higher variance than in the

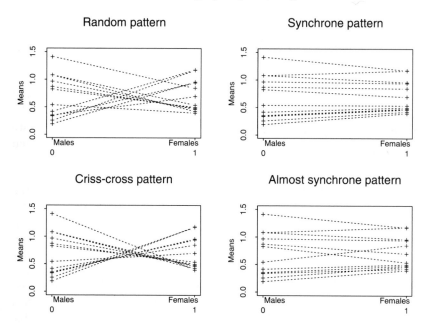

Fig. 4.4. Patterns of variation in within-cluster differences of two categories. Simulated example. The distributions of the means for men and women in each panel are identical, but the joint distributions of the within-cluster means for men and women are not. The means for men and women in a cluster are joined by a dotted line.

other categories is commonly disregarded. Suppose, for illustration, that the category dnk contains units whose classification in the other ('proper') categories has been lost. Then the uneven representation of the proper categories in dnk is reflected by varying differences between dnk and the proper categories, even if the differences among the proper categories are constant across the clusters. The varying differences between dnk and the proper categories may be due to varying composition but may also be due to different within-cluster processes of allocation of observations into the category dnk. Note that in this setting it is not meaningful to define arbitrary linear contrasts involving the category dnk, and so the comments on compatibility of the dummy variables (Section 4.2.2) do not apply.

4.4   Maximum likelihood estimation

All three methods of maximum likelihood estimation described in Sections 2.3–2.5 have extensions for the random coefficient model (4.7). In this section we discuss the Fisher scoring algorithm.

As pointed out in Section 2.3 equations (2.14)–(2.21) apply for any linear regression model given by the log-likelihood

$$l \;=\; -\frac{1}{2}\left\{\log(2\pi) + \log(\det \mathbf{V}) + \mathbf{e}^{\top}\mathbf{V}^{-1}\mathbf{e}\right\}, \tag{4.20}$$

where $\mathbf{e} = \mathbf{y} - \mathbf{X}\boldsymbol{\beta}$ is the vector of residuals. The vector $\mathbf{e}$ depends on the parameters $\boldsymbol{\beta}$; we will also use $\mathbf{e}$ to denote the vector of fitted residuals $\mathbf{y} - \mathbf{X}\hat{\boldsymbol{\beta}}$.

For efficient calculation of the log-likelihood (4.20) and of its scoring vector and information matrix it is essential to avoid numerical inversion of the variance matrices $\mathbf{V}_j$, especially when the data contain clusters with large sample sizes. The following equations express the inverse and the determinant of $\mathbf{V}_j$ in terms of the inverse and determinant of the $r \times r$ matrix $\mathbf{G}_j \;=\; \mathbf{I}_r + \sigma^{-2}\mathbf{Z}_j^{\top}\mathbf{Z}_j\boldsymbol{\Sigma}$:

$$\mathbf{V}_j^{-1} \;=\; \sigma^{-2}\left(\mathbf{I}_{n_j} - \sigma^{-2}\mathbf{Z}_j\boldsymbol{\Sigma}\mathbf{G}_j^{-1}\mathbf{Z}_j^{\top}\right) \tag{4.21}$$

and

$$\det \mathbf{V}_j \;=\; \sigma^{2n_j}\det \mathbf{G}_j. \tag{4.22}$$

The inversion formula is proved by checking that the product of $\mathbf{V}_j$, (4.14), and (4.21) is the identity matrix. Let $\boldsymbol{\Omega} = \sigma^{-2}\boldsymbol{\Sigma}$. We have

$$\sigma^2(\mathbf{I}_{n_j} + \mathbf{Z}_j\boldsymbol{\Omega}\mathbf{Z}_j^{\top})\,\sigma^{-2}(\mathbf{I}_{n_j} - \mathbf{Z}_j\boldsymbol{\Omega}\mathbf{G}_j^{-1}\mathbf{Z}_j^{\top})$$

$$= \mathbf{I}_{n_j} + \mathbf{Z}_j\boldsymbol{\Omega}\left\{\mathbf{I}_{n_j} - (\mathbf{I}_{n_j} + \mathbf{Z}_j^{\top}\mathbf{Z}_j\boldsymbol{\Omega})\mathbf{G}_j^{-1}\right\}\mathbf{Z}_j^{\top} \;=\; \mathbf{I}_{n_j}.$$

To prove (4.22) consider the $(n_j + r) \times (n_j + r)$ matrix $\mathbf{U}$ with the partitioning

$$\mathbf{U} = \begin{pmatrix} \mathbf{I}_r & -\mathbf{Z}_j^\top \\ \mathbf{Z}_j \boldsymbol{\Sigma} & \sigma^2 \mathbf{I}_{n_j} \end{pmatrix}. \tag{4.23}$$

Let $\mathbf{L}_1 = \begin{pmatrix} \mathbf{I}_r & \sigma^{-2}\mathbf{Z}_j^\top \\ 0 & \mathbf{I}_{n_j} \end{pmatrix}$ and $\mathbf{L}_2 = \begin{pmatrix} \mathbf{I}_r & 0 \\ -\mathbf{Z}_j \boldsymbol{\Sigma} & \mathbf{I}_{n_j} \end{pmatrix}$. The determinants of $\mathbf{L}_1$ and $\mathbf{L}_2$ are each equal to unity, and therefore $\det(\mathbf{L}_1 \mathbf{U}) = \det(\mathbf{L}_2 \mathbf{U}) = \det(\mathbf{U})$. But the determinants of $\mathbf{L}_1 \mathbf{U} = \begin{pmatrix} \mathbf{G}_j & 0 \\ \mathbf{Z}_j \boldsymbol{\Sigma} & \sigma^2 \mathbf{I}_{n_j} \end{pmatrix}$ and $\mathbf{L}_2 \mathbf{U} = \begin{pmatrix} \mathbf{I}_r & -\mathbf{Z}_j^\top \\ 0 & \mathbf{V}_j \end{pmatrix}$ are the respective sides of (4.22). Note that the matrix products $\mathbf{L}_1 \mathbf{U}$ and $\mathbf{L}_2 \mathbf{U}$ correspond to partial sweeping of rows of the matrix $\mathbf{U}$.

As in Section 2.3 it is advantageous to consider the parametrization $(\sigma^2, \boldsymbol{\Omega})$, so that the elementary-level variance σ^2 can be separated out from the log-likelihood:

$$l = -\frac{1}{2}\left\{ N \log(2\pi\sigma^2) + \log(\det \mathbf{W}) + \frac{1}{\sigma^2}\mathbf{e}^\top \mathbf{W}^{-1}\mathbf{e} \right\}, \tag{4.24}$$

where $\mathbf{W} = \{\mathbf{W}_j\} \otimes \mathbf{I}_{N_2}$, with $\mathbf{W}_j = \mathbf{I}_{n_j} + \mathbf{Z}_j \boldsymbol{\Omega} \mathbf{Z}_j^\top$ ($= \sigma^{-2}\mathbf{V}_j$); $\mathbf{W}_j$ as a function of $\boldsymbol{\Omega}$ does not depend on σ^2. The scoring equation for σ^2 has an explicit solution (given all the other parameters):

$$\hat{\sigma}^2 = \frac{\mathbf{e}^\top \mathbf{W}^{-1}\mathbf{e}}{N}. \tag{4.25}$$

The identities (2.14) and (2.15) hold for random coefficient models (they do not depend on the form of the variance matrix $\mathbf{V}$), and so the MLE for $\boldsymbol{\beta}$, assuming $\mathbf{V}$ known, coincides with the GLS estimator

$$\hat{\boldsymbol{\beta}} = \left(\mathbf{X}^\top \mathbf{V}^{-1}\mathbf{X}\right)^{-1}\mathbf{X}^\top \mathbf{V}^{-1}\mathbf{y}, \tag{4.26}$$

with $\text{var}(\hat{\boldsymbol{\beta}}) = \left(\mathbf{X}^\top \mathbf{V}^{-1}\mathbf{X}\right)^{-1}$.

We refer to $\boldsymbol{\Omega}$ as the *scaled* (cluster-level) variance matrix, and to its elements as scaled (between-cluster) variances and covariances. The equations for the scoring and information functions, for the elements of $\boldsymbol{\Omega}$,

$$\frac{\partial l}{\partial \omega} = \frac{1}{2}\left\{-\mathrm{tr}\left(\mathbf{W}^{-1}\frac{\partial \mathbf{W}}{\partial \omega}\right) + \sigma^{-2}\mathbf{e}^{\top}\mathbf{W}^{-1}\frac{\partial \mathbf{W}}{\partial \omega}\mathbf{W}^{-1}\mathbf{e}\right\},$$

$$-\mathbf{E}\left(\frac{\partial^2 l}{\partial \omega_1 \omega_2}\right) = \frac{1}{2}\mathrm{tr}\left(\mathbf{W}^{-1}\frac{\partial \mathbf{W}}{\partial \omega_1}\mathbf{W}^{-1}\frac{\partial \mathbf{W}}{\partial \omega_2}\right), \tag{4.27}$$

involve the matrix derivatives $\partial \mathbf{W}_j/\partial \omega = \mathbf{Z}_j(\partial \boldsymbol{\Omega}/\partial \omega)\mathbf{Z}_j^{\top}$. For an element ω of $\boldsymbol{\Omega}$ the matrix $\partial \boldsymbol{\Omega}/\partial \omega$ is the incidence matrix for ω; its elements are equal to zero with the exception of the entries in the same location(s) as ω, which are equal to unity. For example, for a 3×3 variance matrix $\boldsymbol{\Omega}$,

$$\frac{\partial \boldsymbol{\Omega}}{\partial \Omega_{13}} = \begin{pmatrix} 0 & 0 & 1 \\ 0 & 0 & 0 \\ 1 & 0 & 0 \end{pmatrix}$$

and

$$\frac{\partial \boldsymbol{\Omega}}{\partial \Omega_{22}} = \begin{pmatrix} 0 & 0 & 0 \\ 0 & 1 & 0 \\ 0 & 0 & 0 \end{pmatrix}.$$

In a slight departure from our conventions we denote the kth diagonal element of $\boldsymbol{\Omega}$ by Ω_k, and set $\Omega_{kk} = \frac{1}{2}\Omega_k$. We refer to the parametrization of $\boldsymbol{\Omega}$ in terms of Ω_{kk} as the *half-variance* parametrization. Further, let $\boldsymbol{\Delta}_{k,r}$ be the column (indicator) vector which has a 1 in the kth position and zeros elsewhere. Then

$$\frac{\partial \boldsymbol{\Omega}}{\partial \Omega_{kh}} = \boldsymbol{\Delta}_{k,r}\boldsymbol{\Delta}_{h,r}^{\top} + \boldsymbol{\Delta}_{h,r}\boldsymbol{\Delta}_{k,r}^{\top},$$

for both covariances and half-variances. The 'building blocks' for equations (4.25)–(4.27) can be expressed as totals of products of certain quadratic forms in $\mathbf{W}_j^{-1}$:

$$\frac{1}{2}\mathrm{tr}\left(\mathbf{W}_j^{-1}\frac{\partial \mathbf{W}_j}{\partial \Omega_{kh}}\right) = \boldsymbol{\Delta}_{k,r}^{\top}\mathbf{Z}_j^{\top}\mathbf{W}_j^{-1}\mathbf{Z}_j\boldsymbol{\Delta}_{h,r} = (\mathbf{S}_j)_{kh},$$

$$\frac{1}{2}\mathbf{e}_j^{\top}\mathbf{W}_j^{-1}\frac{\partial \mathbf{W}_j}{\partial \Omega_{kh}}\mathbf{W}_j^{-1}\mathbf{e}_j = \mathbf{e}_j^{\top}\mathbf{W}_j^{-1}\mathbf{Z}_j\boldsymbol{\Delta}_{k,r}\boldsymbol{\Delta}_{h,r}^{\top}\mathbf{Z}_j^{\top}\mathbf{W}_j^{-1}\mathbf{e}_j$$

$$= (\mathbf{s}_j)_k(\mathbf{s}_j)_h,$$
$$\tag{4.28}$$

$$\frac{1}{2}\mathrm{tr}\left(\mathbf{W}_j^{-1}\frac{\partial \mathbf{W}_j}{\partial \Omega_{k_1 h_1}}\mathbf{W}_j^{-1}\frac{\partial \mathbf{W}_j}{\partial \Omega_{k_2 h_2}}\right) = (\mathbf{S}_j)_{k_1 k_2}(\mathbf{S}_j)_{h_1 h_2} + (\mathbf{S}_j)_{k_1 h_2}(\mathbf{S}_j)_{h_1 k_2},$$

where $\mathbf{S}_j = \mathbf{Z}_j^{\top}\mathbf{W}_j^{-1}\mathbf{Z}_j$ and $\mathbf{s}_j = \mathbf{Z}_j^{\top}\mathbf{W}_j^{-1}\mathbf{e}_j$; $(\mathbf{S}_j)_{kh}$ and $(\mathbf{s}_j)_k$ stand for the respective elements of the matrix and vector in parentheses. Thus, the elements of the scoring function and of the information matrix for the covariance structure parameters can be obtained from the elements of the matrices $\mathbf{S}_j$ and the vectors $\mathbf{s}_j$. The advantage of the half-variance parametrization is that equations (4.28) hold for both covariance and half-variance parameters. Using (4.21) we obtain the following equation for $\mathbf{S}_j$:

$$\mathbf{S}_j = (\mathbf{I} - \mathbf{Z}_j^\top \mathbf{Z}_j \mathbf{\Omega} \mathbf{G}_j^{-1})\mathbf{Z}_j^\top \mathbf{Z}_j = \mathbf{G}_j^{-1}\mathbf{Z}_j^\top \mathbf{Z}_j, \tag{4.29}$$

and similarly,

$$\mathbf{s}_j = \mathbf{G}_j^{-1}\mathbf{Z}_j^\top \mathbf{e}_j, \tag{4.30}$$

which simplifies the computations somewhat.

The score and information functions for the regression parameters are calculated using the identity

$$\mathbf{u}_1^\top \mathbf{W}^{-1}\mathbf{u}_2 = \mathbf{u}_1^\top \mathbf{u}_2 - \sum_j \mathbf{u}_{1j}^\top \mathbf{Z}_j \mathbf{\Omega} \mathbf{G}_j^{-1}\mathbf{Z}_j^\top \mathbf{u}_{2j}, \tag{4.31}$$

where $\mathbf{u}_1$ and $\mathbf{u}_2$ are arbitrary vectors of length N (e.g., columns of $\mathbf{X}$), and $\mathbf{u}_{1j}$ and $\mathbf{u}_{2j}$ are their respective subvectors corresponding to the cluster j.

In summary, the Fisher scoring algorithm for fitting the random coefficient model (4.7) requires the following data summaries:

- the matrix of sample sums of squares and cross-products

$$(\mathbf{X}, \ \mathbf{y})^\top (\mathbf{X}, \ \mathbf{y}),$$

- the within-cluster sums of squares and cross-products

$$(\mathbf{X}_j, \ \mathbf{y}_j)^\top \mathbf{Z}_j,$$

and $\mathbf{Z}_j^\top \mathbf{Z}_j$ when $\mathbf{Z}_j$ is not a submatrix of $\mathbf{X}_j$.

Note that the corresponding ANCOVA requires the same summaries.

Just as for the random-effects model (2.1) the OLS estimators $\hat{\boldsymbol{\beta}}_0 = (\mathbf{X}^\top \mathbf{X})^{-1}\mathbf{X}^\top \mathbf{y}$ and $\hat{\sigma}^2 = N^{-1}\left(\mathbf{y}^\top \mathbf{y} - \hat{\boldsymbol{\beta}}_0^\top \mathbf{X}^\top \mathbf{X}\hat{\boldsymbol{\beta}}_0\right)$ are suitable starting values for the iterative algorithm. For the matrix $\mathbf{\Omega}$ a non-negative definite diagonal matrix can be chosen as a starting value. For the diagonal elements of $\mathbf{\Omega}$ the following starting values can be used:

$$\hat{\Omega}_k = \frac{\sum_j \left(\mathbf{e}_j^\top \mathbf{z}_j^{(k)}\right)^2}{N_2 \sum_j \left(\mathbf{z}_j^{(k)}\right)^\top \mathbf{z}_j^{(k)}}, \tag{4.32}$$

where the residual vectors $\mathbf{e}_j$ are evaluated for the OLS estimator $\hat{\boldsymbol{\beta}}_0$. This is a generalization of equation (2.30).

4.4.1   CONSTRAINED MAXIMIZATION

Constrained maximization has to be applied for fitting a random coefficient model because the matrix $\boldsymbol{\Omega}$ is non-negative definite, and so should be its estimate $\hat{\boldsymbol{\Omega}}$. In Section 2.3 we dealt with a similar problem, the constraint of non-negativity of the between-cluster variance τ^2, by estimating the square root of τ^2 or by 'temporarily' extending the parameter space for the variance τ^2.

When estimating variance matrices, it does not suffice to insist on non-negativity of the variances. For example, the matrix

$$\begin{pmatrix} 4 & 0 & 3 \\ 0 & 4 & 3 \\ 3 & 3 & 4 \end{pmatrix}$$

is not a variance matrix, even though all its 'correlations', 0, $\frac{3}{4}$, and $\frac{3}{4}$, are smaller than unity in absolute value (the determinant of this matrix is equal to –8). The method of Lagrange multipliers is usually not practicable for maximization of the log-likelihood because the condition of non-negative definiteness of a matrix does not have a suitable algebraic description.

If the estimated variance matrix $\hat{\boldsymbol{\Sigma}}$ does not have any constrained elements then the decomposition $\boldsymbol{\Sigma} = \mathbf{L}\mathbf{L}^{\top}$ into the product of a triangular matrix with its transpose (the Cholesky decomposition) can be estimated. The equations for the Fisher scoring algorithm can be derived by using the chain rule,

$$\frac{\partial \boldsymbol{\Sigma}}{\partial \lambda} = \mathbf{L}\frac{\partial \mathbf{L}^{\top}}{\partial \lambda} + \frac{\partial \mathbf{L}}{\partial \lambda}\mathbf{L}^{\top},$$

for any parameter λ (e.g., an element of $\mathbf{L}$). The matrix $\mathbf{L}$ has $(p^2 + p)/2$ parameters. If $\boldsymbol{\Sigma}$ is singular then some of the elements of $\mathbf{L}$ are not identifiable. Since we do not attach any interpretation to the elements of $\mathbf{L}$ it suffices to constrain the elements of $\mathbf{L}$ to zero one by one, in a suitable order, until all its remaining elements are identified. The ranks of $\mathbf{L}$ and $\boldsymbol{\Sigma}$ coincide; when $\mathbf{L}$ is estimated, finding the rank of $\hat{\boldsymbol{\Sigma}}$ is trivial.

Another approach is to ensure that the estimated covariances do not become too large at the early iterations. We define the *generalized correlation* for a variance matrix $\boldsymbol{\Sigma}$ as the complement to unity of the ratio of its determinant and the product of its diagonal entries:

$$\rho^* = 1 - \frac{\det \boldsymbol{\Sigma}}{\prod_k \Sigma_k}, \tag{4.33}$$

where the matrix $\boldsymbol{\Sigma}$ is assumed to contain no zero variances (otherwise the corresponding row and column of $\boldsymbol{\Sigma}$ are deleted). The starting solution for the variance matrix $\boldsymbol{\Sigma}$ is diagonal, and so it corresponds to $\rho^* = 0$. If the correction for $\hat{\Sigma}$ at the first iteration yields a matrix with a negative eigenvalue, then the correction is halved as many times as necessary for the new solution $\hat{\Sigma}$ to be positive definite. Further halving of the steps is then applied until the generalized correlation is smaller than $\frac{1}{2}$.

Similar step-halving of the solution is then applied at each iteration; at the kth iteration the Fisher scoring corrections are halved until the solution is positive definite and $\rho^* < 1 - 2^{-k}$. This way convergence to a singular solution $\hat{\Sigma}$ may be obtained, although the speed of convergence is somewhat affected by step-halving. The procedure can be fine-tuned to balance the requirements of fast convergence and convergence to a local maximum for a singular variance matrix.

Note that in the EM algorithm each iteration yields a non-negative definite estimated variance matrix, and so the problems of constrained maximization do not arise. However, convergence of the algorithm is very slow when the solution is singular. Also, constraints on the covariances are difficult to implement in the EM algorithm.

4.4.2 CONFOUNDING IN THE VARIATION PART

In general it is difficult to make an informed decision about how complex a description for the variation part of a model can be supported by a given dataset. The analogous problem for regression parameters is usually resolved by reference to familiar results from linear algebra. Covariance structure parameters are, however, not associated with linear spaces, and their possible confounding (correlation structure of the estimates) is in general very difficult to explore.

The condition number of the information matrix for the covariance structure parameters provides some insight into how the variation design and the model parameters affect confounding. Detailed analysis is feasible only in some simple cases. Suppose the variation design is balanced (identical design matrices $\mathbf{Z}_j$), and contains a single quantitative variable. The information matrix for the covariance structure parameters $(\Omega_{11}, \Omega_{12}, \Omega_{22})$ in this case is

$$\mathbf{H} = \frac{1}{2}N_2 \begin{pmatrix} S_{11}^2 & S_{12}^2 & 2S_{11}S_{12} \\ S_{12}^2 & S_{22}^2 & 2S_{12}S_{22} \\ 2S_{11}S_{12} & 2S_{12}S_{22} & 2S_{11}S_{22} + 2S_{12}^2 \end{pmatrix}, \qquad (4.34)$$

where S_{11}, S_{12}, and S_{22} are the elements of $\mathbf{S}$, the common value of all $\mathbf{S}_j$ given by (4.29). The determinant of $\mathbf{H}$ is equal to $\frac{1}{4}N_2^3(S_{11}S_{22} - S_{12}^2)^3$, that is, $\frac{1}{4}N_2^3(\det \mathbf{S})^3$. Conditioning of the information matrix for Ω is related to conditioning of $\mathbf{S}$; if λ is the condition number for $\mathbf{S}$ then λ^3 is the condition number for $\mathbf{H}$. The matrix $\mathbf{H}$ is diagonal if and only if $\mathbf{S}$ is diagonal $(S_{12} = 0)$; then no identification problems arise unless S_{11} or S_{22} is small.

If $\mathbf{Z}_j$ is of full rank then

$$\mathbf{S}_j = \left\{ \left(\mathbf{Z}_j^\top \mathbf{Z}_j \right)^{-1} + \boldsymbol{\Omega} \right\}^{-1}. \tag{4.35}$$

For balanced variation design, $\mathbf{Z}_j$ is of full rank if all the variables in the variation part are elementary-level and are linearly independent. When $\mathbf{Z}$ contains only one variable, this is trivially satisfied. Using (4.35) we can establish when $\mathbf{S}$ is close or equal to a diagonal matrix.

When the entries of $\mathbf{U}_j = \left(\mathbf{Z}_j^\top \mathbf{Z}_j \right)^{-1}$ are small relative to $\boldsymbol{\Omega}$, the properties of $\mathbf{S}$ are influenced primarily by $\boldsymbol{\Omega}$. This appears to be counter-intuitive because, on the whole, the larger the clusters the more influential $\boldsymbol{\Omega}$ is. However, this merely means that small clusters contain little information about $\boldsymbol{\Omega}$, irrespective of the actual value of $\boldsymbol{\Omega}$, whereas for larger clusters the information about $\boldsymbol{\Omega}$ depends on the value of $\boldsymbol{\Omega}$.

The information matrix (4.34) is ill-conditioned when $\mathbf{S}$ is close to a singular matrix. This problem arises when $\boldsymbol{\Omega}$ is (nearly) singular and the entries of $\mathbf{U}$ are small relative to those of $\boldsymbol{\Omega}$. This also contradicts intuition because $\mathbf{U}$ is small when the clusters are large.

Next we explore information about variation when a cluster-level variable is associated with variation in a general unbalanced design. The matrices $\mathbf{Z}_j^\top \mathbf{Z}_j$ are now singular, and so are the matrices $\mathbf{S}_j$ given by (4.29). Then the information matrix (4.34) is singular. Since the association of a cluster-level variable with variation does not have an interpretation in terms of random slopes it is appropriate not to insist on the properties of invariance and to choose a more economic parametrization for $\boldsymbol{\Omega}$, e.g., by constraining the covariance Ω_{12} to zero while estimating the variances Ω_1 and Ω_2.

4.5 Longitudinal analysis

In a typical longitudinal study an outcome variable y is measured at several time-points on each of a sample of subjects. The observations $\{y_{ij}\}_i$ on subject j form a cluster. For each subject we postulate a model for the dependence of the outcome on time:

$$y(t \mid j) = f_j(t) + \varepsilon_t, \tag{4.36}$$

$\varepsilon_t \sim \mathcal{N}(0, \sigma^2)$, i.i.d., and these models (regressions f_j) may vary across the subjects. Polynomial regression in (4.36) is an important example:

$$f_j(t) = \mathbf{t}\boldsymbol{\beta}_j, \tag{4.37}$$

where $\mathbf{t} = (1, t, t^2, \ldots, t^p)$. The usual assumption of multivariate normality of the regression coefficients,

$$\boldsymbol{\beta}_j \sim \mathcal{N}(\boldsymbol{\beta}, \ \boldsymbol{\Sigma}), \tag{4.38}$$

provides a flexible model specification for between-subject variation. Formally, equations (4.36)–(4.38) specify a random coefficient model with a polynomial regression and an unstructured pattern of between-subject variation.

Usually, each subject contributes only a small number of observations, and the resulting dataset contains limited information about each function f_j. Nevertheless, the mean regression $f(\mathbf{t}) = \mathbf{t}\boldsymbol{\beta}$ may be well determined, especially when the dataset contains a large number of subjects. Information about between-subject variation is usually much more scarce than that about regression, and so constraints on the elements of $\boldsymbol{\Sigma}$ have to be considered. In a typical setup $f(t)$ is a polynomial of degree p but coefficients associated with degrees $k + 1, \ldots, p$ are constant across the subjects. In other words, the variation part of the model comprises a polynomial of degree $k \leq p$.

The model given by (4.36)–(4.38) can be supplemented by further explanatory variables. Those recorded for measurement occasions (observation-level variables) are added to the regression part (4.37),

$$\mathbf{f}_j(\mathbf{t}; \mathbf{x}_1) = (\mathbf{t}, \mathbf{x})\boldsymbol{\beta}_j, \tag{4.39}$$

and each of them may be associated with variation, $\boldsymbol{\beta}_j \sim \mathcal{N}(\boldsymbol{\beta}, \ \boldsymbol{\Sigma})$. Subject-level variables can similarly be added to the regression part of the model, although, in general, they should not be associated with variation.

The assumption for between-subject variation, (4.38), can be extended to a subject-level regression model,

$$\boldsymbol{\beta}_j = \mathbf{B}\mathbf{u}_j + \boldsymbol{\delta}_j, \tag{4.40}$$

where $\boldsymbol{\delta}_j \sim \mathcal{N}(\mathbf{0}, \boldsymbol{\Sigma})$, i.i.d., $\mathbf{u}_j$ is a vector of subject-level variables (including the intercept 1 as its first component), and $\mathbf{B}$ is a matrix of parameters.

In many contexts the distinction between observation- and subject-level variables is a natural one because the former are usually subject to temporal variation, whereas the latter represent 'permanent' attributes of the subjects, and, in principle, the distribution of their values can be set either by design or by deliberate selection of subjects.

Equations (4.39) and (4.40) can be combined into a single equation,

$$\mathbf{f}_j(\mathbf{t}) = (\mathbf{t}, \mathbf{x})\mathbf{B}\mathbf{u}_j + (\mathbf{t}, \mathbf{x})\boldsymbol{\delta}_j + \boldsymbol{\varepsilon}_j \tag{4.41}$$

which has the form of a random coefficient model with the regression part consisting of observation- and subject-level variables, and of (some of) their interactions. The interactions allow for consistent differences in time dependence among subjects with different attributes.

The model in (4.41) does not assume any specific distribution of the time-points within the subjects. Subjects may contribute with unequal numbers of unevenly distributed observations. This is an essential advantage over the methods based on multivariate regression (Section 4.6) which assume a balanced (rectangular) design: each subject provides observations at each designated time-point t_k, $k = 1, \ldots, K$.

In observational studies, such a 'rectangular' design is frequently not feasible. In medical studies involving outpatients the subjects may be encouraged to attend at specified time-points, but everyday events, including medical complications, cause changes in the planned appointments. More serious is the effect of informative attrition. Patients may be less likely to attend if the state of their studied condition has improved (worsened). Similar concerns arise in studies of diet of both human and animal subjects. Analysis of studies in which non-participation may be directly linked to the effect of the treatment is particularly problematic. The death of an experimental animal with a short life span, e.g., in a toxicological study, is a case in point. It is usually important to establish whether the cause of death is related to the experimental treatment.

In general, the regression parameters and their estimates in a polynomial regression can be meaningfully interpreted only for low degree polynomials. Otherwise, plots of the polynomials are more informative. However, features of the average regression polynomial $\mathbf{t}\boldsymbol{\beta}$ (such as monotonicity, points of inflexion, concavity, and similar) may not be replicated on each of the within-subject regressions. To obtain an idea of the variety of the functions f_j implied by the model (4.41) it is instructive to plot the polynomials $\{\mathbf{t}(\boldsymbol{\beta} + \boldsymbol{\delta}_j)\}$ for a random sample of $\boldsymbol{\delta}_j$ from $\mathcal{N}(\mathbf{0}, \boldsymbol{\Sigma})$ or $\mathcal{N}(\mathbf{0}, \hat{\boldsymbol{\Sigma}})$. Figure 4.5 presents an example of a cubic regression with variation of the quadratic (and lower-order) coefficients.

The mean polynomial, drawn by a solid bold curve, is

$$\{\mathbf{E}(y) = \} \quad f(x) \;=\; 4 + 12x - 9x^2 + 2x^3$$

$(0 < x < 3.5)$, and the variance matrix for the four regression coefficients is

$$\boldsymbol{\Sigma} \;=\; \begin{pmatrix} 3 & 0 & -1 & 0 \\ 0 & 2 & -1 & 0 \\ -1 & -1 & 1 & 0 \\ 0 & 0 & 0 & 0 \end{pmatrix}.$$

The regression polynomial $f(x)$ has a unique local maximum at $x_{max} = 1$ and a unique local minimum at $x_{min} = 2$; $f(x) \to -\infty$ as $x \to -\infty$ and $f(x) \to +\infty$ as $x \to +\infty$. However, the sample of curves contains a variety of shapes. Several curves do not have a minimum around $x = 2$, and those that do have the minimum for $x > 2$. Several curves do not have a local maximum (other than at $x = 3.5$), and most of those that do attain them at $x < 1$. Features of the regression polynomial $f(x)$ are a poor substitute for the description of the features of curves f_j.

Better insight can be gained by exploring the properties of a typical (conditional) growth curve directly. The derivative of the growth-curve function $f(x) + \delta_0 + \delta_1 x + \delta_2 x^2$ is

$$12 + \delta_1 - (18 - 2\delta_2)x + 6x^2,$$

where δ_1 and δ_2 are the realizations of the deviations in the linear and quadratic terms respectively. The roots of this quadratic function are

$$\frac{(9 - \delta_2) \pm \sqrt{9 - 6\delta_1 - 18\delta_2 + \delta_2^2}}{6}. \tag{4.42}$$

We see that the random terms δ_1 and δ_2 have a dominant influence over the location of the extremes of the growth curve. The discriminant in (4.42) contains the linear combination $\delta = 6\delta_1 + 18\delta_2$, which is normally distribed with zero mean and variance 180. The variance of δ_2^2, a χ_1^2-distributed random variable, is negligible in comparison. Depending on the realization of δ the growth curve is strictly increasing (negative discriminant in (4.42)), or has its maximum for $x < 1$ and minimum for $x > 2$ (with high probability when $\delta < -20$). When the growth curve does have a maximum and minimum, they are symmetric around $x = 1.5$ only when $\delta_2 = 0$.

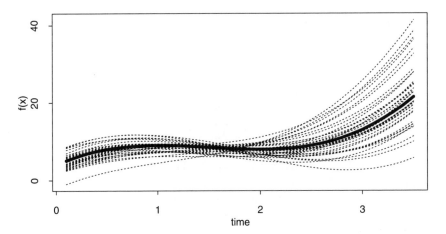

Fig. 4.5. Pattern of variation with random polynomials. Simulated example. The mean of the polynomials is drawn by a bold solid line. The realizations of the polynomials are drawn by dotted lines.

The between-cluster variance of an observation at time-point x is

$$(1, x, x^2) \boldsymbol{\Sigma} (1, x, x^2)^\top = 3 - 2x^3 + x^4.$$

It is easy to check that this variance, as a function of x, has a point of inflexion at $x = 0$ ($f(0) = 3$), a unique minimum at $x = 1.5$ ($f(1.5) = 1.31$), and is increasing for $x > 1.5$, very steeply for $x > 2$ ($f(2) = 3$ and $f(3) = 30$). This pattern of variation can be discerned from Figure 4.5.

4.6   Multivariate regression

In a general multivariate regression model we assume that independent $1 \times K$ (row) vectors of observations $\{\mathbf{y}_j\}$, $j = 1, \ldots, N_2$, are related to $1 \times p$ vectors of explanatory variables $\{\mathbf{x}_j\}$ by linear regression,

$$\mathbf{y}_j = \mathbf{x}_j \mathbf{B} + \boldsymbol{\gamma}_j, \tag{4.43}$$

where $\mathbf{B}$ is a $p \times K$ matrix of parameters and $\boldsymbol{\gamma}_j \sim \mathcal{N}(\mathbf{0}_K^\top, \boldsymbol{\Sigma})$. This allows the components of $\mathbf{y}$ to have arbitrary and unrelated regressions and covariance structures. Longitudinal data with rectangular design are an example of multivariate outcomes. Formally, there are K categories, each corresponding to an experimental condition or a set of circumstances, such as time-points, and each subject has a vector of K observations, $\mathbf{y}$, one for each of these categories.

We denote the matrices $\mathbf{Y} = \{\mathbf{y}_j\} \otimes \mathbf{1}_{N_2}$, $\mathbf{X} = \{\mathbf{x}_j\} \otimes \mathbf{1}_{N_2}$, and $\boldsymbol{\Gamma} = \{\boldsymbol{\gamma}_j\} \otimes \mathbf{1}_{N_2}$, so that

$$\mathbf{Y} = \mathbf{X}\mathbf{B} + \boldsymbol{\Gamma}. \tag{4.44}$$

The log-likelihood for $\mathbf{y}$ is equal to

$$l = -\frac{1}{2} \left\{ N_2 K \log(2\pi) + N_2 \log(\det \boldsymbol{\Sigma}) + \mathrm{tr}(\boldsymbol{\Sigma}^{-1} \mathbf{S}) \right\} \tag{4.45}$$

where $\mathbf{S} = (\mathbf{Y} - \mathbf{X}\mathbf{B})^\top (\mathbf{Y} - \mathbf{X}\mathbf{B})$. The MLE of $\mathbf{B}$ is

$$\hat{\mathbf{B}} = \left(\mathbf{X}^\top \mathbf{X} \right)^{-1} \mathbf{X}^\top \mathbf{Y}, \tag{4.46}$$

and $\hat{\boldsymbol{\Sigma}} = \mathbf{S}/(N_2 - p)$, with $\mathbf{S}$ evaluated for $\mathbf{B} = \hat{\mathbf{B}}$, is the commonly used estimator for $\boldsymbol{\Sigma}$. The estimators $\hat{\mathbf{B}}$ and $\hat{\boldsymbol{\Sigma}}$ are unbiased and have a matrix normal and a Wishart distribution respectively; see, e.g., Mardia *et al.* (1979) for details.

Use of the estimators (4.45) and (4.46) is straightforward when there are no missing data and no constraints (pattern) imposed on $\boldsymbol{\Sigma}$. When there are a small number of missing observations the EM algorithm can be used as outlined in Section 2.5. When the pattern of missingness is varied, so that the EM algorithm is impractical, and subject-wise deletion would lead to the loss of a large proportion of observations, random coefficient models can be used with advantage.

We consider the components of the outcome vectors $\{\mathbf{y}_j\}$ as the elementary-level units and define the clustering of the observations within vectors. Let u_{ij} be the component-level categorical variable indicating the position of component y_{ij} in the vector $\mathbf{y}_j$, that is,

$$u_{ij} = i,$$

and let $\mathbf{z}_{ij}$ be the vector of the corresponding dummy variables,

$$
\begin{aligned}
\mathbf{z}_{1j} &= (1, 0, \ldots, \ldots, 0) \\
\mathbf{z}_{ij} &= (1, 0, \ldots, 0, 1, 0, \ldots, 0)
\end{aligned}
$$

(ones in the first and ith positions). We denote $\mathbf{Z}_j = \{\mathbf{z}_{ij}\}_i \otimes \mathbf{1}_K$. The model in (4.43), with a complete set of $N_2 K$ observations, can now be rewritten as

$$\mathbf{y}_j^\top = \mathbf{X}_j \boldsymbol{\beta} + \mathbf{Z}_j \boldsymbol{\delta}_j, \tag{4.47}$$

where $\mathbf{X}_j$ is the design matrix formed by all the interactions of $\mathbf{x}_j$ and $\mathbf{Z}_j$. Equivalence of (4.43) and (4.47) is clear from the identities $\mathbf{x}_j \mathbf{B} = \mathbf{X}_j \boldsymbol{\beta}$ and $\boldsymbol{\delta}_j = \mathbf{Z}_j^{-1} \boldsymbol{\gamma}_j$. This is a random coefficient model with elementary-level variance $\sigma^2 = 0$. Missing observations in (4.47) are straightforward to accomodate because they correspond to deletion of one row from the design matrices $\mathbf{X}_j$ and $\mathbf{Z}_j$. The design becomes unbalanced but this generates no substantial problems.

Various constraints can be imposed on the regression parameters $\boldsymbol{\beta}$, as well as on the variance matrix $\boldsymbol{\Sigma}$. These include equal regressions for some of the components of $\mathbf{y}_j$, zero elements in $\mathbf{B}$, zero covariances, equal variances, an equicovariance submatrix of $\boldsymbol{\Sigma}$, and the like.

Since the elementary-level variance in (4.47) vanishes, the method of estimation described in Section 4.4 has to be adjusted. We assume that the variance matrix $\boldsymbol{\Sigma}$ is non-singular; otherwise at least one of the components of $\mathbf{y}$ is perfectly described in terms of the others, and could be omitted from the analysis. Let λ_1 be the smallest eigenvalue of $\boldsymbol{\Sigma}$. Then $\boldsymbol{\Sigma} - \sigma^2 \mathbf{I}$ is positive definite for any $\sigma^2 < \lambda_1$, and the variance matrix $\boldsymbol{\Sigma}$ can be formally decomposed into an elementary-level component $\sigma^2 \mathbf{I}$ and the cluster-level component $\boldsymbol{\Sigma}^* = \boldsymbol{\Sigma} - \sigma^2 \mathbf{I}$. To avoid confounding of the covariance structure parameters we constrain σ^2 to a positive value. Since λ_1 is not known (unless $\boldsymbol{\Sigma}$ is known), a guess has to be made. If the estimated variance matrix $\hat{\boldsymbol{\Sigma}}^*$ for a given σ^2 is non-singular, the guess has been appropriate, otherwise the estimating procedure has to be rerun with a smaller value of σ^2.

Usually, we wish to impose constraints on the variance matrix Σ rather than on the matrix $\text{var}(\boldsymbol{\delta}_j)$. This may render the standard parametrization unsuitable.

Fisher scoring equations for maximization of the log-likelihood for the model in (4.47) can be derived by matrix differentiation of the log-likelihood

$$l = -\frac{1}{2}\left\{ N\log(2\pi) + \sum_j \log(\det \mathbf{V}_j) + \sum_j \mathbf{e}_j^\top \mathbf{V}_j^{-1}\mathbf{e}_j \right\}, \qquad (4.48)$$

where N is the total number of non-missing elementary observations in the data ($N_2 K$ for complete data), $\mathbf{e}_j = \mathbf{y}_j^\top - \mathbf{X}_j\boldsymbol{\beta}$ is the vector of residuals for cluster j, and $\mathbf{V}_j$ is the submatrix of Σ corresponding to the available components of $\mathbf{y}_j$. The familiar GLS estimator for $\boldsymbol{\beta}$,

$$\hat{\boldsymbol{\beta}} = \left(\sum_j \mathbf{X}_j^\top \mathbf{V}_j^{-1}\mathbf{X}_j\right)^{-1} \sum_j \mathbf{X}_j^\top \mathbf{V}_j^{-1}\mathbf{y}_j, \qquad (4.49)$$

is obtained, and for elements θ, θ_1, and θ_2 of Σ we have

$$\frac{\partial l}{\partial \theta} = -\frac{1}{2}\sum_j \text{tr}\left(\mathbf{V}_j^{-1}\frac{\partial \mathbf{V}_j}{\partial \theta}\right) + \frac{1}{2}\sum_j \mathbf{e}_j \mathbf{V}_j^{-1}\frac{\partial \mathbf{V}_j}{\partial \theta}\mathbf{V}_j^{-1}\mathbf{e}_j \qquad (4.50)$$

and

$$-\mathbf{E}\left(\frac{\partial^2 l}{\partial \theta_1 \partial \theta_2}\right) = \frac{1}{2}\sum_j \text{tr}\left(\mathbf{V}_j^{-1}\frac{\partial \mathbf{V}_j}{\partial \theta_1}\mathbf{V}_j^{-1}\frac{\partial \mathbf{V}_j}{\partial \theta_2}\right). \qquad (4.51)$$

Equations (4.50) and (4.51) hold for any parametrization of the variance matrices. Note that if the parameter θ is not involved in $\mathbf{V}_j$ then cluster j makes no contribution to the scoring and information functions (4.50) and (4.51).

4.6.1   MULTIVARIATE AND LONGITUDINAL DATA

There are settings in which the distinction between the multivariate and the longitudinal nature of the data is not obvious. When observations are taken at a small number of specific time-points t_k, $k = 1, \ldots, K$, the data collected can be regarded as a set of multivariate data. In such a setting it is often relevant to test the hypothesis that the structure underlying the data is simpler than that provided by the multivariate regression model (4.43). Under this general alternative no prediction can be made about the values of the outcome variable at a time-point t different from all the measurement time-points t_k. In contrast, the longitudinal model given by (4.36) assumes a specific form of dependence on time which is applicable in the entire interval $[t_1, t_K]$ and may be used for modest extrapolation outside this interval.

The longitudinal model (4.36) coincides with the multivariate regression model (4.43) when the dependence on time is specified by a general polynomial of degree $K - 1$, with each degree included in the variation part, because then an arbitrary vector of means $\{\mathbf{E}(\mathbf{y} \,|\, \mathbf{t})\}$ and an arbitrary variance matrix $\mathrm{var}(\mathbf{y} \,|\, \mathbf{t})$ can be fitted. Such a model can be taken as a 'benchmark' against which simpler models could be compared, e.g., by means of the likelihood ratio criterion.

Models for longitudinal data are also applicable when the *metametre* is not time but a different variable. As an example we consider the winning times of the 100, 200, 400, and 800 metre men's races at the Olympic Games 1900–1988. Data from the first Olympic Games (1896) are excluded from the analysis because they are obvious outliers; dramatic improvements in winning times took place between the first two Olympic Games. Generally, the longer the distance of the race the slower the speed, perhaps with the exception of the 200 metre race because reaction at the start and acceleration at the beginning of the 100 metre race are influential factors.

We can consider the data as four time series, one for each of the distances, but also as 20 'time' series, one from each Games. In the latter perspective the distance of the race is the metametre, and the year of the race an explanatory variable defined for the Games. The year of the Games is a proxy variable for the influence of the accumulated expertise and experience in training methods, increasing exposure and popularity of the Games, and the like. Each Olympic Games has its unique atmosphere and specific conditions that influence the performance of the athletes, and these conditions can be represented by a random term, as in the model

$$y_{ij} = \beta_{1i} + x_j \beta_2 + \delta_j + \varepsilon_{ij}, \tag{4.52}$$

where β_{1i}, $i = 1, \ldots, 4$, are the terms specific to the four distances (the theoretical winning times in year 0); and x_j is the year of the Games. The speeds, in metres per second, are given in Table 4.1. They are plotted in Figure 4.6, which also contains a regression fit described below.

A natural parametrization for this variable is to subtract 1900 from the calendar year, so that year 0 refers to the earliest data analysed. Also, we consider subtracting the years covering the two World Wars during which, arguably, no progress in methods of training was made. Then the Games of 1912 and 1920 would be only four 'years' apart, and the Games of 1936 and 1948 only six years apart (excluding the time-periods 1914–18 and 1939–45). The values of this variable, denoted by YR, are given in Table 4.1. In 1968 the Olympic Games took place in Mexico City, at an altitude of 2240 metres, which is deemed to have a substantial positive effect on the speed of the competitors. All the other venues of past Games have

Table 4.1. The speeds of the winners in men's track races for 100, 200, 400, and 800 metres at the Olympic Games 1896–1988. The speeds are in metres per second. YR is the variable calculated from the calendar year starting in 1900, and skipping the years covering the two World Wars. The data from the Olympic Games in 1896 are excluded from the analysis. The 200 metre race was not contested in 1896 (marked by asterisk *) (source of the data: Knight and Troop, 1988)

100 m	200 m	400 m	800 m	YR	Calendar year
8.333	*	7.380	6.107	−4	(1896)
9.259	9.009	8.097	6.590	0	(1900)
9.091	9.259	8.130	6.897	4	(1904)
9.259	8.850	8.000	7.092	8	(1908)
9.259	9.217	8.299	7.149	12	(1912)
9.259	9.091	8.065	7.055	16	(1920)
9.434	9.259	8.403	7.117	20	(1924)
9.259	9.174	8.368	7.156	24	(1928)
9.709	9.434	8.658	7.286	28	(1932)
9.709	9.662	8.602	7.086	32	(1936)
9.709	9.479	8.658	7.326	36	(1948)
9.615	9.662	8.715	7.326	42	(1952)
9.524	9.709	8.565	7.428	46	(1956)
9.804	9.756	8.909	7.526	50	(1960)
10.000	9.852	8.869	7.612	54	(1964)
10.101	10.101	9.132	7.670	58	(1968)
9.862	10.000	8.957	7.554	62	(1972)
9.940	9.886	9.038	7.729	66	(1976)
9.756	9.906	8.909	7.670	70	(1980)
10.010	10.101	9.035	7.767	74	(1984)
10.081	10.127	9.118	7.733	78	(1988)

altitudes in the range 0–200 metres (except Munich, 1972, 520 metres), and so the observations from Mexico City would be highly influential if altitude were accounted for. To keep the illustration simple, we ignore altitude as an explanatory variable. In order to keep the outcomes on a comparable scale, we use the average speed of the winning competitor during the race as the outcome variable.

The mean vector for the four distances is

$$\hat{\mu} = (9.632, \ 9.577, \ 8.626, \ 7.338),$$

and the estimate of the variance matrix is

$$\hat{\Sigma} = \begin{pmatrix} 0.103 & 0.116 & 0.114 & 0.090 \\ 0.116 & 0.157 & 0.142 & 0.113 \\ 0.114 & 0.142 & 0.139 & 0.109 \\ 0.090 & 0.113 & 0.109 & 0.103 \end{pmatrix}. \tag{4.53}$$

A quick and informal way of assessing the dependence structure of the winning performances within a Games is provided by the eigenvalue decomposition of $\hat{\Sigma}$. For the model (4.52) the variance matrix Σ has the eigenvalues $\sigma^2 + 3\tau^2$, σ^2, σ^2, and σ^2, that is, the largest eigenvalue has multiplicity 1 and the rest of the eigenvalues are identical. The eigenvalues of the estimate $\hat{\Sigma}$,

$$0.472, \qquad 0.0152, \qquad 0.0104, \qquad 0.0033,$$

suggest that the model (4.52) is reasonable, and that the venue-level variance τ^2 is much larger than σ^2.

We proceed with the analysis by fitting a linear regression on time; this is essential since the continual improvement of the performances over the years is transparent.

The ML estimate of the matrix of regression parameters (using YR as the explanatory variable) is

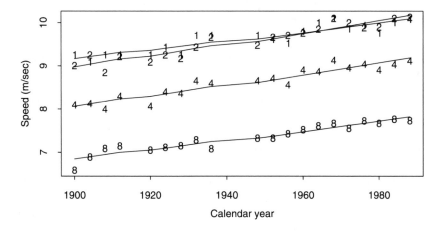

Fig. 4.6. Average speeds of the winners of the men's 100, 200, 400, and 800 metre races at Olympic Games. The solid lines represent linear regression fits using the variable YR which discards the years covering the two World Wars.

$$\hat{\mathbf{B}} = \begin{pmatrix} 9.170 & 8.979 & 8.066 & 6.852 \\ 0.0118 & 0.0153 & 0.0143 & 0.0124 \end{pmatrix}, \tag{4.54}$$

and the unbiased estimate of the variance matrix is

$$\hat{\boldsymbol{\Sigma}} = \begin{pmatrix} 0.019\,52 & 0.007\,93 & 0.012\,64 & 0.001\,11 \\ 0.007\,93 & 0.017\,00 & 0.010\,22 & -0.002\,29 \\ 0.012\,64 & 0.010\,22 & 0.015\,46 & 0.001\,33 \\ 0.001\,11 & -0.002\,29 & 0.001\,33 & 0.010\,32 \end{pmatrix}. \tag{4.55}$$

The ML estimate of $\boldsymbol{\Sigma}$ is the 18/20-multiple of (4.55). However, the correlations in $\hat{\boldsymbol{\Sigma}}$ are not affected by the choice of full or restricted maximum likelihood estimation.

The first row of (4.54) contains the fitted speeds for the four distances in year 0 (1900), and the second line the fitted annual improvement. The standard errors for the parameter estimates in the second row are 0.0012–0.0013 for 100–400 metres and 0.0009 for 800 metres. Clearly, time is a dominant explanatory variable; the winning times have been getting faster over the years. As expected, all the elements of the estimated variance matrix have been reduced substantially; compare the variances in (4.53) with those in (4.55).

We see that the submatrix of (4.55) corresponding to the three shortest races, 100 m, 200 m, and 400 m, is close to an equicovariance matrix, whereas the (adjusted) covariances of the shorter races with the 800 m race are close to zero. The variance for the 800 m races is smaller than the other variances. This pattern can be modelled by the multivariate regression (4.43), with the parameters σ_1^2, σ_2^2, and σ^2 such that the variance matrix, common to all the clusters, is equal to

$$\mathbf{V}_1 = \sum_{k=1}^{3} \sigma_k^2 \mathbf{H}_k, \tag{4.56}$$

where $\mathbf{H}_1 = \mathrm{diag}(\mathbf{h})$, $\mathbf{H}_2 = \mathbf{h}\mathbf{h}^\top$, and $\mathbf{H}_3 = (\mathbf{1}_4 - \mathbf{h})(\mathbf{1}_4 - \mathbf{h})^\top$, with $\mathbf{h} = (1, 1, 1, 0)^\top$.

The ML estimates of the variances σ_1^2, σ_2^2, and σ_3^2 are $\hat{\sigma}_1^2 = 0.007\,07$ (standard error equal to 0.001 58), $\hat{\sigma}_2^2 = 0.010\,26$ (0.004 02), and $\hat{\sigma}_3^2 = 0.010\,32$ (0.003 26). Note that the estimates can be obtained by matching the fitted variance matrix with the observed one. For example,

$$\hat{\sigma}_2^2 = (\hat{\Sigma}_{12} + \hat{\Sigma}_{13} + \hat{\Sigma}_{23})/3,$$

and, trivially, $\hat{\sigma}_3^2 = \hat{\Sigma}_{44}$. This is a consequence of the property of the log-likelihood for distributions from exponential families; maximum likelihood is equivalent to moment matching for functions of sufficient statistics.

The models with unrestricted and patterned covariance structures can be compared by means of the likelihood ratio test statistic. The deviance for the former model is −135.54, whereas the latter, with three covariance structure parameters replacing the original ten, is −128.04, an increase of 7.5 points (null-distribution χ_7^2). The likelihood ratio criterion justifies the model simplification.

The fitted correlation structure corresponds to a bivariate random vector underlying the winning performances; one component represents the sprint events, 100–400 metre races, the other the 800 metre races. This result can be interpreted as follows: deviations of the winners' performances in the 100–400 metre races from the long-term trend can be described by a unidimensional random variable. This variable is independent of the corresponding deviations for the 800 metre races. The (common) fitted correlation of the deviations for the 100–400 metre races is $\hat{\sigma}_2^2/(\hat{\sigma}_1^2 + \hat{\sigma}_2^2) = 0.60$.

4.7   REML estimation

The maximum likelihood estimator of the covariance structure parameters in the Olympic Games example is biased because the estimation procedure ignores the degrees of freedom lost due to the estimated regression parameters. This conflict between unbiased and ML estimation is resolved by applying the maximum likelihood to the *error contrasts*. The general treatment (REML) was presented in Section 2.6, and is also applicable in the current context.

For the Olympic Games data, REML estimates and their estimated standard errors are merely the 20/18-multiples of their ML counterparts, and so the differences between ML and REML are unimportant.

In general, computation for REML is somewhat more complex than for ML. The log-likelihood, and the first-order partial derivatives, have to be adjusted by the term $R = -\log\left\{\det\left(\mathbf{X}^\top \mathbf{V}^{-1} \mathbf{X}\right)\right\}$, see (2.44) and (2.45). In practice, correction of the information matrix for covariance structure parameters is negligible, and can be skipped.

Since REML is based on error contrasts, the fitted deviances can be compared by the likelihood ratio criterion, only for models with the same set of error contrasts. Therefore, the REML deviance can be used only for comparison of models with identical regression parts.

When the focus of the analysis is on the covariance structure (e.g., when the regression parameters are nuisance parameters), the REML approach is preferred. It yields unbiased estimators, and it takes account of uncertainty about *some* of the parameters. One should also consider the symmetric argument: when regression parameters are of primary interest (covariance structure is a nuisance), we could do better by 'integrating out' the covariance structure parameters from the likelihood. The argument is strongly supported by the fact that in most cases information about the covariance structure is very scarce. This point appears not to have been taken up seriously yet.

4.8   Model checking

Thorough model checking involves careful assessment of each element of the model assumptions. In random coefficient models, as in ordinary regression, assessment of the appropriateness of the linear regression formula is essential. In practice this is accomplished by fitting alternative models with regression parts supplemented by one or several variables, and checking that the added variables improve the model fit by less than an adopted criterion. In random coefficient models, essentially the same procedure should be carried out for the variation part, and so a large number of alternative models have to be considered.

Normality of the random model terms is another assumption which has to be subjected to scrutiny. A straightforward procedure involves, essentially, checking for normality (and homogeneity) of the conditional expectations of the random terms. The conditional distributions of the cluster-level random terms are

$$\boldsymbol{\delta}_j \;\sim\; \mathcal{N}(\boldsymbol{\Sigma}\mathbf{G}_j^{-1}\mathbf{Z}_j^{\top}\mathbf{e}_j \,, \boldsymbol{\Sigma}\mathbf{G}_j^{-1}),$$

$$\boldsymbol{\varepsilon}_j \;\sim\; \mathcal{N}(\sigma^2\mathbf{V}_j^{-1}\mathbf{e}_j \,, \sigma^2 - \sigma^4\mathbf{V}_j^{-1}). \tag{4.57}$$

The conditional expectations in (4.57), with the parameters replaced by their ML estimates, are denoted by $\hat{\boldsymbol{\delta}}_j$ and $\hat{\boldsymbol{\varepsilon}}_j$, and are referred to as cluster- and elementary-level *residuals*. It can be shown that $\hat{\boldsymbol{\varepsilon}}_j = \mathbf{e}_j - \mathbf{Z}_j\hat{\boldsymbol{\delta}}_j$. For the elementary-level residuals, checking for normality and homogeneity is analogous to exploration of patterns among OLS residuals, although it should be noted that the conditional expectations of the elementary-level random terms contain another 'layer' of uncertainty because they are extracted from the residuals $\mathbf{e} = \mathbf{y} - \mathbf{X}\hat{\beta}$.

The cluster-level residuals can be checked for univariate normality, component by component. Note that component-wise normality is not equivalent to multivariate normality. It is often more practical and illuminating to plot the linear functions $\mathbf{z}_j\hat{\boldsymbol{\delta}}_j$ within the range of values of $\mathbf{z}$ in cluster j. Of course, this is possible only when the variation part contains a small number of variables.

4.9   General patterns of dependence

We pointed out in Section 4.3 that random coefficient models can be interpreted in terms of varying regressions, (4.5), but they can also be used to model the covariance structure of the observations. In a more general formulation we may assume that

$$\text{var}(\mathbf{y}_j) = \sigma^2 \mathbf{I} + \mathbf{F}(\mathbf{Z}_j; \boldsymbol{\theta}), \tag{4.58}$$

where $\mathbf{F}$ is a matrix function of the variation design matrix $\mathbf{Z}_j$ and a vector of parameters $\boldsymbol{\theta}$. The parameter space for $\boldsymbol{\theta}$ may be given by the condition of non-negative definiteness of either $\mathbf{F}(\mathbf{Z}_j; \boldsymbol{\theta})$ or $\sigma^2 \mathbf{I} + \mathbf{F}(\mathbf{Z}_j; \boldsymbol{\theta})$.

Situations where there is abundant (prior) information about the pattern of variation are rare, and inference about complex patterns of variation is meaningful only for large datasets. There are settings in which a specific pattern (4.58) has a natural interpretation. Longitudinal data with a time-series pattern of dependence is an important example.

For longitudinal data with small clusters the choice of the function $\mathbf{F}$ has little impact on the complexity of the associated Fisher scoring algorithm because the variance matrices $\mathbf{V}_j$ can be inverted numerically. Autoregressive and moving-average-type models for $\mathbf{V}_j$ complement the symmetric patterns of dependence generated by the random coefficient model in (4.5). In some situations it may be opportune to define a parametrization for the *inverse* of $\mathbf{V}_j$. This can simplify the estimation algorithm considerably. Let $\mathbf{C}_j = \mathbf{V}_j^{-1}$ be the inverse of the variance matrix for cluster j; it is often referred to as the *concentration matrix*. Then the log-likelihood (4.24) can be expressed as

$$l = -\frac{1}{2} \left\{ N \log(2\pi) - \sum_j \log(\det \mathbf{C}_j) + \sum_j \mathbf{e}_j^\top \mathbf{C}_j \mathbf{e}_j \right\}. \tag{4.59}$$

The Fisher scoring algorithm leads to the GLS estimator for the regression parameters $\boldsymbol{\beta}$,

$$\hat{\boldsymbol{\beta}} = \left(\mathbf{X}^\top \mathbf{C} \mathbf{X} \right)^{-1} \mathbf{X}^\top \mathbf{C} \mathbf{y},$$

where $\mathbf{C} = \mathbf{C}_j \otimes \mathbf{I}_{N_2}$. Further,

$$\frac{\partial l}{\partial \theta} = \frac{1}{2} \sum_j \text{tr} \left(\mathbf{C}_j^{-1} \frac{\partial \mathbf{C}_j}{\partial \theta} \right) - \frac{1}{2} \sum_j \mathbf{e}_j^\top \frac{\partial \mathbf{C}_j}{\partial \theta} \mathbf{e}_j, \tag{4.60}$$

and using the identity $\mathbf{E}(\mathbf{e}_j \mathbf{e}_j^\top) = \mathbf{C}_j^{-1}$,

$$-\mathbf{E} \left(\frac{\partial^2 l}{\partial \theta_1 \partial \theta_2} \right) = \frac{1}{2} \sum_j \text{tr} \left(\mathbf{C}_j^{-1} \frac{\partial \mathbf{C}_j}{\partial \theta_1} \mathbf{C}_j^{-1} \frac{\partial \mathbf{C}_j}{\partial \theta_2} \right) \tag{4.61}$$

for any parameters θ, θ_1, and θ_2 not involved in the regression part $\mathbf{X}\boldsymbol{\beta}$. The one-step autoregressive type of dependence corresponds to a tridiagonal matrix $\mathbf{C}_j$, and so

$$\frac{\partial \mathbf{C}_j}{\partial \theta} = \mathbf{I}$$

for the common diagonal element of $\mathbf{C}_j$ and

$$\frac{\partial \mathbf{C}_j}{\partial \theta} = \begin{pmatrix} 0 & 1 & 0 & \ldots & & & 0 \\ 1 & 0 & 1 & 0 & \ldots & & \\ 0 & 1 & 0 & 1 & & & \\ \vdots & & \ddots & \ddots & \ddots & & \vdots \\ & & & 0 & 1 & 0 & 1 \\ 0 & \ldots & & & 0 & 1 & 0 \end{pmatrix}$$

for the comon element immediately below or above the diagonal.

Problems of ensuring non-negative definite variance matrices $\mathbf{C}_j$ in the Fisher scoring iterations have analogous solutions to those described in Section 4.4.1. For large clusters recursive equations for the determinant of $\mathbf{C}_j$ can be used with advantage. For illustration, consider the following AR(1)-type dependence structure. Let $\mathbf{R}_n$ be the $n \times n$ tridiagonal matrix with common entries a on the diagonal, and common entries b immediately below the diagonal. Then

$$\det \mathbf{R}_n = a \det \mathbf{R}_{n-1} - b^2 \det \mathbf{R}_{n-2}, \tag{4.62}$$

and $\mathbf{R}_n$ is positive definite if and only if $\det \mathbf{R}_h \geq 0$ for all $h \leq n$. The equation is obtained directly by expanding $\det \mathbf{R}_n$ by its first row.

4.9.1 GENERAL FORM OF ELEMENTARY-LEVEL VARIANCE

When data contain large clusters the assumption of equal elementary-level variances within the clusters may be challenged. It is rarely meaningful to consider unrelated within-cluster variances σ_j^2 but the elementary-level variance may be modelled as a function of some of the cluster-level variables and/or of the cluster size (cluster sample size). Another alternative is to assume that the variances σ_j^2 take on a small number of different values. Such elementary-level variance mixture models can be fitted by embedding the Fisher scoring algorithm in the M-step of the EM algorithm in which the E-step consists of assigning the elementary-level units (or clusters) to the categories according to their within-group variance. Also, the variance σ^2 may be a function of one or several variables. A simple example of the dependence of σ^2 on the subject's gender was considered in Section 4.3.1.

4.10 Bibliographical notes

The issues of predictive validity of standardized educational tests and selection of candidates for enrolment was studied by Novick *et al.* (1972) and
Rubin (1980). Some methodological issues are further developed in Dempster *et al.* (1981). Controversies about 'school effectiveness', that is, assessment of between-school differences in students' academic performances,
stimulated research on the applications of random coefficient methods reported in Aitkin *et al.* (1981), Aitkin and Longford (1986), Raudenbush
and Bryk (1986), and others. These references concentrate on computational issues and issues of statistical efficiency, while skirting or ignoring
the problem of non-random allocation of students to schools. Lack of a
rigorous definition of the studied entity is a niggling deficiency in all applications of random coefficient models to predictive validity and school effectiveness. Any such definition would help to make it clear that the answers
afforded by statistical analyses do not account for an important source of
uncertainty due to departure from the assumptions of random allocation.
Holland (1986) presents a comprehensive framework for problems in which
experimental-design-type inference is sought from the analysis of observational data. Raudenbush (1988) reviews applications of random coefficient
models in educational research.

Hui (1983) and Hui and Berger (1983) use a moment method for the
analysis of a longitudinal study with irregular time-points. Laird and Ware
(1982) and Laird *et al.* (1987) present interesting illustrations of longitudinal analysis with random coefficient models. Goldstein (1986b) discusses a longitudinal analysis of children's growth. Ware (1985) is a very
informative reference on the subject. Lindstrom and Bates (1989) describe
random coefficient models with non-linear regression. Mason *et al.* (1984)
and Wong and Mason (1991) discuss non-standard applications of random
coefficient models for the analysis of data from the World Fertility Survey.

Speeding up the EM algorithm and deriving the information matrix
applicable for the incomplete data are two recurrent topics in research
on improvement of implementations of the EM algorithm. Louis (1982),
Meilijson (1989), and Meng and Rubin (1991) are important contributions
to this area. Wu (1983) and Thompson and Meyer (1986) discuss some
of the undesirable properties of the EM algorithm. As computing power
becomes cheaper and more readily available, computational efficiency of
methods is of lesser concern than the complexity of the programming task.
This point is stressed by Meng and Rubin (1991).

Lange and Ryan (1989) derived formal diagnostic methods for random
coefficient models that take into account uncertainty about the estimated
parameters.

5
Examples using random coefficient models

Random coefficient models are illustrated in this chapter for the datasets introduced in Chapter 2 and analysed in Chapter 3 using random-effects models. The analyses focus on patterns of between-cluster variation.

In all references to statistical significance, 5 per cent level of significance is used.

5.1 Financial ratios

In Section 3.1 we concluded that the within-sector regressions of log-liabilities on log-assets are not identical. Here we carry out a more detailed examination of between-sector variation and further explore the hypothesis that the slope of the log-liabilities on the log-assets is equal to unity. Recall that this hypothesis is linked to the appropriateness of use of the financial ratio of liabilities and assets as a measure of a company's creditworthiness. If in the random-effects model (3.1) the slope $b = 1$, then the conditional expectation of the difference of the log-liabilities and log-assets, $\mathbf{E}\left(y_{ij} - x_{ij} \mid \delta_j^{(1)}\right)$, is equal to $a + \delta_j^{(1)}$, and so $\exp\left(a + \delta_j^{(1)}\right)$ can be regarded as a benchmark for sector j ('average' of the ratios in sector j). In other words, the use of the ratio of liabilities by assets would be warranted if $b = 1$.

The random-effects models applied in Section 3.1 allow for varying intercepts of the within-sector regressions,

$$y_{ij} = a + bx_{ij} + \delta_j^{(1)} + \varepsilon_{ij} \tag{5.1}$$

(y and x are the log-liabilities and log-assets, respectively, for company i in sector j). Note that the arithmetic average of the log-variables corresponds to the geometric average of the original variables. Naturally, the ratio of the liabilities and assets may be appropriate for assessing a company's financial viability in some sectors, but not in others. This hypothesis corresponds

Table 5.1. Record of Fisher scoring iterations. Financial ratios data. Iteration 0 refers to the starting solution based on ordinary least squares. The 0.8-multiple of the OLS residual variance 0.199 was used as the starting solution. The bottom part of the table gives the fitted standard errors from the OLS and maximum likelihood methods

	Regression		Variance		
Iteration	Intercept	Slope	Company	Sector	Deviance
0	−0.101	0.963	0.160 {0.199}	$\begin{pmatrix} 0.2 & 0 \\ 0 & 0 \end{pmatrix}$	2455.54
1	−0.213	0.972	0.172	$\begin{pmatrix} 1.465 & -0.120 \\ -0.120 & 0.0106 \end{pmatrix}$	2280.41
2	−0.269	0.978	0.168	$\begin{pmatrix} 1.679 & -0.140 \\ -0.140 & 0.0127 \end{pmatrix}$	2252.47
3	−0.267	0.978	0.169	$\begin{pmatrix} 1.611 & -0.133 \\ -0.133 & 0.0117 \end{pmatrix}$	2250.74
4	−0.271	0.978	0.169	$\begin{pmatrix} 1.577 & -0.130 \\ -0.130 & 0.0115 \end{pmatrix}$	2250.64
5	−0.270	0.978	0.169	$\begin{pmatrix} 1.589 & -0.131 \\ -0.131 & 0.0116 \end{pmatrix}$	2250.64
6	−0.271	0.978	0.169	$\begin{pmatrix} 1.586 & -0.131 \\ -0.131 & 0.0115 \end{pmatrix}$	2250.64

Method	Standard errors	
ML	(0.0088)	$\left\{ \begin{matrix} (0.535) & (0.050) \\ (0.050) & (0.0048) \end{matrix} \right\}$
OLS	(0.0063)	

to varying slopes on log-assets. We therefore consider the random slopes model

$$y_{ij} = a + bx_{ij} + \delta_j^{(1)} + \delta_j^{(x)} x_{ij} + \varepsilon_{ij}. \qquad (5.2)$$

Table 5.1 contains a record of iterations of the Fisher scoring algorithm for this random slopes model. The starting solution for the regression parameters is based on the OLS fit and the starting sector-level variance matrix is set to have zero variance of the slope and scaled variance of the intercept equal to 0.2. This choice is based on the random-effects model fit in which the scaled variance of the intercept is estimated as 0.176. The starting value for σ^2 is set to the 0.8-multiple of the OLS residual variance. The first iteration homes in on the ML solution, and after the fourth iteration the deviance is reduced only marginally.

Compared to the random-effects model fit the regression parameter estimates are changed only slightly; the regression slope estimate, 0.978, is a bit closer to one, but remains significantly different from one. The fitted company-level variance is almost identical to its counterpart in the random-effects model (0.169 vs. 0.172).

Most remarkable is the increase in the (sector-level) variance of the intercept. Whereas for the random-effects model the fitted variance ratio is constant and equal to 0.176, now the variance ratio appears to be much larger, equal to 1.586. The large fitted intercept variance is a consequence of variance heterogeneity and positivity of the regressors. In fact, the fitted intercept variance is the fitted variance for a hypothetical company with assets of $1000 \times \exp(0) = £1000$. This, of course, is an extreme extrapolation from the data at hand. The three smallest values of assets in the data are £5000, £10 000, and £24 000.

To explore the fitted pattern of variation, and the resulting variance heterogeneity, note that the scaled variance matrix $\hat{\boldsymbol{\Omega}}$ is almost singular. The fitted intercept-by-slope correlation is –0.955; high intercepts are associated with shallow slopes. This implies a pattern of variation close to a left-facing megaphone (see Figure 4.1). The fitted scaled sector-level variance for a company with log-assets x is

$$\tau(x) = (1, \; x)\hat{\boldsymbol{\Omega}}(1, \; x)^{\top},$$

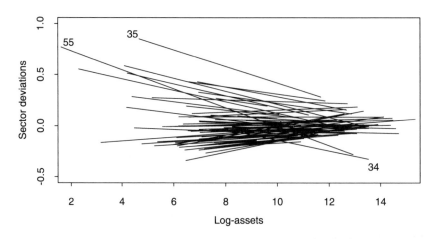

Fig. 5.1. Fitted sector-level deviations in the random slopes model. Financial ratios data. Three outlying sectors are indicated by their order numbers.

and its minimum occurs for $x^* = -\hat{\Omega}_{21}/\hat{\Omega}_2 \doteq 11$. This corresponds to assets of about £60 million, approximately the 85th percentile of the assets in the dataset. The minimum scaled sector-level variance is equal to $\tau(x^*) = 0.11$. The corresponding unconditional variance is $0.169 \times \{1 + \tau(x^*)\} = 0.188$. The fitted scaled variance for smaller companies is much larger. For example, a company with assets of £150 000, the first percentile of the data, has $x \doteq 5$ and its fitted, scaled sector-level variance is $\tau(5) = 0.57$, five times greater than $\tau(x^*)$, and the fitted unconditional variance is $0.169 \times \{1 + \tau(5)\} = 0.265$, 1.4 times greater than the fitted variance for x^*. The fitted covariance structure in the random slopes model at least partly reflects the observed dependence of the variance on assets.

Both the t- and the likelihood ratio test statistics for the hypothesis that the variance of the slopes is equal to zero are significant. The former is equal to $0.0115/0.0048 = 2.42$ and the latter to 23.7 (null-distribution χ_2^2, assuming non-singular Ω).

The fitted deviations of the within-sector regressions, formed as the linear combinations of the conditional expectations, $\hat{\delta}^{(1)} + \hat{\delta}^{(x)}x$, are plotted in Figure 5.1. For each sector represented by more than one company the segment of $\hat{\delta}^{(1)} + x\hat{\delta}^{(x)}$ is plotted in the range of values x that occur in the

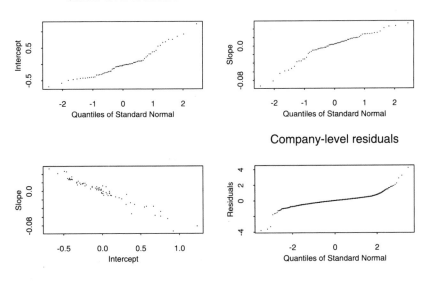

Fig. 5.2. Regression diagnostics for the random slopes model fit. Financial ratios data. The panels contain the normal quantile plots for the sector-level deviations of the intercept $\hat{\delta}^{(1)}$ and the slope $\hat{\delta}^{(x)}$, the scatterplot of these residuals, and the normal quantile plot of the company-level residuals.

data for that sector. The pattern of variation is very close to a left-facing megaphone. This confirms the conclusion based on $\hat{\Omega}$.

Sector 35 (motor components) stands out as the sector with the highest relative liabilities. Sectors 55 (general food manufacturing) and 34 (miscellaneous engineering contractors) counter the trend in that their companies with high assets have relatively low liabilities but small companies have relatively high liabilities. In the case of sector 55 this may be due to some influential observations (note that the sector has a wide range of company sizes). The sector-level residuals appear to be skewed towards shallow slopes and high intercepts. Figure 5.2 contains quantile plots of the components of the sector-level residuals. The company-level residuals, also displayed in the diagram, are distinctly non-normal.

Figure 5.3 contains the plot of the company-level residuals against log-assets. Clearly, the smaller companies have larger company-level variance. Also, there are three distinctly outlying large negative residuals. They belong to the same companies (from sector 58) that were detected as outliers in random-effects analysis.

In the sequel we restrict our analysis to companies with log-assets greater than 7. This corresponds to assets greater than $1000 \times \exp(7) \doteq$ £10^6. For brevity we refer to these as the *large* companies or as the *reduced* dataset. Also, we exclude the three outlying companies from sector 58. This reduced dataset comprises 1881 companies in 71 sectors. The fit for the random slopes model (5.2) is given in the left-hand column of Table 5.2.

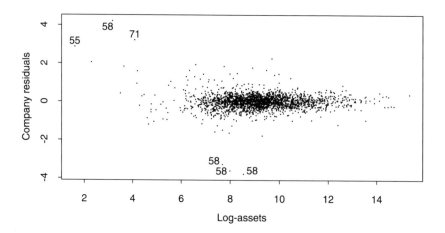

Fig. 5.3. Scatterplot of the company-level residuals and log-assets. Financial ratios data. The sectors of some of the companies with extreme residuals are indicated in the plot.

Table 5.2.   Random coefficient model fits for financial ratios' data, reduced dataset

Parameter	Random slopes		Random effects	
	Estimate	(St. error)	Estimate	(St. error)
Intercept	−0.424		−0.423	
Slope	0.994	(0.0078)	0.994	(0.0060)
Company variance	0.115		0.116	
Scaled sector variance	$\begin{pmatrix} 1.393 & -0.110 \\ -0.110 & 0.0102 \end{pmatrix}$	$\begin{matrix} (0.584) & (0.055) \\ (0.055) & (0.0053) \end{matrix}$	0.224	(0.050)
Deviance	1411.28		1415.30	
Iterations	9		7	

Both the company-level variance and the variance of the slopes on log-assets x have been reduced. The estimate of the latter is $0.0102 \times 0.115 = 0.001\,17$, and the standard deviation $\sqrt{0.001\,17} = 0.034$. Figures 5.4 and 5.5 are the counterparts of Figures 5.1 and 5.2 for the reduced dataset. Judging by the residual plots, the model assumptions are somewhat more

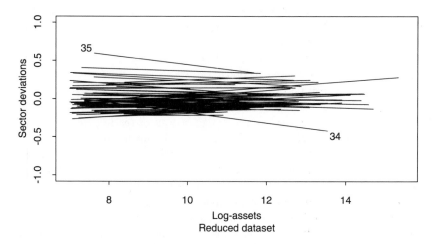

Fig. 5.4.   Fitted sector-level deviations. Financial ratios data. Based on the reduced dataset containing only companies with assets higher than £1 million. Two outlying sectors are indicated by their order numbers.

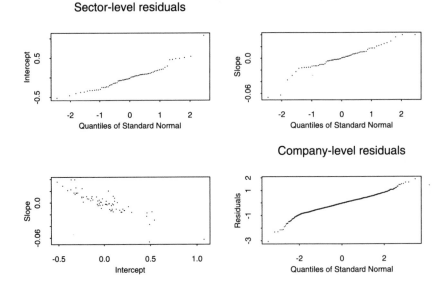

Fig. 5.5.   Diagnostic plots for random slopes model fit. Financial ratios data, reduced dataset. The layout is the same as in Figure 5.2.

acceptable for the reduced dataset, although there are a small number of extreme residuals at both levels.

Large companies exhibit less variation, both within and between sectors. The fitted within-sector variance is about 30 per cent smaller than in the complete dataset, and the entries of the fitted sector-level variance matrix are almost twice as small. The mean regression slope for the large companies is very close to unity, and is not significantly different from it (the corresponding t-ratio is $0.0060/0.0078 \doteq 0.75$). Note that the standard error for the regression slope in the reduced dataset is smaller than its counterpart in the original dataset. This is due to the relative homogeneity of the reduced dataset. An alternative way of testing the hypothesis of appropriateness of the ratios, $b = 1$, is by the likelihood ratio criterion. We refit the random slopes model with the slope constrained to unity. The corresponding deviance is equal to 1141.97; the likelihood ratio test statistic is equal to $1141.97 - 1141.18 = 0.79$ (null-distribution χ_1^2).

Note that similar patterns of sector-level variation are fitted for both the original and reduced datasets (as well as when b is constrained to be equal to unity). For the reduced dataset the fitted minimum variance occurs for $x = 10.75$ (11.0 for the original dataset); smaller companies are associated with larger variance. The fitted variance for a company with log-assets $x = 7$ (the minimum in the reduced dataset) is

$$0.115 \times (1 + 1.393 - 14 \times 0.110 + 49 \times 0.0102) = 0.156,$$

only about 12 per cent higher than the minimum fitted variance 0.139 (at $x = 10.75$). It seems that the model for the reduced dataset could be simplified to constant within-sector slopes on log-assets, that is, to the random-effects model.

The random-effects model fit is given on the right-hand side of Table 5.2. The fitted regression is almost identical to the random slopes model fit. The fitted variance components are $\hat{\sigma}^2 = 0.116$ and $\hat{\sigma}^2\hat{\omega} = 0.0282$. Deletion of the small companies results in substantial reduction of the company-level variance, but only moderate reduction of the sector-level variance. The likelihood ratio test statistics for the comparison of the random slopes and random-effects models is $1415.30 - 1411.28 = 4.02$; the null-distribution of the test statistic is χ_2^2. Thus, the hypothesis of parallel slopes is not rejected. Nevertheless, Figure 5.4 suggests that there are at least two sectors with clearly outlying slopes: Sector 34 has the lowest adjusted liabilities for the largest companies and sector 35 has the highest adjusted liabilities for the smallest companies.

The estimated regression slope is very close to one, and not significantly different from one. Thus, application of the financial ratios is justified for large companies, possibly with the exception of a small number of sectors. The geometric averages of the financial ratios vary across the sectors; the transformed value of the standard deviation from the log-scale is $\exp(\sqrt{0.0282}) \times 100 \doteq 17$ per cent.

We conclude that using the financial ratio of log-liabilities and log-assets as a measure of viability and creditworthiness of the companies is warranted. Nevertheless, there are a number of important qualifications. First, the comparisons are appropriate only for large companies (with assets above a certain threshold, e.g., £1 million). Next, the analysis is based on the assumption that all the companies in the dataset are viable and creditworthy, and that companies that are not have generally higher adjusted liabilities. Further, the sectors differ in their average financial ratios. Also, there are several sectors and single companies for which the financial ratio appears to be inappropriate. Even after reduction of the dataset there are a few outlying companies and sectors. We can pursue model refinement further, e.g., by allowing for within-company heterogeneity, or by identifying further outliers and excluding them from the analysis. The designated outliers should be inspected to ascertain whether there are extraneous factors that justify their exclusion from the dataset.

Financial ratios for the small companies can be explored by approaches similar to those used in this analysis.

5.2 Rat weights

In Section 3.2 we concluded that the weights of newborn rats vary across litters, even after adjustment for litter size. Although the original purpose was to estimate treatment effects associated with the three diets, we use the dataset to illustrate how between-sex differences can be explored.

As a refinement of the description of between-litter differences we consider a pattern of variation, that is dependent on the sex of the rat. The average male–female difference of the rat weights is about 360 mg, or 5.9 per cent. Of course, the within-litter differences may vary around this average. We consider the random coefficient model in which sex is associated with variation; patterns of variation associated with a categorical variable were discussed in Section 4.3.1. The fitted regression for the log-transformed data almost coincides with the regression part for the random-effects model fit, see Table 3.9, and the elementary-level variance has also changed only slightly ($\hat{\sigma}^2 = 0.004\,56$). The fitted scaled variance matrix, ignoring the constraint of non-negative definiteness, is

$$\hat{\Omega} = \begin{pmatrix} 0.637 & -0.151 \\ -0.151 & 0.025 \end{pmatrix}.$$

The corresponding standard errors are

$$\begin{pmatrix} (0.223) & (0.102) \\ (0.102) & (0.124) \end{pmatrix};$$

the t-ratio for the variance of sex differences is about 0.2. The fitted variance matrix $\hat{\Omega}$ has a negative eigenvalue; the fitted correlation is smaller than –1. The value of the deviance associated with this fit is –772.00. If we applied the constraint of non-negative definiteness of $\hat{\Omega}$ the deviance would be higher than –772.00, but lower than the deviance for the random-effects model fit (–769.60). Irrespective of the actual value of the deviance, this would be only a very modest improvement in the model fit. In fact, the fitted scaled variance matrix subject to the constraint of non-negative definiteness is

$$\hat{\Omega} = \begin{pmatrix} 0.505 & -0.139 \\ -0.139 & 0.038 \end{pmatrix}.$$

with $\hat{\sigma}^2 = 0.004\,55$. The deviance corresponding to this fit is –771.98. The log-likelihood appears to be very flat in the neighbourhood of the ML solution. This is also indicated by the slow rate of convergence. Fitting these two models required 13 and 16 iterations for unconstrained and constrained estimation respectively.

The fitted variance for a rat is $\hat{\sigma}^2 \{1 + (1, s)\hat{\Omega}(1, s)^\top\}$, where s is the sex of the rat ($s = 0$ for males, $s = 1$ for females). Thus, the fitted standard deviation for females is 0.0864, and for males 0.0788. Such a difference is trivial in most contexts (note that the standard deviations can be converted to the percentage scale). An alternative way of modelling differential variation is by allowing for a sex-specific, elementary-level variance, while modelling between-litter differences by a univariate random term. The studied dataset provides no evidence of sex-dependent pattern of between-litter variation.

5.3 Pregnancy monitoring

Modelling of varying polynomial regression coefficients for pregnancy monitoring data performed in this section reveals that the pattern of variation fitted by the random-effects models in Section 3.3 is inadequate. We consider random coefficient models with the regression part consisting of cubic polynomials in time before and after day 252 of gestational age. To ensure differentiability of the regression curve at day 252, appropriate constraints on the linear and quadratic adjustment terms are imposed, as in

Table 5.3. Maximum likelihood estimates for pregnancy monitoring data. 'Adj.' stands for adjustment after day 252. The deviance of the model fit is -1521.30

Parameter	Estimate	(St. error)
Intercept	2.949	
Linear	0.004 57	(0.000 77)
Quadratic	$-0.014\,15$	(0.002 77)
Cubic	$-0.001\,07$	(0.000 29)
Adj. Quadratic	$-0.012\,15$	(0.010 13)
Adj. Cubic	0.002 12	(0.001 78)
Size	-0.5692	(0.2126)
Observation variance	0.004 00	

Client-level variance matrix

$$\begin{pmatrix} 14.72 & 0.0626 & 0.0317 & 0.205 \\ 0.0626 & 0.002\,41 & 0.002\,46 & -0.0008 \\ 0.0317 & 0.002\,46 & 0.003\,83 & -0.0047 \\ 0.205 & -0.0008 & -0.0047 & 0.0279 \end{pmatrix}$$

Standard errors for the variances
(2.55) (0.000 68) (0.001 53) (0.0184)

Section 3.3. The variation part contains the absolute, linear, and quadratic terms of the polynomials, that is, four variables since the absolute and linear terms for the adjustment are constrained to be zero. The model fit is summarized in Table 5.3. The differences between the fitted regressions in this random coefficient and the corresponding random-effects models (Table 5.3 and the central part of Table 3.12) are negligible, although there are noticeable differences among the standard errors; paradoxically, most standard errors are smaller for the (more complex) random coefficient model. The fitted client-level variance matrix is almost singular. Its eigenvalues are

$$14.72, \qquad 0.0264, \qquad 0.0042, \qquad 3.64 \times 10^{-4}.$$

The deviance of the model fit is -1521.30, a reduction of 326.70 from the corresponding random-effects model. This is a substantial improvement of the model fit.

The regression slope on the quadratic term for adjustment after day 252, -0.0122 (0.0101), has a small t-ratio, and so does the variation associated with this variable, $0.0279/0.0184 \doteq 1.5$. However, the likelihood ratio test statistic associated with the variable in the variation part is 10.85 (null-distribution χ_4^2). Therefore, the variable is retained in the model.

Inclusion of the cubic terms in the variation part of the model leads to acute confounding, particularly for the term corresponding to adjustment after day 252. Confounding in the variation part is a problem even in the adopted model; the condition number of the information matrix for the

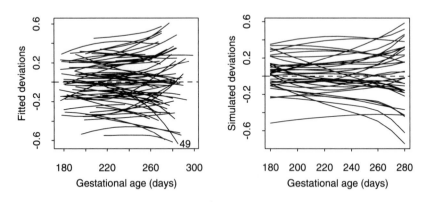

Fig. 5.6. Fitted and simulated client-level deviation curves. Pregnancy monitoring data. The fitted deviation curves are plotted in the left-hand panel. Each curve represents a client. The curves are delineated by the first and last observation of the client. Client 49 is marked by her order number. A set of 35 simulated deviation curves is plotted in the right-hand panel. Each curve represents a 'simulated' client. The coefficients for each curve are generated as a random draw from the fitted client-level distribution.

Table 5.4. Restricted maximum likelihood estimates for pregnancy monitoring data. 'Adj.' stands for adjustment after day 252

Parameter	Estimate	(St. error)
Intercept	2.950	
Linear	0.004 56	(0.000 77)
Quadratic	−0.014 18	(0.002 78)
Cubic	−0.001 08	(0.000 29)
Adj. Quadratic	−0.012 13	(0.010 20)
Adj. Cubic	0.002 13	(0.001 80)
Size	−0.5695	(0.2157)
Observation variance	0.004 01	

Client-level variance matrix

$$\begin{pmatrix} 15.06 & 0.0636 & 0.0331 & 0.207 \\ 0.0636 & 0.002\,48 & 0.002\,58 & -0.0012 \\ 0.0331 & 0.002\,58 & 0.004\,08 & -0.0054 \\ 0.207 & -0.0012 & -0.0054 & 0.0313 \end{pmatrix}$$

Standard errors for the variances
(2.61) (0.000 70) (0.001 58) (0.0192)

covariance structure parameters is 3.3×10^8. However, inclusion of a cubic term raises the condition number to over 10^{12}. Exclusion of any variables in the variation part affects the quality of the fit considerably (as measured by an increase in the deviance).

The REML estimates for the random coefficient model with the same regression and variation parts as above are given in Table 5.4. The differences between ML and REML estimates are trivial. The value of the deviance is omitted from the table so as not to tempt its comparison with the deviance of the ML fit.

Although the client-level residuals $\hat{\boldsymbol{\delta}}_j$ have four components, the fitted deviations $\mathbf{z}\hat{\boldsymbol{\delta}}_j$ are univariate functions of gestational age,

$$f(t, \boldsymbol{\delta}_j) = \mathbf{z}(t)\boldsymbol{\delta}_j,$$

where $\mathbf{z}(t) = (1, t, (t - 252)^2/100, (t - 252)^2/100 \times d)$, and d is the indicator function for $t > 252$. The deviation curves are plotted in the left-hand panel of Figure 5.6. Some of the monotone fitted curves stand out, and would be

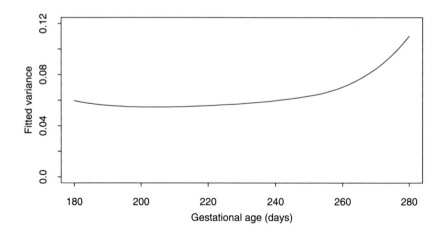

Fig. 5.7. Fitted variance of log-concentration of HPL as a function of gestational age. Pregnancy monitoring data.

the natural candidates for inspection as possible outliers. For comparison, the right-hand panel of Figure 5.6 contains the plot of 35 simulated curves $\mathbf{z}\boldsymbol{\delta}_j$ where $\boldsymbol{\delta}_j$, $j = 1,\ldots 35$, are random draws from $\mathcal{N}(\mathbf{0}, \hat{\boldsymbol{\Sigma}})$. The plot indicates that monotone deviation curves are not unusual, even though

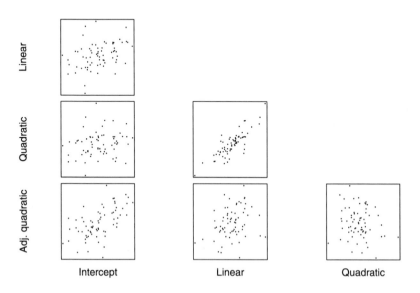

Fig. 5.8. Bivariate plots of the client-level residuals. Pregnancy monitoring data.

none of them decreases as precipitously as the fitted curve for client 49. This client took part in the study during days 228–285, starting with the fitted deviation just under zero, and ending with the lowest fitted deviation, –0.63, on day 285. The plot of simulated curves contains only one curve with value lower than –0.63 for day 285.

The fitted variance of an observation, as a function of gestational age, is plotted in Figure 5.7. The variance is in the range 0.05–0.06 up to about day 240, and then it increases to about 0.11 by day 280. The variance increases about two-fold between days 200 and 280. This is also borne out by the plots in Figure 5.6.

Figure 5.8 contains the pairwise plots of the components of the client-level residuals $\hat{\delta}_j$. These are useful in the search for detailed departures from normality. There appear to be no outliers in either of the plots.

5.4 House prices

Since most of the towns have small numbers of census tracts, modelling of varying within-town regressions is not relevant. However, allowing for varying regression may account for variance heterogeneity as well as for non-linearity of the regression. For example, if the regression with respect to one of the variables is non-linear, but identical across the towns (i.e., not associated with variation), then by fitting linear regression the within-town regression coefficients may become correlated with the average values of the regressor within towns. Figure 5.9 depicts such a situation. Thus, indirectly, variance heterogeneity may provide additional insight into the

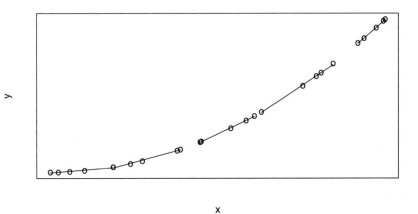

Fig. 5.9.   Non-linear regression and random coefficients. House price data. The observations are marked by points, and the solid lines represent the fitted within-cluster regressions.

Table 5.5.   Fitted slope variances. House price data. The patterns of variation are: $<$ – right-facing megaphone, $>$ – left-facing megaphone, $\times$ – ray pattern or no structure

Variable in random part	Mean St. dev.	Slope var. St. err.	Minimum of variance	Pattern of variation	Deviance
CRIME	3.62 (0.40)	0.0114 (0.0033)	3.97	$\times$	−483.15
ROOM	39.98 (9.09)	0.0066 (0.0013)	55.12	$<$	−528.10
AGE	68.58 (28.14)	0.0026 (0.0005)	67.95	$\times$	−505.79
DIST	1.19 (0.54)	0.197 (0.019)	1.61	$\times$	−532.79
BLACK	0.36 (0.09)	0.399 (0.122)	0.73	$<$	−485.69
LSTAT	−2.23 (0.60)	0.145 (0.025)	−2.38	$\times$	−484.84

choice of transformations for the explanatory variables. Of course, in more clear-cut cases, transformation of variables is better tackled directly by fitting non-linear regression.

The descriptive variables listed in Table 2.5 contain six census-tract level variables: *CRIME, ROOM, AGE, DIST, BLACK,* and *LSTAT.* The variable *DIST,* a measure of the distance from major employment centres, has little within-town variation because the towns are contiguous and most of them are of small area. Associating each census-tract level variable with variation (one at a time) yields substantial reductions of the deviance. Table 5.5 contains summaries for these model fits. The first column contains the sample mean and standard deviation of the variable. The second column gives the estimate of the slope variance and its standard error. In each case the slope variation is significant. For each model the table contains the value of the variable which minimizes the fitted town-level variance. The fitted pattern of variation is described by symbols '$<$', '$\times$', and '$>$', standing, respectively, for (almost) left-facing megaphone (decreasing variance), minimum variance within the range of the data, and (almost) right-facing megaphone (increasing variance). The fitted variance matrix is singular only for the model with *DIST* in the variation part. Since this variable has little within-town variation, variance heterogeneity interpretation of the between-town variance is more appropriate. For more detailed

Table 5.6. Estimates of the regression parameter for *NOXSQ* for random coefficient models

Variable in variation part	Estimate	St. error
CRIME	−0.005 92	(0.001 20)
ROOM	−0.005 27	(0.001 27)
AGE	−0.005 68	(0.001 23)
DIST	−0.005 88	(0.001 23)
BLACK	−0.005 92	(0.001 22)
LSTAT	−0.005 32	(0.001 33)

exploration, town-level variables could also be included in the variation part.

In this example model choice is driven purely by the desire to reduce the value of the deviance by (relatively) parsimonious models, assuming that the estimates of the regression parameters and of their standard errors from 'better fitting' models are more appropriate. More confidence can be attached to the conclusion about the central issue, the association of the variable *NOXSQ* quantifying pollution, if it is only moderately affected by the model choice. The regression parameter estimates for variable *NOXSQ* are given in Table 5.6. The estimates of the slope for *NOXSQ* are in the range −0.0059 to −0.0053, with standard errors 0.0012–0.0013. This is consistent with the estimate from the random-effects analysis, −0.0059 (0.0012), but all these estimates are larger than the OLS estimate, −0.0064 (0.0011). Thus, more detailed modelling of between-cluster variation is not so important for estimation of a regression slope in *this* example.

Further sizeable reduction of the deviance can be achieved by associating several census-tract level variables with variation. For example, the fitted deviance for the model with *ROOM* and *LSTAT* in the variation part is −600.91; the fitted standard deviations for the slopes on *ROOM* and *LSTAT* are 0.0090 (0.000 86) and 0.192 (0.0147), respectively. In comparison with the simpler models (Table 3.13), the standard deviations and their standard errors are only moderately inflated. This suggests that the data contain ample information about variance heterogeneity.

For a model fit with two variables in the variation part, such as *ROOM* and *LSTAT*, detailed model diagnostics can be carried out by plotting the fitted variation parts $\mathbf{x}\hat{\boldsymbol{\delta}}_j$, $\mathbf{x} = (1, ROOM, LSTAT)$, in three dimensions. As an alternative, the linear functions $\mathbf{x}\hat{\boldsymbol{\delta}}_j$ can be plotted for selected values of *ROOM* and *LSTAT*. A set of such plots is drawn in Figure 5.10.

The top left-hand panel contains the fitted deviations

$$(1,\ ROOM,\ LSTAT^*)\hat{\boldsymbol{\delta}}_j\,,$$

as a function of $ROOM$, for $LSTAT^* = -1.93$ (one-half of the standard deviation above the mean of $LSTAT$). The fitted deviation lines are plotted only for the 26 towns for which $LSTAT^*$ lies within the range of their census tracts' values of $LSTAT$. Notable is town 81 (South Boston) because of its wide range of values of $ROOM$. It has the highest fitted deviation for low values of $ROOM$, but the lowest fitted deviation for high values of $ROOM$.

The top right-hand panel of the diagram contains a similar plot for $ROOM$, with the fitted deviations $LSTAT^*$ set to -2.53, half a standard deviation above the mean of $LSTAT$. Deviations for the 38 towns that contain the value $LSTAT^*$ within their ranges of $LSTAT$ are plotted. South Boston is not represented here, because it has only high values of $LSTAT$.

The bottom panels contain the fitted deviation lines for $LSTAT$, with the values of $ROOM$ set to one-half of the standard deviation below and above the mean for $ROOM$. The plots contain 39 and 43 towns respectively.

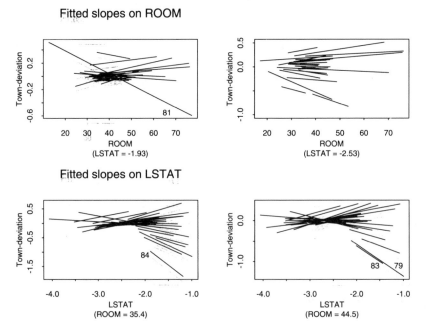

Fig. 5.10. Fitted conditional deviation lines. House price data. The top panels contain fitted town-level deviation lines, functions of $ROOM$, for values of $LSTAT$ given in the subtitles. In the bottom panels the roles of $ROOM$ and $LSTAT$ are interchanged. The order numbers of the towns with outlying deviations are given in the plots; they are: 79 Charlestown, 81 South Boston, 83 Roxbury, and 84 Savin Hill.

The pattern of variation in these two panels is very similar, unlike in the top two panels. The minimum variance occurs around $LSTAT = -2.5$, near the mean. The towns with fitted deviation lines in the lower left-hand corner of the plots are possible outliers; the deviations are distinctly non-normally distributed.

Responding to the model criticism based on Figure 5.10 is, naturally, difficult. However, random coefficient models and such residual plots offer insight into between-town differences, which cannot be gained by methods that ignore the two-level structure of the data. For instance, a more detailed inspection of the data may reveal whether the towns outlying in the bottom panels are adjacent, or have some common characteristics not accounted for by the explanatory variables.

5.5   GRE validity study

An important function of the GRE tests is to assess a student's ability in three designated areas; verbal, quantitative, and analytical. Numerous studies have demonstrated that the three scores do represent three different underlying traits, that is, they are not perfectly correlated, and would not be perfectly correlated even if the tests were perfectly reliable. The departments in the dataset specialize in a variety of academic subjects. Quantitative departments (engineering, physics, mathematics, and the like) prefer and are preferred by students with higher GRE Q scores, and place less emphasis on the GRE V scores. For arts subjects and social sciences (history, sociology, and the like) less emphasis is placed on GRE Q and more on GRE V. The undergraduate grade-point average, GP, is also an important consideration for admission in all departments, even though the exam summary scores, FYA and GP, are on scales specific to each department and undergraduate institution respectively. The admissions procedures of the departments vary in competitiveness, in emphasis on the various aspects of students' background, and in a number of other details. Typically, in small departments more weight is given to informal assessment (interview, letters of recommendation, reputation of the undergraduate school, etc.); in larger departments the standardized scores are more important.

Clearly, more flexibility in modelling of the relationship of FYA on the predictor scores is necessary to capture at least some of the heterogeneity across the departments. This is done by allowing for department-level regressions to vary in all the regression coefficients.

Before proceeding to model fitting we discuss the (potential) use of the fitted within-department regressions. The administrator of the GRE tests, the Educational Testing Service, would like to entice as many departments as possible into cooperation entailing provision of data about their students' FYA and GP scores. To make cooperation more attractive the recruited departments are provided with a fitted regression equation $(1, V, Q, A, GP)\hat{\boldsymbol{\beta}}_j$ and are advised to use it for the selection of applicant students. Such a recommendation is most problematic, though. First, the fitted regression equations describe an amalgam of processes. The selection and self-selection of students into departments is one such process, instruction in the department and the school's environment is another, and, of course, the process of assigning grades is yet another. If a department started using these equations for selection (even only as one of several criteria), the process of selection would be altered, and a different regression equation would then apply. The effects of such a selection procedure are unpredictable. Furthermore, even if the equations did describe the selection process without bias, the standard errors associated with the estimated parameters would be very large for all but the largest departments. Shrinkage estimation of the regression coefficients reduces the mean squared errors of the within-department regressions, but the sampling variation remains considerable. To illustrate this point we omit the details of model specification, which are dealt with later in this section.

The fitted regression for department 1, containing 13 students, is

$$2.65 + \underset{(0.084)}{0.115\,V} + \underset{(0.081)}{0.073\,Q} + \underset{(0.039)}{0.007\,A} + \underset{(0.085)}{0.261\,GP} - 0.220\,\overline{V}.$$

$$(5.3)$$

The standard errors (in parentheses) are derived from the conditional variance matrix for $\hat{\boldsymbol{\delta}}_j$; the estimate of the coefficient on $\overline{V}$ is common to all the departments and its standard error is 0.023. The standard errors are so large that even an equation substantially different from (5.3) is feasible. For comparison, the OLS fit for the department is

$$1.49 + \underset{(0.284)}{0.123\,V} - \underset{(0.472)}{0.082\,Q} + \underset{(0.281)}{0.243\,A} + \underset{(0.194)}{0.374\,GP}; \qquad (5.4)$$

the standard errors are derived using the unbiased estimator of residual variance ($\hat{\sigma}_1^2 = 0.163$). Note that the standard errors in (5.3) ignore the sampling variation associated with the regression and covariance structure parameters, and therefore they are somewhat optimistic. Also, the standard errors in (5.4) are subject to sampling variation since they are a function of $\hat{\sigma}_1^2$, which is based on only 13 observations (8 degrees of freedom).

To underline the lack of information about within-department regression coefficients we give the fitted regression coefficients for department 11

(46 students). Its fitted regression based on the residuals from the random coefficient model is

$$3.40 + \underset{(0.071)}{0.094\,V} - \underset{(0.072)}{0.036\,Q} + \underset{(0.025)}{0.012\,A} + \underset{(0.074)}{0.152\,GP} - 0.220\,\overline{V},$$

(5.5)

while the OLS fit is

$$2.965 + \underset{(0.078)}{0.077\,V} - \underset{(0.082)}{0.018\,Q} + \underset{(0.082)}{0.054\,A} + \underset{(0.095)}{0.112\,GP}$$

(5.6)

($\hat{\sigma}_{11}^2 = 0.037$), so that, judging by the standard errors, OLS is a reasonable competitor to the random coefficient models, but in both (5.5) and (5.6) the standard errors are too large to render the results useful for the department.

One solution to this problem is the aggregation of the departments (graduate programs) within schools or within other meaningful units. This would raise the cluster sizes considerably, although the loss of identification of the department variation may be detrimental to substantive interpretation. Also, the resulting between-cluster variation would be inflated, thus reducing the usefulness of the shrinkage estimators.

Of course, there are numerous other influences on the importance of the GRE scores. On the one hand, all parties involved in the admissions process are well acquainted with the GRE test; its scores appear to be easy to interpret and most undergraduate students with academic ambitions are strongly motivated to score as high as possible on them. On the other hand, the instructors in a department may abandon all expectations based on the students' backgrounds, and spend additional effort, if necessary, to help each student to reach a certain level. Then, of course, the GRE scores would appear to have no 'validity'; all the regression coefficients would be close to zero. A variety of other scenarios can be thought of which give rise to specific patterns of regression coefficients, some of them appearing anomalous if the assumption of random allocation of students to graduate schools were falsely assumed. Also, the assumption of uniform (educational) treatment of the students in a department is not realistic.

For example, when interpreting the department regressions as a description of processes taking place during the first year in graduate school, negative within-department regressions are contradictory. They imply that lower scores on one of the background scores are associated with higher outcome scores – an apparent anomaly. Alternatively, they can be interpreted as an indication of imperfect validity of the GRE tests, a rather undesirable outcome for ETS. Therefore, an assessment of frequency of negative regression coefficients is of interest. Estimating this frequency by counting the departments that have negative *estimated* coefficients is inefficient and biased, since it depends crucially on the distribution of department sizes. Since most departments are small and between-department

Table 5.7. Random coefficients analysis. GRE validity data. The standard errors for the variances of the slopes are in the range 0.024–0.030, for the intercept-by-slope covariances 0.085–0.112, and for the slope-by-slope covariances 0.020–0.022. Twelve Fisher scoring iterations were carried out

Parameter type	Variable					
	1	V	Q	A	GP	$\overline{V}$
Regression	2.883	0.104	0.055	0.032	0.220	−0.220
		(0.0090)	(0.0093)	(0.0087)	(0.0092)	(0.0232)
Student variance			0.0923			
Department variance	$\begin{pmatrix} 3.240 & -0.136 & -0.231 & 0.097 & -0.500 \\ -0.136 & 0.087 & 0.024 & -0.048 & -0.004 \\ -0.231 & 0.024 & 0.087 & -0.042 & 0.014 \\ 0.098 & -0.048 & -0.042 & 0.037 & -0.007 \\ -0.500 & -0.004 & 0.014 & -0.007 & 0.112 \end{pmatrix}$					
Deviance			6158.11			

variation is only moderate, substantial shrinkage towards the pooled regression takes place, and fewer negative estimated coefficients are obtained than there are true negative coefficients.

We start by fitting the model with regression part $(1, V, Q, A, GP, \overline{V})$ and variation part $(1, V, Q, A, GP)$:

$$FYA_{ij} = \mathbf{x}_{ij}\boldsymbol{\beta}_j + \overline{V}_j\gamma + \varepsilon_{ij}$$
$$\boldsymbol{\beta}_j \sim \mathcal{N}(\boldsymbol{\beta}, \boldsymbol{\Sigma}).$$

(5.7)

The parameter estimates are given in Table 5.7. Compared to the random-effects model the deviance is substantially reduced ($6370.56 - 6158.11 = 212.45$, for 14 additional parameters). Note that the fitted variance matrix is singular. Therefore, the likelihood ratio test cannot be used in its standard form (hypothesis testing is not of interest in this context, though).

The regression parameter estimates are similar to those obtained in random-effects analysis (Section 3.5); the standard errors are only moderately inflated. The fitted department-level variances are much larger than expected. Based on these variances we would estimate, for example, that

$$\Phi(-0.104/\sqrt{0.0923 \times 0.087}) \times 100 = 12$$

per cent of the departments have a negative coefficient on V. Similarly, the fitted proportion of negative coefficients for the scores Q, A, and GP are 27, 36, and 1.5 per cent respectively. Note that this method does not identify the actual departments that have negative coefficients. In contrast, the numbers of negative estimated coefficients are only 6, 42, 162, and none (out of 748), for the respective scores V, Q, A, and GP, and only 204 departments (27 per cent) have at least one negative estimated coefficient. This, nevertheless, seems to be a sizeable proportion of the departments.

The model (5.7) is a special case of the contextual model

$$FYA_{ij} = \mathbf{x}_{ij}\boldsymbol{\beta}_j + \varepsilon_{ij}, \tag{5.8}$$

in which the random coefficients $\boldsymbol{\beta}_j$ are related to the context variable $\overline{V}$ by linear regressions,

$$\boldsymbol{\beta}_j = \boldsymbol{\Gamma}(1, \overline{V}_j)^\top + \boldsymbol{\delta}_j. \tag{5.9}$$

We denote the elements of $\boldsymbol{\Gamma}$ by $\Gamma_{k,h}$ ($k = 1, V, Q, A, GP$, $h = 1, 2$). We consider first the consequences of the model in (5.8)–(5.9) on the occurrence of negative coefficients. The regression slope of FYA on a student-level variable, say A, is

$$r_{A,\overline{V}} = \Gamma_{A,1} + \Gamma_{A,2}\overline{V}_j,$$

which is a linear function of $\overline{V}_j$. If both $\Gamma_{A,1}$ and $\Gamma_{A,2}$ are positive then $r_{A,\overline{V}}$ is positive for all $\overline{V}$, and so in any narrow band of $\overline{V}$ more than half of the departments are likely to have positive coefficients on A. More detailed discussion of the frequency of negative regression coefficients (on A) can be based on the distribution of the regression coefficients on A for the departments in a narrow band around $\overline{V}$,

$$\beta_{A,j} \sim \mathcal{N}(r_{A,\overline{V}}, \Sigma_{AA}).$$

Typically, either departments with low values or those with high values of $\overline{V}$ contain the largest proportion of negative coefficients. Nevertheless, if the variance Σ_{AA} is large, negative coefficients are likely, even when the expectation of the slope, $r_{A,\overline{V}}$, is positive.

The unconditional expectation of the outcomes FYA is

$$FYA = \mathbf{x}\boldsymbol{\Gamma}(1, \overline{V})^\top = \mathbf{x}^*\boldsymbol{\gamma}, \tag{5.10}$$

where $\mathbf{x}^* = (1, V, Q, A, U, \overline{V}, V\overline{V}, Q\overline{V}, A\overline{V}, U\overline{V})$. The GRE scores V, Q, and A are highly correlated; their correlation matrix is

$$\begin{pmatrix} 1.0 & 0.475 & 0.559 \\ 0.475 & 1.0 & 0.707 \\ 0.559 & 0.707 & 1.0 \end{pmatrix}.$$

Table 5.8. Random coefficients analysis with the context variable $\overline{V}$ associated with all scores. GRE validity data

Parameter type	Variable				
	1	V	Q	A	GP
Regression	2.291	0.108	0.142	0.099	0.250
		(0.0092)	(0.0095)	(0.0090)	(0.0094)
Context $(\times\overline{V})$	0.004	−0.002	−0.032	−0.025	−0.012
	(0.144)	(0.034)	(0.035)	(0.033)	(0.035)
Student variance	0.0923				
Department variance	$\begin{pmatrix} 6.209 & -0.133 & -0.714 & 0.108 & -0.914 \\ -0.133 & 0.087 & 0.024 & -0.049 & -0.005 \\ -0.714 & 0.024 & 0.087 & -0.043 & 0.015 \\ 0.108 & -0.049 & -0.043 & 0.037 & -0.007 \\ -0.914 & -0.005 & 0.015 & -0.007 & 0.112 \end{pmatrix}$				
Deviance	6151.57				

Addition of further regressors which are related to GRE scores would contribute to the problem of confounding of the regression parameters. Also, most departments are small and, since they tend to contain students with a narrow band of V, V and $\overline{V}$ are also highly correlated; $\mathrm{cor}(V,\overline{V}) = 0.572$. On the other hand, the large dataset at hand allows for fitting regression with highly correlated explanatory variables. Of interest is how complex a regression formula involving these intercorrelated regressors should be used that would balance the requirements of identification and adequacy of the model.

The estimates for the random coefficient model (5.8)–(5.9) are given in Table 5.8. The deviance for this model fit is 6151.67, a reduction of only 6.44 for four regression parameters. Thus the 'context' for the slope deviations is not significant. Nevertheless, the regression parameter estimates, if taken at face value, have important consequences on the frequency of negative coefficients. The mean fitted slope on A is $0.099 - 0.025\overline{V}$, which is very close to zero for the highest values of $\overline{V}$. About half the departments with high values of $\overline{V}$, $\overline{V} \doteq 4.0$, are likely to have negative within-department coefficients on A. A similar situation arises for the coefficients on Q, whereas the mean coefficients for V and GP depend on $\overline{V}$ much less critically. The fitted variance matrix $\hat{\boldsymbol{\Sigma}}$ differs from its counterpart for the model (5.7) only in the terms involving the intercept.

Thus, fitting of more complex regression models results in more variation of the fitted slopes, and, as a consequence, in a higher proportion of negative fitted coefficients. Note that estimation of the proportion of negative (theoretical) coefficients with 'contextual' models is more complex because the mean slopes depend on $\overline{V}$.

The standard errors for the regression parameter estimates in the model (5.8)–(5.9), as compared to the model (5.7), are somewhat inflated. Ill-conditioning of the regression can be explored by an eigenvalue analysis of the information matrix $\mathbf{H} = \mathbf{X}^\top \mathbf{V}^{-1} \mathbf{X}$. Since most departments are small, little precision is lost by considering the matrix $\mathbf{X}^\top \mathbf{X}$ instead of $\mathbf{H}$. This has the advantage that ill-conditioning does not depend on the unknown matrix $\mathbf{V}$ and can be explored even before model fitting. The eigenvalues for the regressors $(1, V, Q, A, GP, \overline{V})$ are

$$443\,000, \quad 3010.1, \quad 1715.1, \quad 1033.2, \quad 710.9 \quad 86.45;$$

the condition number (the ratio of the largest and smallest eigenvalues) is 5124.3. In contrast, the eigenvalues for the regressors in the contextual model with $\overline{V}$, $(1, V, Q, A, U, GP, \overline{V}, V\overline{V}, Q\overline{V}, A\overline{V}, GP\overline{V})$, are

$$3.142 \times 10^6, \quad 23\,518, \quad 14\,582, \quad 8324, \quad 3531.1$$
$$759.0, \quad 338.0, \quad 18.22, \quad 11.02, \quad 1.627.$$

The condition number is now 1.93×10^6, a 375-fold increase, confirming acute ill-conditioning in the regression part of the contextual model.

We conclude that modelling of the context by a within-department mean does not improve the model fit for the FYA scores. On the contrary, it exacerbates the problem of confounding of the regression parameters.

Some further model simplification can be achieved by simplifying the parametrization for between-department variation. Since linear combinations of the three GRE scores are meaningful, the covariances involving pairs of GRE scores should not be constrained to zero, even though they are (nominally) statistically not significant. However, combining GP with the GRE scores is not meaningful, and therefore the constraints

$$\Sigma_{25} = \Sigma_{35} = \Sigma_{45} = 0 \tag{5.11}$$

can be imposed. Note that when estimated using the models in (5.7) and (5.8), these covariances have low t-ratios.

The model fit for (5.7) with the constraints (5.11) is summarized in Table 5.9. The numbers of negative coefficients are essentially the same as before. Compared to the model fit with no constraints on Σ, only two fitted coefficients on A (out of 162) have changed sign from negative to positive.

Table 5.9. Random coefficients analysis. GRE validity data. The standard errors for the slope variances are in the range 0.024–0.030, for the intercept-by-slope covariances 0.085–0.112, and for the slope-by-slope covariances, 0.020–0.022. Twelve Fisher scoring iterations were carried out

Parameter type	Variable					
	1	V	Q	A	GP	$\overline{V}$
Regression	2.883	0.104	0.055	0.031	0.220	−0.220
		(0.0090)	(0.0093)	(0.0087)	(0.0092)	(0.0232)
Student variance			0.0923			
Department variance	$\begin{pmatrix} 3.188 & -0.151 & -0.192 & 0.077 & -0.489 \\ -0.151 & 0.086 & 0.025 & -0.049 & 0.000 \\ -0.192 & 0.025 & 0.088 & -0.041 & 0.000 \\ 0.077 & -0.049 & -0.041 & 0.036 & 0.000 \\ -0.489 & 0.000 & 0.000 & 0.000 & 0.112 \end{pmatrix}$					
Deviance			6158.53			

The variance of the coefficients on A is not significantly different from zero, and could be constrained to zero (together with all the covariances in the same row and column of Σ). This would eliminate another four covariance structure parameters. This is contentious, though, because the A score can be combined with the other GRE scores.

The fitted variance matrices $\hat{\Sigma}$ for (5.7) and (5.8) each have a negative eigenvalue. The eigenvalues for the estimates of $\hat{\Sigma}$ based on (5.7) and (5.8) are

$$3.343, \quad 0.136, \quad 0.061, \quad 0.027, \quad -0.0040$$

and

$$3.285, \quad 0.143, \quad 0.063, \quad 0.024, \quad -0.0040$$

respectively. They are essentially of rank 4, and so their parametrization contains an element of redundancy. This is of little importance for model parsimony, and is not pursued further.

Figure 5.11 contains the pairwise plots of the department-level residuals. Correlations of the deviations, e.g., between GRE A and GRE V are transparent, and confirm the pattern in the fitted department-level variance matrix. In particular, the assumption of independence of the GP-deviations with those for the GRE scores is supported, with the possible exception of the correlation between GP and GRE A scores.

Finally, we return to the issue of how to use the fitted department-level regressions (arguably, this is what the analysis should have started with). As we pointed out earlier, these equations describe an amalgam of processes associated with students' selection of the graduate school, criteria for admissions, instruction in the department, and the academic environment in general. If each student is achieving the same (or similar) results at the end of the first year, so that FYA has little variation within the school, the undergraduate scores are not associated with FYA. This does not mean that the GRE scores have poor predictive validity; rather, it simply means that the academic staff devoted more time and effort to students who needed more assistance to achieve a certain level. This would be essential if the curriculum in the first year were a prerequisite for the rest of the course. If students are given inflated grades in a school, then, obviously, the school appears to 'convert' GRE scores much more favourably than do other schools.

Students with the same undergraduate scores V, Q, A, and GP have a variety of interests and ambitions, and so, even if they entered the same

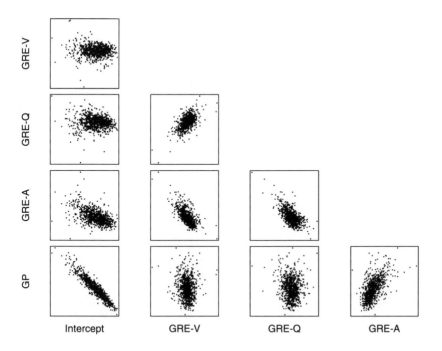

Fig. 5.11. Pairwise plots of the fitted department-level residuals. GRE validity data. Each panel contains the scatterplot of a pair of components of the vector of department-level residuals.

graduate school, they might achieve different FYA scores. Admissions staff probably attempt to select the most suitable students conditionally on their background scores. The administrator of the GRE validity study claimed that the fitted equations could provide the schools with information about how to improve the admissions process (e.g., by selecting students with the highest predicted FYA scores. Such a claim is difficult to substantiate. If the admissions process (policy, criteria, etc.) is altered, so is the department regression, because the equation is influenced by these processes. Also, the department may select the best students, but most of them may be offered places in more competitive departments and enrol there.

Meaningful interpretation of these equations is possible only if account of the admission and instruction processes can be taken. This is not a straightforward task, even when one is well acquainted with the 'local' circumstances of the graduate school. Comparison of these equations across time, or with competitor departments (which have similar selection and instruction processes) is meaningful, but differences may reflect phenomena of limited interest.

The administrator of the GRE tests has a vested interest in the assessment of the predictive validity of the tests. If the non-random nature of

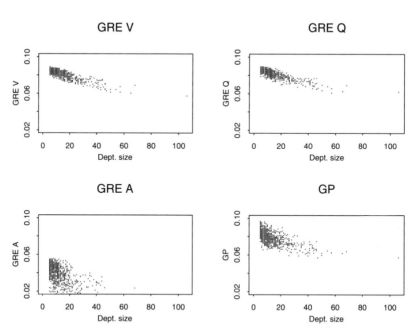

Fig. 5.12. Dependence of the conditional (posterior) standard deviation of the department-level residuals on department size. All four plots have the same scale for each axis.

assignment of students to departments were ignored, predictive validity should demonstrate itself in positive regression coefficients. However, it is not clear how large these parameters have to be for predictive validity to be satisfactory. Monitoring the parameter estimates over years would help to determine whether the validity is improving or not, but this is hampered by the changing composition of the participating departments.

We have concluded that, apart from problems of interpretation, the fitted within-department regression equations are of limited use, because their coefficients are associated with large standard errors. It is therefore of interest to determine the smallest department size for which the fitted standard errors would be smaller than a preset benchmark. Figure 5.12 contains the plots of the conditional standard deviations for the components of the residuals $\hat{\boldsymbol{\delta}}_j$ against the department size. The plots are bearers of bad news; for the V, Q, and GP scores the standard deviations decrease very slowly, and the standard deviations would be smaller than, say, 0.05, only for unrealistically large departments. The source of this problem appears to be that the within-department design matrices $\mathbf{Z}_j$ are almost singular even for the largest departments. This is so because the students in larger departments are more tightly (self-)selected on GRE and GP scores than in small departments, with the possible exception of the GRE A scores.

6
Multiple levels of nesting

6.1   Models

A natural extension of the one-way (two-level) nesting design studied in Chapters 2 and 4 is obtained when the clusters, say families or, generally, units at level 2, are further clustered into districts, areas, or, in general, into units at level 3. Such designs arise in clustered sampling. For instance, in an educational survey students are sampled within schools, schools within districts, and districts are sampled from the districts of the country.

The levels of nesting can be generated by a combination of data structures that on their own would give rise to a two-level design. As an example, consider a longitudinal study with replicated measurements. Since each observation in the longitudinal study is subject to measurement error or some other source of uncertainty, it is repeated independently (replicated) several times at each time-point. Without replication we would consider the random coefficient model

$$y_{jh} = \mathbf{x}_{jh}\boldsymbol{\beta}_h + \gamma_{jh}, \qquad (6.1)$$

with the usual assumptions of normality and independence, $\boldsymbol{\beta}_h \sim \mathcal{N}(\boldsymbol{\beta}, \boldsymbol{\Sigma})$, $\gamma_{jh} \sim \mathcal{N}(0, \sigma_2^2)$, where j indexes the observations and h the subjects. Temporal variation and measurement error are confounded in the random term γ; they are components of the variance σ_2^2. With replicated observations we consider the model

$$y_{ijh} = \mathbf{x}_{jh}\boldsymbol{\beta}_h + \gamma_{jh} + \varepsilon_{ijh}, \qquad (6.2)$$

with the same assumptions as in (6.1), supplemented by $\varepsilon_{ijh} \sim \mathcal{N}(0, \sigma_1^2)$ and the usual assumptions of independence. The variance σ_1^2 describes uncertainty due to the measurement error (replication within subject and within time-point), and σ_2^2 describes the departures of the 'observations' that are devoid of the measurement error (the 'true' values of y) from the subject's growth curve $\mathbf{x}_{jh}\boldsymbol{\beta}_h$. The matrix $\boldsymbol{\Sigma} = \mathrm{var}(\boldsymbol{\beta}_h)$ describes the pattern of between-subject variation.

The conditional expectation of the outcome y_{ijh}, given the subject's coefficients $\boldsymbol{\beta}_h$, is $\mathbf{x}_{jh}\boldsymbol{\beta}_h$, and the unconditional expectation (the population mean growth curve) is $\mathbf{x}_{jh}\boldsymbol{\beta}$. More generally, the linear regression formula $\mathbf{x}\boldsymbol{\beta}$ may involve variables defined at each of the three levels:

$$\mathbf{E}(y_{ijh}) = \beta^{(0)} + \mathbf{x}_{ijh}^{(1)}{}'\boldsymbol{\beta}^{(1)} + \mathbf{x}_{jh}^{(2)}{}'\boldsymbol{\beta}^{(2)} + \mathbf{x}_{h}^{(3)}{}'\boldsymbol{\beta}^{(3)}$$
$$= \mathbf{x}_{ijh}\boldsymbol{\beta}, \tag{6.3}$$

where $\boldsymbol{\beta} = \left(\beta^{(0)}, \boldsymbol{\beta}^{(1)^\top}, \boldsymbol{\beta}^{(2)^\top}, \boldsymbol{\beta}^{(3)^\top}\right)^\top$ and $\mathbf{x}_{ijh} = \left(1, \mathbf{x}_{ijh}^{(1)}, \mathbf{x}_{jk}^{(2)}, \mathbf{x}_{h}^{(3)}\right)$. We refer to the levels of clustering as elementary level, level 2, and level 3, and use the term 'level k' both as a noun and as an adjective.

We assume that there are N_3 units at level 3, the units $h = 1, \ldots, N_3$ comprising $n_{2,h}$ level 2 clusters each and each of these clusters jh, $j = 1, \ldots, n_{2,h}$, comprising $n_{1,jh}$ elementary units. The number of level 2 clusters in the sample is $N_2 = \sum_h n_{2,h}$, and the number of elementary-level units is $N = \sum_h \sum_j n_{1,jh}$. By $n_{1,h}$ we denote the number of elementary-level units in the level 3 unit h: $n_{1,h} = \sum_j n_{1,jh}$. All these cluster sizes are defined for the sample; analogous definitions and notation can be introduced for the sampling frame (target population).

We denote $\mathbf{X}_{jh} = \{\mathbf{x}_{ijh}\}_i \otimes \mathbf{1}_{n_{1,jh}}$ and $\mathbf{y}_{jh} = \{y_{ijh}\}_i \otimes \mathbf{1}_{n_{1,jh}}$, the design matrix and the vector of outcomes, respectively, for the level 2 cluster jh. By further stacking we define the level 3 versions, $\mathbf{X}_h = \{\mathbf{X}_{jh}\} \otimes \mathbf{1}_{n_{2,h}}$ and $\mathbf{y}_h = \{\mathbf{y}_{jh}\} \otimes \mathbf{1}_{n_{2,h}}$, and the sample versions, $\mathbf{X} = \{\mathbf{X}_h\} \otimes \mathbf{1}_{N_3}$ and $\mathbf{y} = \{\mathbf{y}_h\} \otimes \mathbf{1}_{N_3}$, of the regression design matrix and the vector of outcomes. We use the analogous notation for other stacked objects.

The simplest non-trivial covariance structure in the three-level design corresponds to the random-effects model

$$\mathbf{y}_{jh} = \mathbf{X}_{jh}\boldsymbol{\beta} + \delta_h + \gamma_{jh} + \varepsilon_{ijh}, \tag{6.4}$$

where $\delta_h \sim \mathcal{N}(0, \sigma_3^2)$, $\gamma_{jh} \sim \mathcal{N}(0, \sigma_2^2)$, and $\varepsilon_{ijh} \sim \mathcal{N}(0, \sigma_1^2)$ form three mutually independent random samples. The variance of an observation in (6.4),

$$\operatorname{var}(y_{ijh}) = \sigma_3^2 + \sigma_2^2 + \sigma_1^2, \tag{6.5}$$

decomposes into the elementary-, second-, and third-level variances (variance components). The covariance of two observations in the same level 2 unit is

$$\operatorname{cov}(y_{ijh}, y_{i'jh}) = \sigma_3^2 + \sigma_2^2 \quad (i \neq i'), \tag{6.6}$$

and the covariance of two observations in the same level 3 but different level 2 units is

$$\operatorname{cov}(y_{ijh}, y_{i'j'h}) = \sigma_3^2 \quad (j \neq j'). \tag{6.7}$$

Zero variances, $\sigma_3^2 = 0$ and $\sigma_2^2 = 0$, correspond to trivial submodels of (6.4). When $\sigma_3^2 = 0$ (6.4) becomes a two-level model; level 2 clusters are mutually independent, and their further clustering is therefore irrelevant. Similarly, when $\sigma_2^2 = 0$ (6.4) becomes a two-level model with the nesting structure of elementary units within level 3 units.

The model in (6.4) can be generalized to allow arbitrary within-cluster regressions, at both levels 2 and 3:

$$\mathbf{y}_{jh} = \mathbf{X}_{jh}\boldsymbol{\beta}_{jh} + \boldsymbol{\varepsilon}_{jh}, \qquad (6.8)$$

where $\boldsymbol{\beta}_{jh}$ has the decomposition

$$\boldsymbol{\beta}_{jh} = \boldsymbol{\beta} + \boldsymbol{\delta}_{jh}^{(2)} + \boldsymbol{\delta}_h^{(3)} \qquad (6.9)$$

into the vector of regression parameters and level 2 and level 3 random terms. We assume that $\boldsymbol{\delta}_{jh}^{(2)} \sim \mathcal{N}(\mathbf{0}, \boldsymbol{\Sigma}_2)$ and $\boldsymbol{\delta}_h^{(3)} \sim \mathcal{N}(\mathbf{0}, \boldsymbol{\Sigma}_3)$ form two independent random samples, mutually independent of the random sample $\{\varepsilon_{ijh}\}_{ijh}$.

Flexibility of the model (6.8)–(6.9) is achieved by suitable constraints on the variance matrices $\boldsymbol{\Sigma}_2$ and $\boldsymbol{\Sigma}_3$. Much of the discussion from Chapter 4 applies directly to three-level models, because (6.8)–(6.9) becomes a two-level model by setting either $\boldsymbol{\Sigma}_2$ or $\boldsymbol{\Sigma}_3$ to a zero matrix.

In a different perspective each level 3 cluster can be viewed as satisfying a two-level model. For example, the model in (6.8), applied solely for level 3 cluster $h = 1$, becomes a two-level random coefficient model:

$$\mathbf{y}_{j1} = \mathbf{X}_{j1}(\boldsymbol{\beta}_1^* + \boldsymbol{\delta}_{j1}^{(2)}) + \boldsymbol{\varepsilon}_{j1},$$

with regression parameters $\boldsymbol{\beta}_1^* = \boldsymbol{\beta} + \boldsymbol{\delta}_1^{(3)}$, and level 2 variation given by $\boldsymbol{\delta}_{j1}^{(2)} \sim \mathcal{N}(\mathbf{0}, \boldsymbol{\Sigma}_2)$. This can be an important device for preliminary analysis of the data. Often, several level 3 clusters contain sufficiently extensive data to permit meaningful cluster-wise analyses. These analyses may provide important clues about model selection for the entire three-level dataset, even though the resulting estimators are much less efficient than those obtained by a single (three-level) analysis of the entire dataset. When each level 2 unit contains a single elementary observation, or each level 3 cluster contains a single level 2 unit, two-level data structures result. In each case a pair of sources of variation becomes indistinguishable. For example, in the latter case, variation at levels 2 and 3 cannot be separated. Another case of singularity arises when the data contain only one unit at level 3. Then the regression parameters are confounded with the realized random terms at level 3.

When sampling within units at a level is 'thin', that is, most units contain small numbers of subunits, modelling for each level of aggregation may not be possible. Then it is more useful to consider only the aggregate of the two sources of variation.

An alternative representation of the model in (6.8)–(6.9) is

$$\mathbf{y}_{jh} = \mathbf{X}_{jh}\boldsymbol{\beta} + \mathbf{Z}_{jh}^{(3)}\boldsymbol{\delta}_h^{(3)} + \mathbf{Z}_{jh}^{(2)}\boldsymbol{\delta}_{jh}^{(2)} + \boldsymbol{\varepsilon}_{jh}, \qquad (6.10)$$

which is a direct extension of the two-level model in (4.7). We define the design matrices $\mathbf{Z}_h^{(m)} = \mathbf{Z}_{jh}^{(m)} \otimes \mathbf{1}_{n_{2,h}}$, $\mathbf{Z}^{(m)} = \mathbf{Z}_h^{(m)} \otimes \mathbf{1}_{N_3}$ for $m = 2, 3$.

The matrix $\mathbf{X}_{jh}$ $(\mathbf{X}_h, \mathbf{X})$ is referred to as the regression part design matrix for level 2 cluster jh (for level 3 cluster h, for the entire sample), $\mathbf{Z}_{jh}^{(3)}$ $\left(\mathbf{Z}_h^{(3)}, \mathbf{Z}^{(3)}\right)$ as the level 3 variation design matrix for cluster jh (cluster h, entire sample), and similarly for $\mathbf{Z}_{jh}^{(2)}$ $\left(\mathbf{Z}_h^{(2)}, \mathbf{Z}^{(2)}\right)$. The linear combinations $\mathbf{Z}_{jh}^{(3)}\boldsymbol{\delta}_h^{(3)}$ and $\mathbf{Z}_{jh}^{(2)}\boldsymbol{\delta}_{jh}^{(2)}$ represent the departures of the regression for level 3 cluster h from the population (universal) regression $\mathbf{x}\boldsymbol{\beta}$ and of the regression for level 2 cluster jh from the regression for the level 3 cluster h, $\mathbf{x}\boldsymbol{\beta} + \mathbf{z}^{(3)}\boldsymbol{\delta}_h^{(3)}$.

We adopt the convention that the variance matrices $\boldsymbol{\Sigma}_3$ and $\boldsymbol{\Sigma}_2$ do not contain variances that have been set to zero, and that each covariance is either a free parameter or is set to zero. We denote the dimensions of $\boldsymbol{\Sigma}_3$ and $\boldsymbol{\Sigma}_2$ by r_3 and r_2 respectively.

The interpretation of the three-level model in terms of varying within-cluster regressions implies that the level 2 variation design should contain only elementary-level variables, while the level 3 variation design may contain elementary-level and level 2 variables for which it is meaningful to consider coefficients varying among the clusters of level 3. For modelling of variance heterogeneity it is meaningful to consider models which do not conform to these conventions. Further development in this chapter does not rely on these conventions.

6.1.1   LEVEL-WISE EQUATIONS

The level-wise equations discussed in Section 4.5 have a straightforward extension for three levels. We consider the model (6.8)–(6.9) with additional equations for the random terms:

$$\boldsymbol{\delta}_h^{(3)} = \mathbf{B}^{(3)}\mathbf{x}_h^{(3)} + \boldsymbol{\delta}_{h*}^{(3)},$$

$$\boldsymbol{\delta}_{jh}^{(2)} = \mathbf{B}^{(2)}\mathbf{x}_{jh}^{(2)} + \boldsymbol{\delta}_{jh*}^{(2)}, \qquad (6.11)$$

where $\mathbf{B}^{(2)}$ and $\mathbf{B}^{(3)}$ are matrices of (regression) parameters, and $\{\boldsymbol{\delta}_{h*}^{(3)}\}$ and $\{\boldsymbol{\delta}_{jh*}^{(2)}\}$ are mutually independent random samples from $\mathcal{N}(\mathbf{0}, \boldsymbol{\Sigma}_3)$ and $\mathcal{N}(\mathbf{0}, \boldsymbol{\Sigma}_2)$ respectively. It is easy to see that after substituting (6.11) in

(6.8) another model of the form (6.8) is obtained, with various cross-level interactions of the variables included in the regression part of the model:

$$\mathbf{y}_{jh} = \mathbf{X}_{jh}\boldsymbol{\beta} + \mathbf{X}_{jh}\mathbf{B}^{(3)}\mathbf{x}_h^{(3)} + \mathbf{X}_{jh}\mathbf{B}^{(2)}\mathbf{x}_{jh}^{(2)} + \mathbf{X}_{jh}\boldsymbol{\delta}_{h*}^{(3)} + \mathbf{X}_{jh}\boldsymbol{\delta}_{jh*}^{(2)} + \boldsymbol{\varepsilon}_{jh}.$$

$$(6.12)$$

Since the expectation of (6.12) is linear in the elements of the matrices $\mathbf{B}^{(2)}$ and $\mathbf{B}^{(3)}$, (6.12) can be expressed in the standard form (6.8), with the regression part containing interactions of $\mathbf{x}$ with $\mathbf{x}^{(2)}$ and $\mathbf{x}^{(3)}$.

In many applications it is necessary to impose constraints on the matrices $\mathbf{B}^{(2)}$ and $\mathbf{B}^{(3)}$ to avoid having an excessive number of regression parameters. Also, when the variables $\mathbf{x}$, $\mathbf{x}^{(2)}$, and $\mathbf{x}^{(3)}$ are close to being linearly dependent, so that problems with confounding of the regression parameters would arise, addition of numerous interaction terms usually exacerbates these problems.

6.2   Estimation

The log-likelihood for the model in (6.10) is equal to

$$l = -\frac{1}{2}\left\{ N\log(2\pi) + \log(\det \mathbf{V}) + \mathbf{e}^\top \mathbf{V}^{-1}\mathbf{e} \right\},$$

$$(6.13)$$

where $\mathbf{V} = \{\mathbf{V}_{h,3}\} \otimes \mathbf{I}_{N_3}$,

$$\mathbf{V}_{h,3} = \{\mathbf{V}_{jh,2}\}_j \otimes \mathbf{I}_{n_{2,h}} + \mathbf{Z}_h^{(3)} \boldsymbol{\Sigma}_3 \left(\mathbf{Z}_h^{(3)}\right)^\top,$$

$$(6.14)$$

and

$$\mathbf{V}_{jh,2} = \sigma^2 \mathbf{I}_{n_{1,jh}} + \mathbf{Z}_{jh}^{(2)} \boldsymbol{\Sigma}_2 \left(\mathbf{Z}_{jh}^{(2)}\right)^\top.$$

$$(6.15)$$

To derive the equations for the Fisher scoring algorithm we follow the outline of Section 1.9, see also Section 4.4. The first step is to derive equations for the inverse and determinant of $\mathbf{V}$ ($\mathbf{V}_{h,3}$), which are computationally feasible for large clusters (many level 2 clusters, or many elementary-level observations in a level 3 cluster).

For level 2 variance matrix $\mathbf{V}_{jh,2}$ we have the inversion formula (4.21):

$$\mathbf{V}_{jh,2}^{-1} = \sigma^{-2}\left\{ \mathbf{I}_{n_{1,jh}} - \sigma^{-2}\mathbf{Z}_{jh}^{(2)} \boldsymbol{\Sigma}_2 \mathbf{G}_{jh,2}^{-1} \left(\mathbf{Z}_{jh}^{(2)}\right)^\top \right\},$$

$$(6.16)$$

where $\mathbf{G}_{jh,2} = \mathbf{I}_{r_2} + \sigma^{-2}\left(\mathbf{Z}_{jh}^{(2)}\right)^\top \mathbf{Z}_{jh}^{(2)} \boldsymbol{\Sigma}_2$. Also,

$$\det \mathbf{V}_{jh,2} = \left(\sigma^2\right)^{n_{1,jh}} \det \mathbf{G}_{jh,2}.$$

$$(6.17)$$

The inverse of the variance matrix $\mathbf{V}_{h,3}$ can be expressed in terms of the inverse matrices $\mathbf{V}_{jh,2}^{-1}$. Let $\mathbf{U}_h = \{\mathbf{V}_{jh,2}^{-1}\} \otimes \mathbf{I}_{n_{2,h}}$; then

$$\mathbf{V}_{h,3}^{-1} = \mathbf{U}_h - \mathbf{U}_h \mathbf{Z}_h^{(3)} \boldsymbol{\Sigma}_3 \mathbf{G}_{h,3}^{-1} \left(\mathbf{Z}_h^{(3)}\right)^\top \mathbf{U}_h, \qquad (6.18)$$

where

$$\mathbf{G}_{h,3} = \mathbf{I}_{r_3} + \left(\mathbf{Z}_h^{(3)}\right)^\top \mathbf{U}_h \mathbf{Z}_h^{(3)} \boldsymbol{\Sigma}_3$$

$$= \mathbf{I}_{r_3} + \sum_j \left(\mathbf{Z}_{jh}^{(3)}\right)^\top \mathbf{V}_{jh,2}^{-1} \mathbf{Z}_{jh}^{(3)} \boldsymbol{\Sigma}_3.$$

Equation (6.18) is a generalization of (6.16); it does not rely on a specific form of the matrices $\mathbf{U}_h$. In particular, when the blocks $\mathbf{V}_{jh,2}$ are arbitrary positive scalars the (diagonal) variance matrices $\mathbf{U}_h$ correspond to the models with unequal elementary-level variances (see Section 4.9.1).

The identity in (6.18) is proved by direct multiplication:

$$\left\{\mathbf{U}_h^{-1} + \mathbf{Z}_h^{(3)} \boldsymbol{\Sigma}_3 \left(\mathbf{Z}_h^{(3)}\right)^\top\right\} \left\{\mathbf{U}_h - \mathbf{U}_h \mathbf{Z}_h^{(3)} \boldsymbol{\Sigma}_3 \mathbf{G}_{h,3}^{-1} \left(\mathbf{Z}_h^{(3)}\right)^\top \mathbf{U}_h\right\}$$

$$= \mathbf{I}_{n_{1,h}} + \mathbf{Z}_h^{(3)} \boldsymbol{\Sigma}_3 \left[\mathbf{I}_{r_3} - \left\{\mathbf{I}_{r_3} + \left(\mathbf{Z}_h^{(3)}\right)^\top \mathbf{U}_h \mathbf{Z}_h^{(3)} \boldsymbol{\Sigma}_3\right\} \mathbf{G}_{h,3}^{-1}\right] \left(\mathbf{Z}_h^{(3)}\right)^\top \mathbf{U}_h,$$

which is equal to the identity matrix.

For the determinants we have the identity

$$\det \mathbf{V}_{h,3} = \prod_j \det \mathbf{V}_{jh,2} \det \mathbf{G}_{h,3}, \qquad (6.19)$$

which is proved by a simple extension of the similar identity used for the variance matrix of two-level data, (4.22). Let $\mathbf{T}$ be the partitioned matrix

$$\mathbf{T} = \begin{pmatrix} \mathbf{I}_{r_3} & -\left(\mathbf{Z}_h^{(3)}\right)^\top \\ \mathbf{Z}_h^{(3)} \boldsymbol{\Sigma}^{(3)} & \mathbf{V}_{jh,2} \otimes \mathbf{I}_{n_{2,h}} \end{pmatrix},$$

and let $\mathbf{L}_1 = \begin{pmatrix} \mathbf{I}_{r_3} & \left(\mathbf{Z}_h^{(3)}\right)^\top \mathbf{U}_h \\ \mathbf{0} & \mathbf{I}_{n_{2,h}} \end{pmatrix}$ and $\mathbf{L}_2 = \begin{pmatrix} \mathbf{I}_{r_3} & \mathbf{0} \\ -\mathbf{Z}_h^{(3)} \boldsymbol{\Sigma}_3 & \mathbf{I}_{n_{1,h}} \end{pmatrix}$.

The determinants of $\mathbf{L}_1$ and $\mathbf{L}_2$ are each equal to unity, and so $\det(\mathbf{T}) = \det(\mathbf{L}_1\mathbf{T}) = \det(\mathbf{L}_2\mathbf{T})$. But the determinants of $\mathbf{L}_1\mathbf{T}$ and $\mathbf{L}_2\mathbf{T}$ are the sides of (6.19).

Now the log-likelihood (6.13) can be expressed in terms of the matrices $\mathbf{U}_h$, $\mathbf{G}_{h,3}$, and $\mathbf{G}_{jh,2}$:

$$l = -\frac{1}{2}\{N\log(2\pi\sigma^2) + \sum_h\sum_j\log(\det\mathbf{G}_{jh,2}) + \sum_h\log(\det\mathbf{G}_{h,3})$$
$$+ \sum_h\mathbf{e}_h^\top\mathbf{U}_h\mathbf{e}_h - \sum_h\mathbf{e}_h^\top\mathbf{U}_h\mathbf{Z}_h^{(3)}\boldsymbol{\Sigma}_3\mathbf{G}_{h,3}^{-1}\left(\mathbf{Z}_h^{(3)}\right)^\top\mathbf{U}_h\mathbf{e}_h\} , \tag{6.20}$$

where $\mathbf{e}_h = \{\mathbf{e}_{jh}\} \otimes \mathbf{1}_{n_{2,h}}$ and $\mathbf{e}_{jh} = \mathbf{y}_{jh} - \mathbf{X}_{jh}\boldsymbol{\beta}$. The quadratic forms $\mathbf{e}_h^\top\mathbf{U}_h\mathbf{Z}_h^{(3)}$ can be expressed in terms of $\mathbf{G}_{jh,2}^{-1}$ using (6.16).

The MLE for the regression parameters $\boldsymbol{\beta}$ coincides with the GLS estimator

$$\hat{\boldsymbol{\beta}} = \left(\mathbf{X}^\top\mathbf{V}^{-1}\mathbf{X}\right)^{-1}\mathbf{X}^\top\mathbf{V}^{-1}\mathbf{y}; \tag{6.21}$$

its derivation in Section 2.3 is applicable for an arbitrary variance matrix $\mathbf{V}$, so long as $\mathbf{V}$ is functionally independent of $\boldsymbol{\beta}$. Computation of (6.21) is facilitated by (6.18) and (6.16): for arbitrary $N \times 1$ vectors $\mathbf{u}_1$ and $\mathbf{u}_2$

$$\mathbf{u}_1^\top\mathbf{V}^{-1}\mathbf{u}_2 = \frac{1}{\sigma^2}\left\{\mathbf{u}_1^\top\mathbf{u}_2 - \frac{1}{\sigma^2}\sum_h\sum_j\mathbf{u}_{1,jh}^\top\mathbf{Z}_{jh}^{(2)}\boldsymbol{\Sigma}_2\mathbf{G}_{jh,2}^{-1}\left(\mathbf{Z}_{jh}^{(2)}\right)^\top\mathbf{u}_{2,jh}\right\}$$
$$- \sum_h\mathbf{u}_{1,h}^\top\mathbf{U}_h\mathbf{Z}_h^{(3)}\boldsymbol{\Sigma}_3\mathbf{G}_{h,3}^{-1}\left(\mathbf{Z}_h^{(3)}\right)^\top\mathbf{U}_h\mathbf{u}_{2,h} , \tag{6.22}$$

where $\mathbf{u}_{1,jh}$, $\mathbf{u}_{2,jh}$, $\mathbf{u}_{1,h}$, and $\mathbf{u}_{2,h}$ are the subvectors of $\mathbf{u}_1$ and $\mathbf{u}_2$ corresponding to units jh and h, so that $\{\mathbf{u}_{k,jh}\} \otimes \mathbf{1}_{N_2} = \mathbf{u}_k = \{\mathbf{u}_{k,h}\} \otimes \mathbf{1}_{N_3}$ for $k = 1, 2$. The quadratic forms in $\mathbf{U}_h$ are evaluated using (6.16). Usually, $\mathbf{V}$ is unknown and is replaced in (6.21) by its ML estimate.

For the derivation of the scoring vector and the information matrix with respect to the covariance structure parameters we follow the outline of Section 4.4. We use the parametrization with scaled variance matrices $\boldsymbol{\Omega}_2 = \sigma^{-2}\boldsymbol{\Sigma}_2$ and $\boldsymbol{\Omega}_3 = \sigma^{-2}\boldsymbol{\Sigma}_3$, so that the matrices $\mathbf{G}_{jh,2}$ and $\mathbf{G}_{h,3}$, as functions of $\boldsymbol{\Omega}_2$ and $\boldsymbol{\Omega}_3$, do not depend on σ^2:

$$\mathbf{G}_{jh,2} = \mathbf{I}_{r_2} + \left(\mathbf{Z}_{jh}^{(2)}\right)^\top\mathbf{Z}_{jh}^{(2)}\boldsymbol{\Omega}_2,$$
$$\mathbf{G}_{h,3} = \mathbf{I}_{r_3} + \sum_j\left(\mathbf{Z}_{jh}^{(3)}\right)^\top\mathbf{W}_{jh,2}^{-1}\mathbf{Z}_{jh}^{(3)}\boldsymbol{\Omega}_3, \tag{6.23}$$

where $\mathbf{W}_{jh,2} = \sigma^{-2}\mathbf{V}_{jh,2}$. It is easy to show that the scoring function for σ^2 has the unique root

$$\hat{\sigma}^2 = \frac{\mathbf{e}^\top\mathbf{W}^{-1}\mathbf{e}}{N}, \tag{6.24}$$

where the scaled variance matrix $\mathbf{W} = \sigma^{-2}\mathbf{V}$, as a function of $\boldsymbol{\Omega}_2$ and $\boldsymbol{\Omega}_3$, does not depend on σ^2. The scoring function for a covariance structure parameter θ is

$$\frac{\partial l}{\partial \theta} = -\frac{1}{2}\left\{ \operatorname{tr}\left(\mathbf{W}^{-1}\frac{\partial\mathbf{W}}{\partial\theta}\right) - \frac{1}{\sigma^2}\mathbf{e}^\top\mathbf{W}^{-1}\frac{\partial\mathbf{W}}{\partial\theta}\mathbf{W}^{-1}\mathbf{e}\right\}. \tag{6.25}$$

The partial derivative $\partial\mathbf{W}/\partial\theta$ is block-diagonal with patterned blocks:

$$\frac{\partial\mathbf{W}}{\partial\theta_2} = \left\{\mathbf{Z}_{jh}^{(2)}\frac{\partial\boldsymbol{\Omega}_2}{\partial\theta_2}\left(\mathbf{Z}_{jh}^{(2)}\right)^\top\right\}_{jh} \otimes \mathbf{I}_{N_2} \tag{6.26}$$

and

$$\frac{\partial\mathbf{W}}{\partial\theta_3} = \left\{\mathbf{Z}_{h}^{(3)}\frac{\partial\boldsymbol{\Omega}_3}{\partial\theta_3}\left(\mathbf{Z}_{h}^{(3)}\right)^\top\right\}_{h} \otimes \mathbf{I}_{N_3} \tag{6.27}$$

for parameters θ_2 and θ_3 involved in $\boldsymbol{\Omega}_2$ and $\boldsymbol{\Omega}_3$ respectively. The partial derivatives $\partial\boldsymbol{\Omega}_k/\partial\theta_k$ are incidence matrices, analogous to their counterparts defined in Section 4.4. For the half-variance parametrization applied to both $\boldsymbol{\Omega}_2$ and $\boldsymbol{\Omega}_3$ we have

$$\frac{\partial\boldsymbol{\Omega}_k}{\partial\theta_k} = \boldsymbol{\Delta}_{p_1,k}\boldsymbol{\Delta}_{p_2,k}^\top + \boldsymbol{\Delta}_{p_2,k}\boldsymbol{\Delta}_{p_1,k}^\top,$$

where $\boldsymbol{\Delta}_{p,k} = (0,\ldots,1,\ldots 0)$ is the $r_k \times 1$ indicator vector for the component p, consisting of zeros with the exception of the pth element, which is equal to unity. Then

$$\operatorname{tr}\left(\mathbf{W}^{-1}\frac{\partial\mathbf{W}}{\partial\theta_2}\right) = \sum_h\sum_j\left(\mathbf{S}_{jh}^{(2)}\right)_{p_1,p_2}, \tag{6.28}$$

where $\mathbf{S}_{jh}^{(2)} = \left(\mathbf{Z}_{jh}^{(2)}\right)^\top(\mathbf{W}^{-1})_{[jh]}\mathbf{Z}_{jh}^{(2)}$, $(\mathbf{S})_{p_1,p_2}$ denotes the (p_1,p_2) element of $\mathbf{S}$, and $(\mathbf{W})_{[jh]}$ is the submatrix of $\mathbf{W}$ corresponding to the level 2 unit jh (an $n_{2,jh} \times n_{2,jh}$ matrix). Application of (6.18) yields

$$\mathbf{S}_{jh}^{(2)} = \left(\mathbf{Z}_{jh}^{(2)}\right)^\top\mathbf{W}_{jh,2}^{-1}\mathbf{Z}_{jh}^{(2)} - \left(\mathbf{Z}_{jh}^{(2)}\right)^\top\mathbf{W}_{jh,2}^{-1}\mathbf{Z}_{jh}^{(3)}\boldsymbol{\Omega}_3\mathbf{G}_{h,3}^{-1}\left(\mathbf{Z}_{jh}^{(3)}\right)^\top\mathbf{W}_{jh,2}^{-1}\mathbf{Z}_{jh}^{(2)}$$

$$= \mathbf{G}_{jh,2}^{-1}\left\{\left(\mathbf{Z}_{jh}^{(2)}\right)^\top\mathbf{Z}_{jh}^{(2)} - \left(\mathbf{Z}_{jh}^{(2)}\right)^\top\mathbf{Z}_{jh}^{(3)}\boldsymbol{\Omega}_3\mathbf{G}_{h,3}^{-1}\left(\mathbf{Z}_{jh}^{(3)}\right)^\top\mathbf{Z}_{jh}^{(2)}\left(\mathbf{G}_{jh,2}^{-1}\right)^\top\right\}. \tag{6.29}$$

For the second term in (6.25) we have

$$\mathbf{e}^\top \mathbf{W}^{-1} \frac{\partial \mathbf{W}}{\partial \theta_2} \mathbf{W}^{-1} \mathbf{e} = \sum_h \sum_j \left(\mathbf{s}_{jh}^{(2)} \right)_{p_1} \left(\mathbf{s}_{jh}^{(2)} \right)_{p_2}, \qquad (6.30)$$

where $\mathbf{s}_{jh}^{(2)} = \left(\mathbf{Z}_{jh}^{(2)} \right)^\top (\mathbf{W}^{-1})_{[jh]} \, \mathbf{e}_{jh}$ and $(\mathbf{s})_p$ stands for the pth component of $\mathbf{s}$. The vectors $\mathbf{s}_{jh}^{(2)}$ are evaluated using the identity (6.18):

$$\mathbf{s}_{jh}^{(2)} = \mathbf{G}_{jh,2}^{-1} \left(\mathbf{Z}_{jh}^{(2)} \right)^\top \left\{ \mathbf{e}_{jh} - \mathbf{Z}_{jh}^{(3)} \boldsymbol{\Omega}_3 \mathbf{G}_{h,3}^{-1} \left(\mathbf{Z}_{jh}^{(3)} \right)^\top \mathbf{W}_{jh,2}^{-1} \mathbf{e}_{jh} \right\}. \qquad (6.31)$$

Similar calculations for the parameter θ_3 yield

$$\mathrm{tr} \left(\mathbf{W}^{-1} \frac{\partial \mathbf{W}}{\partial \theta_3} \right) = \sum_h \left(\mathbf{S}_h^{(3)} \right)_{p_1, p_2}, \qquad (6.32)$$

with

$$\mathbf{S}_h^{(3)} = \left(\mathbf{Z}_h^{(3)} \right)^\top \mathbf{W}_h^{-1} \mathbf{Z}_h^{(3)} = \mathbf{G}_{h,3}^{-1} \sum_j \left(\mathbf{Z}_{jh}^{(3)} \right)^\top \mathbf{W}_{jh}^{-1} \mathbf{Z}_{jh}^{(3)}; \qquad (6.33)$$

the quadratic forms $\left(\mathbf{Z}_{jh}^{(3)} \right)^\top \mathbf{W}_{jh}^{-1} \mathbf{Z}_{jh}^{(3)}$ are evaluated using (6.16). Further,

$$\mathbf{e}^\top \mathbf{W}^{-1} \frac{\partial \mathbf{W}}{\partial \theta_3} \mathbf{W}^{-1} \mathbf{e} = \sum_h \left(\mathbf{s}_h^{(3)} \right)_{p_1} \left(\mathbf{s}_h^{(3)} \right)_{p_2}, \qquad (6.34)$$

where

$$\mathbf{s}_h^{(3)} = \left(\mathbf{Z}_h^{(3)} \right)^\top \mathbf{W}_h^{-1} \mathbf{e}_h = \mathbf{G}_{h,3}^{-1} \sum_j \left(\mathbf{Z}_{jh}^{(3)} \right)^\top \mathbf{W}_{jh}^{-1} \mathbf{e}_{jh}. \qquad (6.35)$$

The information function for a pair of covariance structure parameters θ and θ' has the general form

$$-\mathbf{E} \left(\frac{\partial^2 l}{\partial \theta \, \partial \theta'} \right) = \frac{1}{2} \mathrm{tr} \left(\mathbf{W}^{-1} \frac{\partial \mathbf{W}}{\partial \theta} \mathbf{W}^{-1} \frac{\partial \mathbf{W}}{\partial \theta'} \right). \qquad (6.36)$$

The parameters θ and θ' are associated with elements of the variance matrices $\boldsymbol{\Omega}_2$ and $\boldsymbol{\Omega}_3$; we denote the corresponding rows and columns by p_1 and p_2, and p_1' and p_2' respectively. We distinguish three cases:

1. Both θ and θ' are involved in $\boldsymbol{\Omega}_2$. Then

$$\mathrm{tr}\left(\mathbf{W}^{-1}\frac{\partial\mathbf{W}}{\partial\theta}\mathbf{W}^{-1}\frac{\partial\mathbf{W}}{\partial\theta'}\right)$$

$$= \sum_h\sum_j\left\{\left(\mathbf{S}_{jh}^{(2)}\right)_{p_1,p_2}\left(\mathbf{S}_{jh}^{(2)}\right)_{p_1',p_2'} + \left(\mathbf{S}_{jh}^{(2)}\right)_{p_1,p_2'}\left(\mathbf{S}_{jh}^{(2)}\right)_{p_1',p_2}\right\}.$$

$$(6.37)$$

2. Both θ and θ' are involved in $\mathbf{\Omega}_3$. Then

$$\mathrm{tr}\left(\mathbf{W}^{-1}\frac{\partial\mathbf{W}}{\partial\theta}\mathbf{W}^{-1}\frac{\partial\mathbf{W}}{\partial\theta'}\right)$$

$$= \sum_h\left\{\left(\mathbf{S}_h^{(3)}\right)_{p_1,p_2}\left(\mathbf{S}_h^{(3)}\right)_{p_1',p_2'} + \left(\mathbf{S}_h^{(3)}\right)_{p_1,p_2'}\left(\mathbf{S}_h^{(3)}\right)_{p_1',p_2}\right\}.$$

$$(6.38)$$

3. θ is involved in $\mathbf{\Omega}_2$ and θ' in $\mathbf{\Omega}_3$. Then

$$\mathrm{tr}\left(\mathbf{W}^{-1}\frac{\partial\mathbf{W}}{\partial\theta}\mathbf{W}^{-1}\frac{\partial\mathbf{W}}{\partial\theta'}\right)$$

$$= \sum_h\sum_j\left\{\left(\mathbf{S}_{jh}^{(23)}\right)_{p_1,p_2}\left(\mathbf{S}_{jh}^{(23)}\right)_{p_1',p_2'} + \left(\mathbf{S}_{jh}^{(23)}\right)_{p_1,p_2'}\left(\mathbf{S}_{jh}^{(23)}\right)_{p_1',p_2}\right\},$$

$$(6.39)$$

where

$$\mathbf{S}_{jh}^{(23)} = \left(\mathbf{Z}_{jh}^{(3)}\right)^\top\left(\mathbf{W}^{-1}\right)_{[jh]}\mathbf{Z}_{jh}^{(2)}$$

$$= \mathbf{G}_{h,3}^{-1}\left(\mathbf{Z}_{jh}^{(3)}\right)^\top\mathbf{Z}_{jh}^{(2)}\left(\mathbf{G}_{jh,2}^{-1}\right)^\top.$$

$$(6.40)$$

6.2.1 ORGANIZING COMPUTATIONS

From equations (6.20)–(6.39) we see that the Fisher scoring algorithm requires the following data summaries:

- sample totals of cross-products $(\mathbf{X},\ \mathbf{y})^\top(\mathbf{X},\ \mathbf{y})$;
- level 3 totals of cross-products $(\mathbf{X}_h,\ \mathbf{y}_h)^\top\mathbf{Z}_h^{(3)}$, and $\left(\mathbf{Z}_h^{(3)}\right)^\top\mathbf{Z}_h^{(3)}$ if $\mathbf{Z}_h^{(3)}$ is not contained in $\mathbf{X}_h$;
- level 2 totals of cross-products $(\mathbf{X}_{jh},\ \mathbf{y}_{jh})^\top\mathbf{Z}_{jh}^{(2)}$, and $\left(\mathbf{Z}_{jh}^{(2)}\right)^\top\mathbf{Z}_{jh}^{(2)}$ and $\left(\mathbf{Z}_{jh}^{(3)}\right)^\top\mathbf{Z}_{jh}^{(2)}$ if $\mathbf{Z}_{jh}^{(2)}$ and $\mathbf{Z}_{jh}^{(3)}$ are not contained in $\mathbf{X}_{jh}$.

For a large dataset it is advantageous to calculate these summaries and store them, so that the original data do not have to be accessed during the Fisher scoring iterations. The summaries often form an object much smaller than the original data. Fitting a model comprises the following steps:

1. Loop over the level 3 units. For each level 3 unit h, loop over its level 2 subunits jh and calculate the cross-products

$$(\mathbf{X}_{jh},\ \mathbf{y}_{jh})^\top(\mathbf{X}_{jh},\ \mathbf{y}_{jh}) \ = \ \sum_{i=1}^{n_{1,jh}} (\mathbf{x}_{ijh},y_{ijh})^\top(\mathbf{x}_{ijh},y_{ijh}).$$

Accumulate these matrices of cross-products, and store their submatrices

$$(\mathbf{X}_{jh},\ \mathbf{y}_{jh})^\top\mathbf{Z}_{jh}^{(2)}.$$

Accumulate the level 3 totals

$$(\mathbf{X}_h,\ \mathbf{y}_h)^\top(\mathbf{X}_h,\ \mathbf{y}_h) \ = \ \sum_{j=1}^{n_{2,h}} (\mathbf{X}_{jh},\ \mathbf{y}_{jh})^\top(\mathbf{X}_{jh},\ \mathbf{y}_{jh}),$$

and store their submatrices

$$(\mathbf{X}_h,\ \mathbf{y}_h)^\top\mathbf{Z}_h^{(3)}.$$

Store the sample totals

$$(\mathbf{X},\ \mathbf{y})^\top(\mathbf{X},\ \mathbf{y}) \ = \ \sum_{h=1}^{N_3} (\mathbf{X}_h,\ \mathbf{y}_h)^\top(\mathbf{X}_h,\ \mathbf{y}_h).$$

2. Ordinary least squares. The starting solution is calculated:

$$\hat{\beta}_0 \ = \ \left(\mathbf{X}^\top\mathbf{X}\right)^{-1}\mathbf{X}^\top\mathbf{y},$$

$$\hat{\sigma}_0^2 \ = \ N^{-1}\left\{\mathbf{y}^\top\mathbf{y} - \hat{\beta}_0^\top\left(\mathbf{X}^\top\mathbf{X}\right)^{-1}\hat{\beta}_0\right\},$$

$$\hat{\Omega}_{2,0} \ = \ \mathbf{0},$$

$$\hat{\Omega}_{3,0} \ = \ \mathbf{0},$$

or other guesses for Ω_2 and Ω_3 (diagonal matrices, though).

3. Fisher scoring iteration.

4. Terminate iterations or return to 3.

The calculations for a Fisher scoring iteration can be organized in such a way that each level 3 unit is accessed only once. However, its level 2 subunits have to be accessed twice because, for instance, calculation of level 3 terms $\mathbf{G}_{h,3}$ requires level 2 terms $\mathbf{G}_{jh,2}$, but (6.40) requires both $\mathbf{G}_{jh,2}$ and $\mathbf{G}_{h,3}$. The following steps describe an economic algorithm for the level 3 calculations:

1. Loop over level 2 subunits. For each subunit calculate and store $\mathbf{G}_{jh,2}^{-1}$ and

$$(\mathbf{X}_{jh}, \ \mathbf{Z}_{jh}, \ \mathbf{y}_{jh})^{\top} \mathbf{W}_{jh,2}^{-1} \mathbf{Z}_{jh},$$

where $\mathbf{Z}_{jh} = \left(\mathbf{Z}_{jh}^{(2)}, \ \mathbf{Z}_{jh}^{(3)} \right)$. Accumulate $\log(\det \mathbf{W}_{jh,2})$ and

$$\left(\mathbf{X}_{jh}, \ \mathbf{Z}_{jh}^{(3)}, \ \mathbf{y}_{jh} \right)^{\top} \mathbf{Z}_{jh}^{(2)} \mathbf{\Omega}_2 \mathbf{G}_{jh,2}^{-1} \left(\mathbf{Z}_{jh}^{(2)} \right)^{\top} \left(\mathbf{X}_{jh}, \ \mathbf{Z}_{jh}^{(3)}, \ \mathbf{y}_{jh} \right),$$

for (6.20) and (6.21), using (6.19) and (6.22).

2. Calculate $\left(\mathbf{X}_h, \ \mathbf{Z}_h^{(3)}, \ \mathbf{y}_h \right)^{\top} \mathbf{U}_h \mathbf{Z}_h^{(3)}, \ \mathbf{G}_{h,3}^{-1}, \ \mathbf{S}_h^{(3)}$, and $\mathbf{s}_h^{(3)}$, and their contributions to (6.20), (6.21), (6.32), (6.34), and (6.38).

3. Calculate $\mathbf{S}_{jh}^{(2)}, \ \mathbf{s}_{jh}^{(2)}$, and $\mathbf{S}_{jh}^{(23)}$, and their contributions to (6.20), (6.21), and (6.28), (6.30), (6.37), and (6.39).

The discussion of criteria for convergence in Section 2.3.1 applies also to this algorithm. Adjustments to ensure that the estimates $\hat{\mathbf{\Omega}}_2$ and $\hat{\mathbf{\Omega}}_3$ are positive definite throughout the iterations, and that the algorithm finds a maximum on the boundary of the parameter space, have to be carried out separately for $\hat{\mathbf{\Omega}}_2$ and $\hat{\mathbf{\Omega}}_3$.

Substantial simplification of the algorithm takes place when no variables are included in either variation part, that is, for the random-effects model in (6.2). The corresponding algorithm is described in Section 6.7 for a somewhat more general case. Also, when $\mathbf{Z}^{(3)}$ is a submatrix of $\mathbf{Z}^{(2)}$ then $\left(\mathbf{Z}_h^{(3)} \right)^{\top} \mathbf{U}_h \mathbf{Z}_h^{(3)}$ is a submatrix of

$$\left(\mathbf{Z}_h^{(2)} \right)^{\top} \mathbf{U}_h \mathbf{Z}_h^{(2)} = \sum_j \mathbf{G}_{jh,2}^{-1} \left(\mathbf{Z}_{jh}^{(2)} \right)^{\top} \mathbf{Z}_{jh}^{(2)}.$$

This is considerably easier to evaluate than using (6.22).

6.3 Model choice

We have seen in Chapters 4 and 5 that for two-level data it is meaningful to make an inference about complex patterns of variation only when there are many large clusters. This applies to an even greater extent to inference with three-level data. In general, it is difficult to obtain an idea of how much information about certain patterns of variation a dataset (a design) contains.

Obviously, the variance matrices $\Sigma_3 = \text{var}(\delta_h^{(3)})$ and $\Sigma_2 = \text{var}(\delta_{jh}^{(2)})$ cannot be estimated more efficiently than if the realizations of $\{\delta_h^{(3)}\}$ and $\{\delta_{jh}^{(2)}\}$ were observed directly. Since these are two normal random samples, the properties of the MLEs based on these realizations can be established from standard normal theory. The factors influencing sampling variation of these estimators are the respective level 3 and level 2 sample sizes, N_3 and N_2, but also the (unknown) variance matrices Σ_3 and Σ_2. If the variance matrices Σ_3 and Σ_2 are of comparable size then estimation of Σ_3 is less accurate because the sample contains fewer realizations from $\mathcal{N}(0, \Sigma_3)$ than from $\mathcal{N}(0, \Sigma_2)$ (since $N_3 < N_2$). Note that complexity of the pattern of variation (dimensionality of the variance matrices Σ_3 and Σ_2) does not influence the sampling variation of the covariance structure parameters in this hypothetical scenario.

In a realistic situation it is illuminating to consider two sources of uncertainty about a variance (matrix), say Σ_3. In addition to the finiteness of the 'sample', N_3, each random term $\delta_h^{(3)}$ is observed indirectly. The vector $\delta_h^{(3)}$ is realized on $n_{1,h}$ elementary observations – higher $n_{1,h}$ and smaller elementary-level variance σ^2 are associated with more information about $\delta_h^{(3)}$. However, even if the elementary-level terms ε_{ijh} and the regression parameters β were known, $\delta_h^{(3)}$ could not be uniquely separated from the linear combinations $\mathbf{Z}_{jh}^{(2)}\delta_{jh}^{(2)} + \mathbf{Z}_h^{(3)}\delta_h^{(3)}$, $j = 1, \ldots, n_{2,h}$. This problem is more acute when the counts $n_{2,h}$ are small. Even when β is known, each variance component σ^2, Σ_2, and Σ_3 contributes to uncertainty about the variance matrices Σ_2 and Σ_3.

When the level 2 sample size N_2 is large relative to the level 3 sample size N_3, that is, most level 3 units have few level 2 subunits, the random terms $\delta_h^{(3)}$ and $\delta_{jh}^{(h)}$ are almost or completely confounded. In such a nesting design the parameters in the variance matrices Σ_2 and Σ_3 are also confounded, more so when the dimensions of these variance matrices are large. It is then meaningful to posit a complex pattern of variation at most for one of these variance matrices.

In selecting the variables for the variation design at one level (e.g., for Σ_2) it is useful to consider the trivial variation design at the other level ($\Sigma_3 = 0$). Then the discussion in Section 4.4.2 about confounding of covariance structure parameters is relevant. In particular, high correlation of the variables in the variation part promotes ill-conditioning of the covariance structure parameters.

6.4 Restricted maximum likelihood

The *full* log-likelihood (6.13) for three-level data may involve a large number of regression parameters $\boldsymbol{\beta}$. When they are nuisance parameters, inference about the covariance structure parameters can be strengthened by eliminating $\boldsymbol{\beta}$ from the log-likelihood. This is of particular importance when the number of regression parameters, p, is a substantial fraction of the number of elementary observations N.

Restricted maximum likelihood estimation, REML, described in Section 2.6, can be employed in this context. Briefly, instead of the N observations y we consider $N - p$ linearly independent error contrasts of y and maximize their log-likelihood. Harville's (1974) result is applicable in this context; the log-likelihood (6.13) is adjusted by the term

$$R = -\frac{1}{2} \log \left\{ \det \left(\mathbf{X}^\top \mathbf{V}^{-1} \mathbf{X} \right) \right\}, \qquad (6.41)$$

so that the REML estimator is the maximizer of $l_R = l + R$. Since R does not depend on $\boldsymbol{\beta}$ the ML and REML estimators of $\boldsymbol{\beta}$ coincide with the generalized least squares estimator (6.21), although they use different fitted variance matrices $\hat{\mathbf{V}}$. The elementary-level variance σ^2 can be extracted from (6.41):

$$R = -\frac{1}{2} p \log \sigma^2 - \frac{1}{2} \log \left\{ \det \left(\mathbf{X}^\top \mathbf{W}^{-1} \mathbf{X} \right) \right\}, \qquad (6.42)$$

where $\mathbf{W} = \sigma^{-2} \mathbf{V}$. It is easy to see from (6.13) that the REML estimator of the elementary-level variance is equal to

$$\hat{\sigma}^2 = \frac{\mathbf{e}^\top \mathbf{W}^{-1} \mathbf{e}}{N - p}. \qquad (6.43)$$

The scoring vector for the covariance structure parameters has to be adjusted by the partial derivatives of (6.41). We have

$$\frac{\partial R}{\partial \theta} = \frac{1}{2} \mathrm{tr} \left\{ \mathbf{W}^{-1} \mathbf{X} \left(\mathbf{X}^\top \mathbf{W}^{-1} \mathbf{X} \right)^{-1} \mathbf{X}^\top \mathbf{W}^{-1} \frac{\partial \mathbf{W}}{\partial \theta} \right\}, \qquad (6.44)$$

and by substituting (6.26) and (6.27) we obtain

$$\frac{\partial R}{\partial \theta_2} = \frac{1}{2} \sum_h \sum_j \left\{ \left(\mathbf{z}_{jh}^{(2)} \right)^\top \left(\mathbf{W}^{-1} \right)_{[jh]} \mathbf{X}_{jh} \right.$$

$$\left. \times \left(\mathbf{X}^\top \mathbf{W}^{-1} \mathbf{X} \right)^{-1} \mathbf{X}_{jh}^\top \left(\mathbf{W}^{-1} \right)_{[jh]} \mathbf{z}_{jh}^{(2)} \right\}_{p_1, p_2} \qquad (6.45)$$

for a level 2 parameter $\theta_2 = (\mathbf{\Omega}_2)_{p_1,p_2}$, and

$$
\frac{\partial R}{\partial \theta_3} = \frac{1}{2} \sum_h \left\{ \left(\mathbf{Z}_h^{(3)} \right)^\top \mathbf{W}_h^{-1} \mathbf{X}_h \left(\mathbf{X}^\top \mathbf{W}^{-1} \mathbf{X} \right)^{-1} \mathbf{X}_h^\top \mathbf{W}_h^{-1} \mathbf{Z}_h^{(3)} \right\}_{p_1,p_2}
$$

$$(6.46)$$

for a level 3 parameter.

The adjustments for the information matrix are obtained by further differentiation of (6.45) and (6.46). In practice the exact values of the standard errors of the estimated parameters are of little importance, especially in the presence of a complex dependence structure of the associated estimators, and the derivatives of (6.45) and (6.46) can be ignored. Note that for REML the quadratic forms $\left(\mathbf{Z}_h^{(3)} \right)^\top \mathbf{W}_h^{-1} \mathbf{X}_h$ are required; for the ML algorithm only terms $\left(\mathbf{Z}_h^{(3)} \right)^\top \mathbf{W}_h^{-1} \mathbf{Z}_h^{(3)}$ are required. For evaluation of (6.45) and (6.46) it is advantageous to use the value of $\mathbf{X}^\top \mathbf{W}^{-1} \mathbf{X}$ from the previous iteration (its OLS version, $\mathbf{X}^\top \mathbf{X}$, at the first iteration).

Several comparisons of ML and REML estimates in Chapter 5 imply that whether $\boldsymbol{\beta}$ is assumed known or estimated matters only in small datasets with relatively complex regression designs. For preliminary assessment of accuracy of estimation of the covariance structure parameters, $\boldsymbol{\beta}$ can in most cases be assumed known.

6.5   Model diagnostics

Each model fit should be accompanied by a check of the model assumptions. When searching for a suitable regression part $\mathbf{X}\boldsymbol{\beta}$ it often suffices to use ordinary regression methods. This entails searching for patterns of the OLS residuals with respect to the variables included in the regression part (e.g., to ascertain whether polynomial regression should be fitted), or with respect to other 'candidate' regression variables.

When checking the assumptions of normality, thin-tailedness of the distribution of the residuals is of principal interest. The residuals, that is, the conditional expectations of the random terms, are:

$$
\hat{\boldsymbol{\delta}}_h^{(3)} = \mathbf{E}\left(\boldsymbol{\delta}_h^{(3)} \,|\, \mathbf{y}; \boldsymbol{\beta}, \sigma^2, \mathbf{\Omega}_2, \mathbf{\Omega}_3 \right) = \mathbf{\Omega}_3 \mathbf{s}_h^{(3)} ,
$$

$$
\hat{\boldsymbol{\delta}}_{jh}^{(2)} = \mathbf{E}\left(\boldsymbol{\delta}_{jh}^{(2)} \,|\, \mathbf{y}; \boldsymbol{\beta}, \sigma^2, \mathbf{\Omega}_2, \mathbf{\Omega}_3 \right) = \mathbf{\Omega}_2 \left(\mathbf{Z}_{jh}^{(2)} \right)^\top \left(\mathbf{W}^{-1} \right)_{[jh]} \mathbf{e}_{jh} ,
$$

$$
\hat{\boldsymbol{\varepsilon}}_h = \mathbf{E}\left(\boldsymbol{\varepsilon}_h^{(3)} \,|\, \mathbf{y}; \boldsymbol{\beta}, \sigma^2, \mathbf{\Omega}_2, \mathbf{\Omega}_3 \right) = \mathbf{W}_h^{-1} \mathbf{e}_h ,
$$

$$(6.47)$$

with respective conditional variance matrices

$$\text{var}\left(\hat{\boldsymbol{\delta}}_{jh}^{(2)} \mid \mathbf{y}; \, \boldsymbol{\beta}, \sigma^2, \boldsymbol{\Omega}_2, \boldsymbol{\Omega}_3\right) = \sigma^2 \boldsymbol{\Omega}_2(\mathbf{I}_{r_2} - \sigma^2 \boldsymbol{\Omega}_2 \mathbf{S}_{jh}^{(2)} \boldsymbol{\Omega}_2),$$

$$\text{var}\left(\hat{\boldsymbol{\delta}}_{h}^{(3)} \mid \mathbf{y}; \, \boldsymbol{\beta}, \sigma^2, \boldsymbol{\Omega}_2, \boldsymbol{\Omega}_3\right) = \sigma^2 \boldsymbol{\Omega}_3(\mathbf{I}_{r_3} - \sigma^2 \mathbf{S}_{h}^{(3)} \boldsymbol{\Omega}_3)$$

$$= \sigma^2 \boldsymbol{\Omega}_3 \left(\mathbf{Z}_{h}^{(3)}\right)^{\top} \mathbf{U}_h \mathbf{Z}_{h}^{(3)},$$

$$\text{var}\left(\hat{\boldsymbol{\varepsilon}}_h \mid \mathbf{y}; \, \boldsymbol{\beta}, \sigma^2, \boldsymbol{\Omega}_2, \boldsymbol{\Omega}_3\right) = \sigma^2 \left(\mathbf{I}_{n_{1,h}} - \sigma^2 \mathbf{W}_{h}^{-1}\right).$$

It can be shown that

$$\mathbf{e}_{jh} = \mathbf{Z}_{jh}^{(3)} \hat{\boldsymbol{\delta}}^{(3)} + \mathbf{Z}_{jh}^{(2)} \hat{\boldsymbol{\delta}}^{(2)} + \hat{\boldsymbol{\varepsilon}}_{jh}, \tag{6.48}$$

so that $\hat{\boldsymbol{\varepsilon}}_{jh}$ can be calculated more easily than by (6.47).

Exploration of the residuals often serves two purposes: apart from checking for normality, outlying (extreme) residuals may be of considerable importance since they identify clusters with regressions that are substantially different from the rest. Small clusters contain little information about their own regression coefficients, and therefore their residuals are shrunk with a large weight to zero; they are therefore less likely to have extreme residuals. Rarely can the residuals be clustered meaningfully so that they would provide further feedback for model improvement.

A straightforward check for normality is accomplished by constructing histograms, normal quantile plots, or the like, for each component of the residuals. When a variation part is a simple regression $(\delta_1 + z\delta_2)$, it is more informative to plot these regression lines over the ranges of realized values of z.

6.6   Likelihood ratio testing

The likelihood ratio test is applicable for the comparison of models based on the same dataset. In this section we describe a few typical settings in three-level analysis. Likelihood ratio test compares two models, say, H_A and H_B, where H_A is a submodel of H_B:

$$H_A : (\boldsymbol{\beta}, \, \sigma^2, \, \boldsymbol{\Omega}_2, \, \boldsymbol{\Omega}_3) \in \boldsymbol{\Theta}_A,$$
$$H_B : (\boldsymbol{\beta}, \, \sigma^2, \, \boldsymbol{\Omega}_2, \, \boldsymbol{\Omega}_3) \in \boldsymbol{\Theta}_B,$$

with $\boldsymbol{\Theta}_A \subset \boldsymbol{\Theta}_B$. Typically, the model H_A is constructed from the model H_B by imposing linear constraints on some of the parameters. Thus the model H_B has p_B, m_{1B}, m_{2B}, and m_{3B} unknown and functionally independent parameters for the regression part, level 1 variance, and level 2 and level 3 variance matrices, respectively, and the corresponding numbers for the model H_A are $p_A \leq p_B$, $m_{1A} \leq m_{1B}$, $m_{2A} \leq m_{2B}$, and $m_{3A} \leq m_{3B}$. Let d be the number of linear constraints, that is,

$$d = p_B - p_A + m_{1B} - m_{1A} + m_{2B} - m_{2A} + m_{3A} - m_{3B}.$$

The likelihood ratio test statistic is the difference of the deviances of the models H_A and H_B, $2(l_B - l_A)$. Under the null-hypothesis H_A, if the parameter vector lies in the interior of the parameter space Θ, and under certain regularity conditions deemed to be satisfied here, this statistic has, asymptotically, the χ^2 distribution with d degrees of freedom.

When $p_A = p_B$ either ML or REML can be used. When $p_A < p_B$, that is, when testing for some of the regression parameters, REML cannot be used because essentially different datasets (error contrasts) correspond to the models H_A and H_B.

When the parameter vector lies on the boundary of the parameter space the likelihood ratio test statistic does not have a χ^2 distribution. The likelihood ratio test statistic can still be used as an informal criterion. The larger the difference of the deviances, the more evidence for the model H_B. Often, the likelihood ratio test is used in this informal way, especially when the search for an appropriate model involves a larger number of tests.

The problems related to non-openness of the parameter space can be dealt with effectively by considering the extended parameter space, that is, by allowing negative within-cluster covariances, and more generally, negative eigenvalues of the matrices Σ_2 and Σ_3.

6.6.1   INDEPENDENT DATA?

If the three-level data did not have the anticipated within-cluster covariance structure, substantial simplification of their description would be possible, and ordinary regression for their description would be applicable. Under this hypothesis the membership of an elementary unit in level 2 and level 3 clusters would be immaterial because the same conditional (within-cluster) regressions, equal to the common unconditional regression, would apply.

Assuming the same regression part of the model under both hypothesis and alternative, the likelihood ratio test for independence ($\Omega_2 = \mathbf{0}$ and $\Omega_3 = \mathbf{0}$) has, asymptotically, the χ^2 distribution with $d = m_{2B} + m_{3A}$ degrees of freedom. Note that the alternative in this test is $\Omega_k \neq \mathbf{0}$, $k = 2, 3$, not that these (variance) matrices are non-negative definite. In particular, when one of these matrices is a scalar, then the corresponding component of the alternative is that the 'variance' is non-zero ('two-sided' alternative). Negative cluster-level 'variance' can be interpreted as negative within-cluster covariance.

The test for independence of clusters has ML and REML versions. It would appear that the REML version is more powerful, although for large samples the distributions of the corresponding test statistics almost coincide.

6.6.2   AN IRRELEVANT VARIABLE

It may be of interest to test the hypothesis that an explanatory variable makes no contribution to the description of the outcome variable. The variable may be included not only in the regression part but also in one or both variation parts of the model. Exclusion of the variable corresponds to constraints on one regression parameter ($K-1$ parameters for a categorical variable), one ($K-1$) constraint on variance at each level where the variable is considered under the alternative, and constraints on all the covariances associated with this variable. Since a constraint on a regression parameter is involved, REML cannot be used.

6.7   More than three levels of nesting

Of course, levels of nesting beyond the third can arise. Level 3 units may be clustered within level 4 units, and so forth. Usually, sampling of the subunits at each level is thin, otherwise the data size is bound to be enormous. For such data only the simple covariance structures are meaningful, assuming a univariate random term associated with each cluster,

$$y_{i_k,i_{k-1},\ldots,i_2,i_1} =$$
$$\mathbf{x}_{i_k,i_{k-1},\ldots,i_2,i_1}\boldsymbol{\beta} + \delta_{i_k} + \delta_{i_k,i_{k-1}} + \ldots + \delta_{i_k,i_{k-1},\ldots,i_2} + \varepsilon_{i_k,i_{k-1},\ldots,i_1}.$$
$$(6.49)$$

The expectation of an outcome is equal to $\mathbf{x}\boldsymbol{\beta}$ and the variance matrix of the vector of outcomes $\mathbf{y}$ is

$$\mathbf{V} = \sigma^2\mathbf{I}_N + \sum_{m=2}^{k} \sigma_m^2\mathbf{J}^{(m)}, \qquad (6.50)$$

where $\mathbf{J}^{(m)}$ is the incidence matrix for level m; $\mathbf{J}^{(m)}$ is a block-diagonal matrix, with blocks of ones corresponding to the clusters of level m:

$$\mathbf{J}^{(m)} = \sum_{j=1}^{N_m} \mathbf{z}_{j,m}\mathbf{z}_{j,m}^{\top}, \qquad (6.51)$$

where $\mathbf{z}_{j,m}$ is the $N \times 1$ incidence vector for a level m cluster j; its components are equal to unity if they correspond to observations in cluster j, and to zero otherwise. The summation is over all N_m units at level m. Note that

$$\frac{\partial \mathbf{V}}{\partial \sigma_m^2} = \mathbf{J}^{(m)}. \qquad (6.52)$$

We define the partial matrix totals

$$\mathbf{V}_M = \sigma^2 \mathbf{I}_N + \sum_{m=2}^{M} \sigma_m^2 \mathbf{J}^{(m)}. \tag{6.53}$$

The inverse of the matrix $\mathbf{V}$ can be derived recursively using the identity

$$\mathbf{V}_M^{-1} = \mathbf{V}_{M-1}^{-1} - \sigma_M^2 \mathbf{V}_{M-1}^{-1} \sum_{j=1}^{N_M} \frac{\mathbf{z}_{j,M} \mathbf{z}_{j,M}^\top}{1 + \sigma_M^2 C_{j,M}} \mathbf{V}_{M-1}^{-1}, \tag{6.54}$$

where $C_{j,M} = \mathbf{z}_{j,M}^\top \mathbf{V}_{M-1}^{-1} \mathbf{z}_{j,M}$. For completeness we define $C_{i,1} \equiv 1$. Since $\mathbf{z}_{j,M}$ is the sum of the incidence vectors $\mathbf{z}_{i,M-1}$ for all its subunits i at level $M-1$, and the blocks of $\mathbf{V}_{M-1}$ correspond to units of level $M-1$,

$$C_{j,M} = \sum_{i \subset j} \frac{C_{i,M-1}}{1 + \sigma_M^2 C_{i,M-1}}, \tag{6.55}$$

where the summation is over all units i at level $M-1$ that belong to unit j at level M. Calculation of the various quadratic forms in $\mathbf{V}^{-1}$ can be carried out recursively using the inversion formula (6.54) for $M = 2, \ldots, k$. For a general $N \times 1$ vector $\mathbf{u}$ we have

$$
\begin{aligned}
\mathbf{u}^\top \mathbf{V}_M^{-1} \mathbf{z}_{j,M} &= \\
\mathbf{u}^\top \mathbf{V}_{M-1}^{-1} \mathbf{z}_{j,M} &- \sigma_M^2 \sum_{i \subset j} \frac{\mathbf{u}_i^\top \mathbf{V}_{M-1}^{-1} \mathbf{z}_{i,M-1} \mathbf{z}_{i,M-1}^\top \mathbf{V}_{M-1}^{-1} \mathbf{z}_{i,M-1}}{1 + \sigma_M^2 C_{j,M}} \\
&= \sum_i \frac{\mathbf{u}_i^\top \mathbf{V}_{M-1}^{-1} \mathbf{z}_{i,M-1}}{1 + \sigma_M^2 C_{j,M}}, \tag{6.56}
\end{aligned}
$$

where $\mathbf{u}_i$ is the vector of length N containing the subvector of $\mathbf{u}$ corresponding to unit i at level $M-1$ in the same position as in $\mathbf{u}$ and zeros elsewhere, and all the summations are over subunits i of unit j at level M. For example, for four-level data the information matrix for (σ_2^2, σ_3^2) is equal to

$$\frac{1}{2} \sum_{h=1}^{N_4} \sum_{j \subset h} \sum_{i \subset j} \left(\mathbf{z}_{j,3}^\top \mathbf{V}_{4,h}^{-1} \mathbf{z}_{i,2} \right)^2,$$

where the inner summation is over all level 2 subunits i of level 3 unit j, and the middle summation is over all level 3 subunits j of level 4 unit h. To evaluate the quadratic form $\mathbf{z}_{j,3}^\top \mathbf{V}_{4,h}^{-1} \mathbf{z}_{i,2}$ we use (6.54):

$$\mathbf{z}_{j,3}^{\top}\mathbf{V}_{4,h}^{-1}\mathbf{z}_{i,2} = \left(1 - \mathbf{z}_{j,3}^{\top}\mathbf{V}_{3,j}^{-1}\mathbf{z}_{j,3}\frac{\sigma_4^2}{1 + C_{4,h}\sigma_4^2}\right)\mathbf{z}_{j,3}^{\top}\mathbf{V}_{3,j}^{-1}\mathbf{z}_{i,2}$$

$$= \frac{C_{2,i}}{1 + C_{3,j}\sigma_3^2}\left(1 - \frac{\sigma_4^2}{1 + C_{4,h}\sigma_4^2}\frac{1}{1 + C_{3,j}\sigma_3^2}\sum_{l \subset j} C_{2,l}\right),$$

if h is the unit of level 4 which contains both unit j at level 3 and unit i at level 2; otherwise the expression is equal to zero. The summation is over all subunits l of unit j.

The determinant of the variance matrix $\mathbf{V}$ can be derived by recursive application of the identity in (6.19):

$$\det \mathbf{V} = \sigma^{2N} \prod_{M=2}^{k} \prod_{j=1}^{N_M} (1 + \sigma_M^2 C_{j,M}). \tag{6.57}$$

The identities (6.56) and (6.57) can be used for evaluation of all the items required for the Fisher scoring algorithm: the log-likelihood, (6.13), the GLS estimator given by (6.21), the equation for the elementary-level variance, (6.24), and the scoring vector and the information matrix for the other variance parameters, (6.25) and (6.36).

When multilevel data do contain abundant information about patterns of variation, it is usually only at one level. An efficient algorithm for fitting models relevant to this situation can be constructed by combining the recursive scheme described in this section with the method of Section 6.2.

6.8 Hearing loss data

This section describes an analysis of a three-level dataset originating from a survey on hearing loss in Great Britain in 1981. We focus on identifying the sources of variation and assessing their relative size and importance, as well as on the association of hearing loss with age and sex.

It is generally believed that the hearing of human subjects deteriorates with age. Hearing loss for a given frequency is defined as the loudest acoustic signal which is not registered sensually by the subject. Hearing loss is measured separately for each ear and is defined as a function of frequency. Loudness of an impulse is measured as the logarithm of the energy per area generated by the impulse; the units of measurement are decibels (dB). Hearing loss of up to 20 dB at any frequency is of little consequence; hearing loss of 70 dB and higher is generally classified as severe. Hearing loss is specific to frequencies; a subject may be deaf or hard of hearing at a certain range of frequencies, but not at others. The hearing loss is likely to be

Table 6.1.   Data for the first subject. Hearing loss data. The subject is a 73-year-old man

	Frequencies in Hz							
Ear side	100	200	300	500	1000	2000	5000	10 000
Left	70	64	62	87	125	113	125	125
Right	60	64	63	97	125	125	125	125

highly correlated, both across the various frequencies for the same ear and between the two ears of the same subject.

The Institute of Hearing Research in Nottingham, England, conducts regular surveys in which a sample of adult subjects in Great Britain is given a standard audiological test as a result of which, among other measurements, hearing loss at eight different frequencies that span almost the entire range of audible sound, and for both ears, is measured. Here we analyse a dataset containing these sets of 16 measurements for 99 subjects. This is a subset of the complete dataset collected by the National Hearing Survey in 1981. Table 6.1 contains the data for the first subject in the sample. He is a 73-year-old man, deaf at frequencies of 1000 Hz and above, and with severe hearing loss at lower frequencies. In fact, 125 Hz is the highest signal that could be recorded in the survey.

The dataset comprises 55 male and 44 female adults, aged 18–87 years. The stem-and-leaf plot of the ages for each sex is given in Table 6.2. The

Table 6.2.   Stem-and-leaf plot of the ages of the subjects in hearing loss data. Ages are in years; the stems are in multiples of 10

Women		Men
99	: 1 :	8
2234	: 2 :	046
0145559	: 3 :	11556777
08	: 4 :	23344557899
0244566799	: 5 :	2333444489
0333445555	: 6 :	00123344466679
0122346	: 7 :	011227
17	: 8 :	00

two sexes have comparable distributions of ages. For simplicity of illustration we disregard all sampling issues (the survey employed a stratified probability sampling scheme).

There are numerous hypothesized causes of hearing loss. Genetic predisposition, environment, and accidents (e.g., sudden exposure to loud noise, such as presence at an explosion) are three well-identified categories of causes. Some of these causes affect hearing in both ears uniformly, others affect hearing only at certain frequencies, or only in one ear. Obviously, we cannot distinguish between these causes of hearing loss, but the survey data enable the sources of variation to within- and between-subject components to be assigned, so that aggregates of some of the causes of hearing loss can be assessed.

Figure 6.1 contains plots of the mean hearing loss for the eight frequencies for each ear on the linear scale (the left-hand panel) and on the logarithmic scale (the right-hand panel). The means for the left and right ears are denoted by the respective symbols 'L' and 'R'. It is not meaningful to have the frequencies on a linear scale of Hz, and therefore the frequencies at which measurements are carried out are coded as 1, 2, ..., 8. The mean hearing loss appears to be an increasing function of the frequency, with the exception of the lowest and highest frequencies. The means for the left and right ears differ insubstantially.

In essence, each subject's data are a pair of 'time' series, with the frequency in the role of a metametre. The hearing loss is a function of

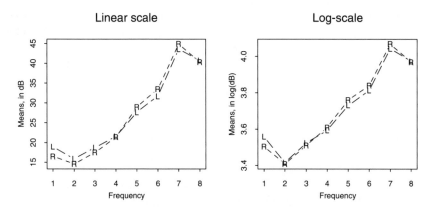

Fig. 6.1. Mean hearing loss for the eight frequencies. Hearing loss data. The right-hand panel contains the plot of the arithmetic means of hearing loss by frequency and ear, and the left-hand panel contains the same plot for the means of the logarithms of the hearing loss. The means for the left and right ears are denoted by 'L' and 'R' respectively. The frequencies on the horizontal axis are coded as 1–100 Hz, 2–200 Hz, 3–300 Hz, 4–500 Hz, 5–1000 Hz, 6–2000 Hz, 7–5000 Hz, and 8–10 000 Hz.

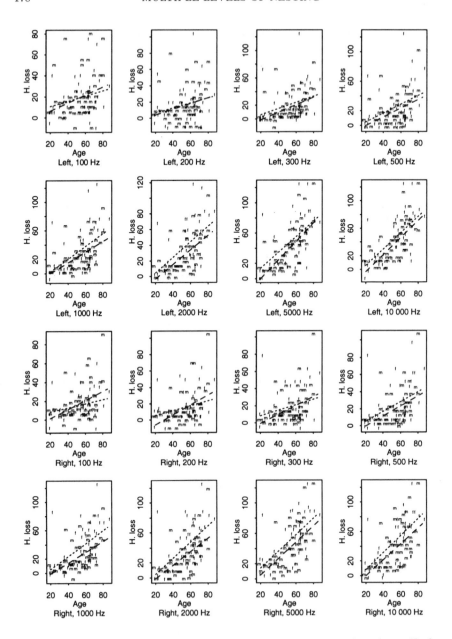

Fig. 6.2. Hearing loss for each ear and frequency. Hearing loss data. Each plot is for a frequency and either the left or right ear. Observations are marked 'f' for women and 'm' for men. The fitted regressions are drawn by a dashed line (women) and a dotted line (men).

frequency (as a continuous argument), specific for each ear. The values
of this function are observed at the designated frequencies and are subject to measurement error. Thus, the data have the nesting structure of
observations within ears, and ears within subjects.

For initial data exploration it is informative to plot the data for each
frequency, as is done in Figure 6.2. Observations for men and women
are marked by different symbols, and the fitted regressions are also drawn
with different line types. No systematic pattern of differences among these
regressions can be observed, except that men have uniformly higher fitted
hearing loss than women for high frequencies. Note that the regressions on
age are not linear.

As a crude model for the data we consider the subjects' hearing loss
functions to be parallel:

$$y_{ijk} = \mathbf{x}_{ijk}\boldsymbol{\beta} + \delta_k^{(3)} + \delta_{jk}^{(2)} + \varepsilon_{ijk}, \tag{6.58}$$

where i indexes the frequencies, $i = 1, 2, \ldots, 8$, j the ears of a subject,
$j = 1, 2$, and k the subjects, $k = 1, \ldots, 99$, and $\left\{\delta_k^{(3)}\right\}$, $\left\{\delta_{jk}^{(2)}\right\}$, and $\{\varepsilon_{ijk}\}$
are mutually independent random samples from respective distributions
$\mathcal{N}(0, \sigma_3^2)$, $\mathcal{N}(0, \sigma_2^2)$, and $\mathcal{N}(0, \sigma_1^2)$. The variance σ_3^2 describes between-
subject variation not accounted for by the available subject-level variables
(age and sex); σ_2^2 describes the variation between a subject's two ears,
and σ_1^2 is the amalgam of measurement error, temporal variation, and in-
adequacy of the model description for differences between the frequencies.

For the regression parameters $\boldsymbol{\beta}$ we first consider unrelated regressions
on age for each frequency and sex. This amounts to defining interactions
of age, sex, and age-by-sex with the frequency (as a categorical variable),
involving 32 regression parameters, or, equivalently, a 4×8 matrix of
regression parameters, with rows corresponding to intercept, age, sex, and
age×sex interaction, and columns corresponding to the frequencies.

The regression parameters and their standard errors are displayed in the
top panel of Table 6.3 for models with and without the age×sex×frequency
interactions. For illustration, the fitted hearing loss for a 50-year-old
woman is $-14.41 + (0.433 - 0.199) \times 50 + 9.62 = 6.90$. Owing to the
balanced design the regression estimates coincide with the least squares
estimates. The standard errors are common to all the parameter estimates
in a row of the table, and are given in the rightmost column.

The most prominent feature of the estimates is the association of hearing loss with age, especially for high frequencies. For example, at a frequency of 2000 Hz, 40 years' age difference is associated with hearing loss of
$40 \times 0.91 = 36.4$ dB, the difference between satisfactory hearing (e.g., 10 dB)
and noticeably impaired hearing (46 dB). Although the within-frequency

Table 6.3. Regression parameter estimates. Hearing loss data. The top panel contains results for the outcome variable on the original scale, the bottom panel for the log-transformed variable, $\log(y+20)$. The deviances for the models with age × sex × frequency interactions and without them are 12 532.95 and 12 537.69 respectively. The deviance for the model fit for $\log(y+20)$ is -1116.31

Variable	Frequencies								St. errors
	1	2	3	4	5	6	7	8	
Intercept	-14.41	-21.31	-22.94	-17.47	-16.87	-24.95	-34.30	-33.56	(19.90)
age	0.433	0.481	0.538	0.591	0.757	0.914	1.157	1.166	(0.164)
sex	9.62	10.57	11.14	53.94	1.22	4.39	13.67	8.57	(12.45)
age × sex	-0.199	-0.177	-0.149	-0.001	0.147	0.123	-0.096	0.009	(0.225)
	Without age × sex interactions								
Intercept	1.60	-6.98	-10.90	-17.42	-28.81	-34.93	-26.54	-34.31	(8.267)
age	0.327	0.387	0.459	0.590	0.836	0.980	1.106	1.171	(0.112)
sex	-0.82	1.23	3.29	5.36	9.00	10.90	8.61	9.06	(3.76)
	Analysis for $\log(y+20)$								
Intercept	3.126	2.797	2.799	2.748	2.593	2.514	2.890	2.628	(0.164)
age	0.0077	0.0102	0.0121	0.0136	0.0181	0.0201	0.0186	0.0219	(0.0022)
sex	-0.002	0.054	0.056	0.095	0.137	0.174	0.125	0.127	(0.075)

sex differences appear to be substantial (around 10 dB), they are not significant (the corresponding standard errors are in the range 10–14).

The estimates of the elementary-level variance σ_1^2 and of the variance ratios $\omega_2 = \sigma_2^2/\sigma_1^2$ and $\omega_3 = \sigma_3^2/\sigma_1^2$ are:

$$\hat{\sigma}_1^2 = 114.91,$$
$$\hat{\omega}_2 = 0.47 \qquad (0.11),$$
$$\hat{\omega}_3 = 2.27 \qquad (0.37).$$

The fitted correlation of measurements of hearing loss for a subject's ears is $(0.47+2.27)/(1+0.47+2.27) = 0.73$. As a measure of similarity of hearing loss at different frequencies (for an ear) it is more appropriate to consider the ratio $0.47/1.47 = 0.32$. If we assume that the measurement error variance is $100a$ per cent of the elementary-level variance σ_1^2, then the correlation of 'true' hearing loss for frequencies within an ear is $0.47/(1-a+0.47)$. This correlation is an increasing function of a, and so 0.32 is an underestimate of the within-ear correlation of hearing loss. For example, if $a = 0.3$ then the correlation is 0.40. Similarity of hearing loss in a subject's two ears is characterized by the ratio $\omega_3/(\omega_2+\omega_3)$, estimated as 0.83. This implies that hearing is to a large extent a subject-level characteristic, as opposed to an ear-level one.

The standard deviation of the person mean hearing loss is $\sqrt{114.9 \times 2.27}$ $\doteq 16.1$. Combined with the regression fit we can estimate the percentage of population of certain age and sex that have a serious hearing loss. For example, if we regard the hearing loss of 60 dB as severe, then the fitted percentage of 30-year-old men with serious hearing loss at 1000 Hz is $100 - 100\Phi\{(60-5.84)/16.1\} = 3.8 \times 10^{-2}$ per cent; the percentage of 60-year-old men is 2.54 per cent.

The regression part of the model contains 32 parameters, but the age-by-sex interaction is negligible. Exclusion of these eight parameters is associated with reduction of deviance 4.74 (from 12 537.69 to 12 532.95). The changes in the variance components and their standard errors are also negligible. Further, a trend of the slopes on age across the frequencies is obvious. Differences in hearing loss between young and old subjects are an increasing (distinctly non-linear) function of the frequency. The age × frequency and sex × frequency interactions are important; if either set of eight parameters is excluded from the regression part the deviance increases by more than 100. For example, when both sets of interactions are excluded the model fit has the deviance 12 916.

Since age is the dominant explanatory variable, it is useful to explore whether the relationship of hearing loss to age is linear. The quadratic term for the polynomial regression is important, as can be guessed from Figure 6.2. As an alternative to polynomial regression we consider the transformation $\log(20 + y)$ for the hearing loss.

The estimates of the elementary-level variance and the variance ratios are

$$\hat{\sigma}_1^2 = 0.068,$$
$$\hat{\omega}_2 = 0.196 \qquad (0.059),$$
$$\hat{\omega}_3 = 1.411 \qquad (0.225).$$

Details of the model fit are given in bottom panel of Table 6.3. The variance ratios are changed substantially: both $\hat{\omega}_2$ and $\hat{\omega}_3$ are much smaller than for the analysis with the original response variable. Other alternatives for the transformation of variables also produce substantially different results. Although this is somewhat disconcerting, in all model fits the level 3 variance is much larger than the level 2 variance (3–8 times). Also, a different pattern of (frequency-specific) sex differences emerges. Now men have a higher hearing loss for all frequencies except 100 Hz, and the adjusted difference increases in the range 200–5000 Hz. Thus, transformation(s) of the outcomes is an important component of model choice.

The log-transformation of hearing loss, $\log(20 + y)$, is useful in that interactions of sex and age remain non-significant, and the level 2 and level 3 residuals are no longer distinctly non-normal. Figure 6.3 contains the quantile plots of the residuals. Level 2 residuals are not aligned along the straight line, but the departure is less distinct than for other models. We do not attempt any parsimonious modelling of the age × frequency interaction parameters because of dependence on the scale for frequency.

Next we explore alternatives for the covariance structure of the observations. If we associate the frequency (as a factor with eight levels) with variation at both ear- and subject-level, then we essentially have a two-level multivariate dataset: a pair of 8-variate outcomes for a subject's ears.

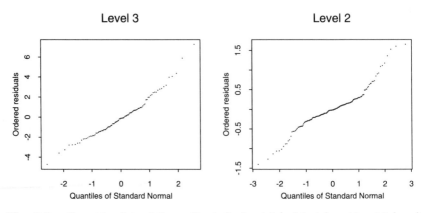

Fig. 6.3. Quantile plots of the residuals for level 3 (subjects) and level 2 (ears). Hearing loss data. Both sets of residuals are multiplied by 100.

The between-person variation is described by an 8×8 variance matrix, and the within-person variation is described as $\sigma_1^2 I + \Sigma_2$, also 8×8. Since data within an ear contain no replication, only the sum of the observation- and ear-level variance matrices can be identified. Thus the covariance structure is given by two 8×8 matrices. Their estimates (for the log-transformed hearing loss) are

$$\hat{\Sigma}_2^* = \begin{pmatrix}
0.140 & 0.142 & 0.102 & 0.088 & 0.080 & 0.074 & 0.062 & 0.064 \\
0.142 & 0.213 & 0.137 & 0.119 & 0.094 & 0.092 & 0.074 & 0.066 \\
0.102 & 0.137 & 0.127 & 0.117 & 0.093 & 0.090 & 0.070 & 0.065 \\
0.088 & 0.119 & 0.117 & 0.130 & 0.111 & 0.100 & 0.085 & 0.080 \\
0.080 & 0.094 & 0.093 & 0.111 & 0.125 & 0.116 & 0.095 & 0.087 \\
0.074 & 0.092 & 0.090 & 0.100 & 0.116 & 0.121 & 0.094 & 0.089 \\
0.062 & 0.074 & 0.070 & 0.085 & 0.095 & 0.094 & 0.091 & 0.089 \\
0.064 & 0.066 & 0.065 & 0.080 & 0.087 & 0.089 & 0.089 & 0.101
\end{pmatrix}$$

and

$$\hat{\Sigma}_1^* = \begin{pmatrix}
0.039 & 0.024 & 0.023 & 0.022 & 0.020 & 0.013 & 0.013 & 0.011 \\
0.024 & 0.066 & 0.022 & 0.014 & 0.016 & 0.008 & 0.012 & 0.010 \\
0.023 & 0.022 & 0.062 & 0.032 & 0.026 & 0.018 & 0.016 & 0.017 \\
0.022 & 0.014 & 0.032 & 0.050 & 0.030 & 0.022 & 0.011 & 0.014 \\
0.020 & 0.016 & 0.026 & 0.030 & 0.042 & 0.024 & 0.016 & 0.023 \\
0.013 & 0.008 & 0.018 & 0.022 & 0.024 & 0.036 & 0.013 & 0.016 \\
0.013 & 0.012 & 0.016 & 0.011 & 0.016 & 0.013 & 0.034 & 0.022 \\
0.011 & 0.010 & 0.017 & 0.014 & 0.023 & 0.016 & 0.022 & 0.050
\end{pmatrix}.$$

Such a description for variation of hearing loss is not very informative. Some features of Σ_1^* and Σ_2^* can be inferred from the eigenvalue decompositions of $\hat{\Sigma}_1^*$ and $\hat{\Sigma}_2^*$. Their respective sets of eigenvalues are:

$$\begin{array}{llllllll}
(\hat{\Sigma}_1^*) & 0.179 & 0.058 & 0.046 & 0.030 & 0.021 & 0.018 & 0.015 & 0.012, \\
(\hat{\Sigma}_2^*) & 0.787 & 0.148 & 0.047 & 0.026 & 0.024 & 0.010 & 0.004 & 0.002.
\end{array}$$

Both matrices have a dominant first eigenvalue. It comes as no surprise that the elements of the corresponding eigenvectors have the same sign and similar magnitude:

$$\begin{array}{llllllll}
(\hat{\Sigma}_1^*) & 0.345 & 0.436 & 0.368 & 0.376 & 0.359 & 0.346 & 0.293 & 0.282, \\
(\hat{\Sigma}_2^*) & 0.327 & 0.338 & 0.458 & 0.408 & 0.397 & 0.292 & 0.257 & 0.307.
\end{array}$$

The natural interpretation of this finding is that the most important component underlying the adjusted hearing loss is uniform across the frequencies both for between-subject and between-ear variation. It is somewhat intriguing that even the eigenvectors corresponding to the second largest eigenvectors of $\hat{\boldsymbol{\Sigma}}_1^*$ and $\hat{\boldsymbol{\Sigma}}_2^*$ are quite similar:

$$
\begin{pmatrix} \hat{\boldsymbol{\Sigma}}_1^* \\ \hat{\boldsymbol{\Sigma}}_2^* \end{pmatrix}
\begin{matrix} -0.376 & -0.591 & -0.193 & 0.065 & 0.315 & 0.334 & 0.342 & 0.375, \\ -0.210 & -0.832 & -0.018 & 0.168 & 0.196 & 0.266 & 0.123 & 0.334. \end{matrix}
$$

They can be loosely interpreted as contrasts of the hearing loss at the lowest and highest frequencies.

The eigenvalue decompositions suggest that the variance matrices at the ear and subject levels have some common features. It would therefore be useful to explore models in which covariance structure parameters are 'shared' across levels. We reanalyse the hearing loss data in Chapter 7 to explore patterns of variation common to the two levels of nesting.

The differences between left and right ears are negligible for each frequency and for all adjustments considered (age, sex, and age × sex), and so the corresponding estimates are not given.

6.9 Model based estimation in surveys

The National Assessment of Educational Progress (NAEP) is a large scale survey of US primary and secondary education. It employs a stratified three-stage clustered probability sampling design for students in various age/grade groups, and a complex partially balanced incomplete block design (spiralling) for the administered items. Spiralling enables collection of information about a large number of items without administering each item to every student in the sample.

The survey items represent content areas (such as reading, history, and mathematics), and for each content area an underlying proficiency scale is defined. The items of a content area are grouped into blocks. For example, in the 1983–4 survey of 13-year-olds there were 13 'reading' blocks and most students were administered a set of three blocks. Twenty–four different sets of blocks of reading items were 'rotated' between the students. NAEP is administered at regular intervals (e.g., every three years for a specific age/grade), and details of the design change from one administration to next. See Johnson and Zwick (1990) for further background.

The scores on each proficiency scale are estimated from the responses to the cognitive items for all the students in the sample who were administered at least one block of items. For a detailed background for these methods, see Lord (1980).

Results of each survey are published in the form of 'summary tables'. They contain the sample (weighted) means of the proficiency scores, and the estimated standard errors for these means, for each combination of attitude item and response to it, cross-classified by the demographic background

Table 6.4. Weighted means of reading proficiency scores for a reading item. NAEP survey data. The entries in columns A, B, C, D, and 'dnk' (do not know) are the estimates of the mean proficiencies and of their standard errors for the students who gave these responses to item 14 in block N for the subpopulation indicated in the leftmost column. For example, the mean proficiency of the Hispanic students who gave response A is 227.6 with standard error 3.9. The correct response, 'C', is marked with an asterisk *. The weighted percentages within the subpopulations for the responses are: 17–25 (response A), 6–10 (B), 45–62 (C), 9–16 (D), 4–8 (dnk)

| Population | N | Response | | | | |
		A	B	C*	D	dnk
Total	2341	242.1 (1.5)	229.6 (3.3)	271.1 (1.1)	243.4 (2.3)	238.7 (3.1)
Male	1154	240.1 (2.5)	229.5 (4.0)	267.1 (1.9)	238.6 (2.9)	236.0 (3.7)
Female	1187	244.4 (2.4)	229.8 (5.6)	274.5 (1.3)	247.0 (3.5)	242.3 (3.8)
White	1670	248.4 (2.1)	232.6 (3.9)	274.7 (1.2)	250.3 (2.7)	240.3 (4.2)
Black	320	232.0 (4.7)	220.8 (6.9)	254.4 (2.9)	224.8 (6.3)	231.3 (7.7)
Hispanic	263	227.6 (3.9)	223.5 (7.3)	253.8 (3.9)	228.1 (5.1)	240.2 (6.7)
Other	88	229.1 (7.5)	249.4 (12.4)	280.6 (6.3)	249.8 (9.9)	223.5 (21.2)

variables. Table 6.4 contains an example from the 1983–4 survey of 13-year-olds. The full sample comprised approximately 31 000 students, each of whom was administered at least one of the 13 reading blocks of items. The reading block N was administered to 3078 students. Of these, 2139 (about 70 per cent) chose the response option A to item no. 1. One entry in the summary tables contains an estimate of the mean proficiency of these pupils, and an estimate of the associated standard error. Most items have five response options, and so the estimate of a typical entry in the summary tables is based on a small proportion of the total sample.

The sampling scheme involved $H = 32$ strata, and within each stratum a pair of primary sampling units (PSU) was drawn. Within each selected PSU a sample of schools, and within each selected school a sample of students was drawn. The sampling procedures at the levels of clustering (PSU, school, student) were conditionally independent, given selection of the units at the higher level. The conditional probabilities of selection were unequal, so as to oversample certain minority groups. We ignore the issue of missingness (which is dealt with by an adjustment of weights for non-response), and the fact that the proficiency scores are estimated subject to error.

We describe first the methods of estimation of the means and associated standard errors that are used currently, and then propose a more efficient alternative based on variance component models.

We denote by Y_{ijkh} the proficiency score of student i in school j in PSU k in stratum h, and by N_{jkh} the number of students in the school (j, i, k). The population mean is defined as

$$\bar{Y} = \frac{\sum_{ijkh} Y_{ijkh}}{\sum_{jkh} N_{jkh}} ; \qquad (6.59)$$

the summations are over all the units (students i, schools j) in the sampling frame. The mean for a subpopulation is defined similarly, with the summation over students (index i) in that subpopulation, and the counts N_{jkh} replaced by the corresponding subpopulation counts.

In NAEP the traditional ratio estimator for the (sub-) population means is used:

$$\bar{y} = \frac{\sum_{ijkh} w_{ijkh} y_{ijkh}}{\sum_{ijkh} w_{ijkh}} , \qquad (6.60)$$

where w_{ijkh} are the sampling weights, and the summations are over all the students in the sample, and, if applicable, in the given subpopulation.

The jackknife is a general method for the reduction of bias of an estimator, see Wolter (1985) for a detailed background. In NAEP for each stratum $h = 1, \ldots, H = 32$, the h-pseudosample is defined by replacing the data for the first PSU of stratum h in the original sample by the data from the other PSU in the stratum. The h-pseudosample estimator is defined as the estimator (6.60) for this pseudosample, and is denoted by $\bar{y}_h$. The jackknife estimator of $\bar{Y}$ is

$$\bar{y}^* = \frac{\sum_h \bar{y}_h}{H} . \qquad (6.61)$$

The jackknife estimator of the sampling variance of $\bar{y}^*$ is defined as the corrected sum of squares of the pseudosample means:

$$\sum_h (\bar{y}_h - \bar{y})^2, \qquad (6.62)$$

In practice, $\bar{y}$ and $\bar{y}^*$ are almost indistinguishable; in NAEP $\bar{y}$ is used throughout.

6.9.1   A MODEL BASED ALTERNATIVE

If the sampling stages (student, school, and PSU) are considered as sources of variation, the student's proficiency scores can be represented by the variance component model

$$y_{ijkh} = \mu_h + \delta_{kh}^{(3)} + \delta_{jkh}^{(2)} + \varepsilon_{ijkh}, \qquad (6.63)$$

where μ_h, $h = 1, \ldots, H$, are the stratum means (unknown parameters), and $\left\{\delta_{kh}^{(3)}\right\}$, $\left\{\delta_{jkh}^{(2)}\right\}$, and $\{\varepsilon_{ijkh}\}$ are three mutually independent random samples from respective distributions $\mathcal{N}(0, \sigma_3^2)$, $\mathcal{N}(0, \sigma_2^2)$, and $\mathcal{N}(0, \sigma_1^2)$. We denote the vector of stratum means by $\boldsymbol{\mu} = \{\mu_h\} \otimes \mathbf{1}_H$, and the corresponding MLE by $\hat{\boldsymbol{\mu}}$.

The estimator of the population mean is the weighted mean of the stratum mean estimators,

$$\hat{\mu} = \frac{\mathbf{w}^\top \hat{\boldsymbol{\mu}}}{\mathbf{w}^\top \mathbf{1}_H}, \qquad (6.64)$$

where $\mathbf{w}$ is the (column) vector of the within-stratum totals of weights. The sampling variance of $\hat{\mu}$ is estimated as $\mathbf{w}^\top \mathbf{H}^{-1} \mathbf{w}$, where $\mathbf{H}$ is the estimated information matrix for $\hat{\boldsymbol{\mu}}$.

The analysis has to accomodate unequal sampling weights associated with the students in the sample. The ML and REML algorithms for estimation with (6.63) are based on within-school totals of proficiencies. We replace these by their sample-weighted versions, with the sampling weights normalized (multiplied by a constant) so that their mean would be equal to one. Potthoff et al. (1992) propose a different normalization of the weights.

The estimates of the means and standard deviations for a few item-response combinations are given in Table 6.5. The results for the jackknife and variance component methods are not substantially different; the largest difference in the means is 0.4 and the largest difference in standard errors is a mere 0.04.

The jackknife estimator of the sampling variance of $\bar{y}$, (6.62), is based on the 32 pseudosample means $\bar{y}_h$. Although its distribution is not known, it is conjectured that for smaller datasets it is a scalar multiple of the χ^2 distribution with fewer than 32 degrees of freedom. Since minority students tend to be distributed very unevenly across the strata, whole PSUs or even strata may not be represented in a dataset for a minority. Therefore, efficient estimation of the standard errors for such datasets is of added importance.

To demonstrate the advantage of the ML method based on the model in (6.63) we briefly describe a simulation study and report a small subset of its results. Consider the nesting designs of PSUs, schools, and students for a small dataset. For these students we simulate a set of proficiency

Table 6.5.   Jackknife and variance component analyses. NAEP survey example. The column $\hat{\sigma}_1$ contains the estimates of the within-school standard deviations, $\hat{\omega}_1$ and $\hat{\omega}_2$ are the estimated variance ratios

Item and response	Students	Jackknife $\bar{y}$ (St. error)	Variance components $\hat{\mu}$ (St. error)	$\hat{\sigma}_1$	$\hat{\omega}_2$	$\hat{\omega}_3$
6 B	309	250.8 (1.69)	251.0 (1.69)	9.83	0.143	0.000
9 C	560	257.2 (1.77)	257.2 (1.73)	10.28	0.220	0.008
5 E	1038	264.8 (1.26)	264.4 (1.27)	10.45	0.157	0.000
8 A	1312	252.4 (1.25)	252.1 (1.24)	10.39	0.141	0.008
4 A	2240	254.8 (1.26)	255.1 (1.30)	10.94	0.146	0.033
Sample	3076	253.4 (1.05)	253.5 (1.05)	11.55	0.122	0.023

scores according to (6.63), with unit weights, constant means, $\mu_h \equiv \mu$, and realistic values of the variance components σ_m^2, $m = 1, 2, 3$. The sampling distributions of the jackknife and MLE estimators of the mean and of their (asymptotic) sampling variances can then be compared in a straightforward manner. Table 6.6 contains the results for such a dataset. For each estimator (jackknife or ML, of the mean or of the standard error) we consider

Table 6.6.   Simulation study using a small dataset. NAEP survey data. The dataset consists of 368 students from 98 schools. The schools are from 52 PSUs in 30 strata. The variance components are $\sigma_1^2 = 1$, $\sigma_2^2 = 0.10$, and $\sigma_3^2 = 0.025$. The estimators are $\bar{y}$, its jackknife mean squared error, JMSE, its square root, JSTE, $\hat{\mu}$, its asymptotic variance, MLV, and the asymptotic standard error, MLSE

Jackknife			Maximum likelihood		
Estimator	Mean	Observed st. dev.	Estimator	Mean	Observed st. dev.
$\bar{y}$	0.0027	0.0748	$\hat{\mu}$	−0.0078	0.0711
JMSE	0.0060	0.0022	MLV	0.0053	0.0010
JSTE	0.0760	0.0139	MLSE	0.0730	0.0065

the mean of the simulated values, the observed mean squared error, and for $\bar{y}$ and $\hat{\mu}$ the mean of the estimates of the sampling variance.

The difference of the observed sampling variances of $\bar{y}$ and $\hat{\mu}$ is negligible. The estimate of the sampling variance for $\hat{\mu}$ is biased but its mean squared error is several times smaller than its counterpart for the jackknife estimator. Since the difference between the estimators $\bar{y}$ and $\hat{\mu}$ can be ignored, the estimator of the sampling variance of $\hat{\mu}$ is more efficient than the jackknife estimator. It is highly unlikely that the difference between the estimators $\bar{y}$ and $\hat{\mu}$ is the sole reason for the differences in efficiency. In any case, the estimator $\hat{\mu}$ is also slightly more efficient.

The artificial datasets were generated according to the model in (6.63), which may be a less than perfect model for the real data. Therefore, the simulation study is somewhat unfair to the jackknife method. It is unlikely, though, that this would explain the apparent gain in efficiency.

6.10   Bibliographical notes

Ahrens (1965) constructed an algorithm based on the moment method for variance component analysis of three-level data. Sahai (1976) compared a number of estimators of the variance components in a three-level balanced design. The methods of Hartley and Rao (1967), Patterson and Thompson (1971), and the EM algorithm, can in principle be extended to three or more levels of nesting, or even to data involving cross-classified random effects, although practical implementations in most cases are very complex. The problems of slow convergence are severely compounded in the EM algorithm. The iteratively reweighted least squares method of Goldstein (1986a), as well as the Fisher scoring algorithm, afford relatively simple and computationally efficient extensions for multilevel analysis. Organization of the computations is important; using the new generation of statistical languages (Splus, GAUSS or matlab) gives the qualified analyst a distinct advantage over the modules in general purpose statistical software (e.g., in BMDP or SAS), or specialized software for multilevel analysis (HLM, Bryk et al. 1988, ML3, Rasbash et al. 1991, or VARCL, Longford 1988b).

The residuals, i.e., estimates of the random effects, are essentially estimates of certain conditional means. In the perspective of *BLUP* (best linear unbiased predictor), particularly popular in animal research, the residuals are unbiased estimators of the realizations of the random effects. See Robinson (1991) for a review of this approach.

7
Factor analysis and structural equations

7.1 Introduction

The problem of within-cluster correlation of observations is by no means restricted to linear regression. Random sampling design, or independence of the observations, is a central assumption in many other routinely applied statistical methods, and independence of the observations is often the prime target of model criticism and diagnostic procedures. This chapter describes the extensions of factor analysis and of structural equation models for two-level data.

Factor analysis can be regarded as a regression analysis with unknown regressors, and structural equation models are essentially regression models with indirectly observed regressors. Extensions of these models and of the associated computational methods for two- (multi-) level design closely resemble multilevel regression models. In the developments, we follow the outlines of Chapters 3 and 5, which dealt with regression models.

Sections 7.2 and 7.3 give a minimal background for factor analysis with independent observations. Methods of estimation for the two-level factor analysis, discussed in Section 7.4, share several features with the methods in the previous chapters. Methods of estimation with structural equation models are somewhat more complex. In general, the problems of model choice and identification carry over from the unilevel case but are further compounded by modelling of within-cluster correlation. The development in Sections 7.8 and 7.9 is motivated by measurement error models, but its generalization presents few conceptual problems.

The chapter is concluded with examples.

7.2 Factor analysis

In classical factor analysis we have a normally distributed random sample of N $p \times 1$ outcome vectors $\mathbf{y}_i$, and assume that, apart from a vector of independent normal variates, each vector $\mathbf{y}_i$ is generated as a set of linear combinations of a small number r $(r < p)$ of unobservable (latent) variables $\boldsymbol{\delta}_i$ called *factors*:

$$\mathbf{y}_i = \boldsymbol{\mu} + \boldsymbol{\Lambda}\boldsymbol{\delta}_i + \boldsymbol{\varepsilon}_i, \qquad (7.1)$$

where $\boldsymbol{\mu}$ is a vector of expectations, $\boldsymbol{\Lambda}$ is a $p \times r$ matrix of parameters called *factor loadings*, $\{\boldsymbol{\delta}_i\}$ is a random sample from $\mathcal{N}_r(\mathbf{0}, \boldsymbol{\Psi})$, and $\{\boldsymbol{\varepsilon}_i\}$ is a random sample from $\mathcal{N}_p(\mathbf{0}, \boldsymbol{\Theta})$, where $\boldsymbol{\Theta}$ is a diagonal matrix. These two random samples are mutually independent. The variances in the matrix $\boldsymbol{\Theta}$ are referred to as the *uniquenesses*. The matrices $\boldsymbol{\Lambda}$, $\boldsymbol{\Psi}$, and $\boldsymbol{\Theta}$ require some additional specification to ensure their identifiability. This is discussed below.

Since the distribution of a normal random vector is uniquely determined by its mean vector and variance matrix, the model in (7.1) is equivalent to

$$\begin{aligned} \mathbf{E}(\mathbf{y}_i) &= \boldsymbol{\mu}, \\ \boldsymbol{\Sigma} = \mathrm{var}(\mathbf{y}_i) &= \boldsymbol{\Lambda}\boldsymbol{\Psi}\boldsymbol{\Lambda}^\top + \boldsymbol{\Theta}. \end{aligned} \qquad (7.2)$$

Instead of the constant mean $\boldsymbol{\mu}$, multivariate regression can be considered: $\mathbf{E}(\mathbf{y}_i) = \mathbf{B}\mathbf{x}_i^\top$, where $\mathbf{B}$ is a $p \times K$ matrix of parameters and $\mathbf{x}_i$ a $1 \times K$ vector of regressors. The linear predictor function $\mathbf{B}\mathbf{x}_i^\top$ can be interpreted as an adjustment for explanatory variables.

The variance matrix $\boldsymbol{\Sigma}$ in (7.2) decomposes into a diagonal matrix and a matrix of (small) rank r. Obviously, such a decomposition is unique only in some trivial cases. Even if the matrix $\boldsymbol{\Theta}$ were known, the difference matrix $\boldsymbol{\Sigma} - \boldsymbol{\Theta}$ would have a continuum of decompositions of the form $\boldsymbol{\Lambda}\boldsymbol{\Psi}\boldsymbol{\Lambda}^\top$. To see this, let $\mathbf{A}$ be an arbitrary $r \times r$ non-singular matrix; then $\boldsymbol{\Lambda}\boldsymbol{\Psi}\boldsymbol{\Lambda}^\top = \boldsymbol{\Lambda}^*\boldsymbol{\Psi}^*\boldsymbol{\Lambda}^{*\top}$ for $\boldsymbol{\Lambda}^* = \boldsymbol{\Lambda}\mathbf{A}^{-1}$ and $\boldsymbol{\Psi}^* = \mathbf{A}\boldsymbol{\Psi}\mathbf{A}^\top$. Thus, replacement of $\boldsymbol{\Lambda}$ by $\boldsymbol{\Lambda}^*$ corresponds to replacing $\boldsymbol{\delta}_i$ with $\mathbf{A}\boldsymbol{\delta}_i$ in equation (7.1).

This ambiguity is usually resolved by adopting one of the following assumptions, referred to as *modes* of analysis:

- *Exploratory mode*: the latent vectors $\boldsymbol{\delta}_i$ have the multivariate standard normal distribution, $\boldsymbol{\delta}_i \sim \mathcal{N}_r(\mathbf{0}, \mathbf{I})$, and the matrix $\boldsymbol{\Lambda}^\top\boldsymbol{\Theta}^{-1}\boldsymbol{\Lambda}$ is diagonal;
- *Confirmatory mode*: the factors (components of $\boldsymbol{\delta}_i$) have unit variances and arbitrary correlation structure, and the matrix $\boldsymbol{\Lambda}$ has a specified pattern (e.g., of zeros).

See Jöreskog (1979), Bartholomew (1987), or Mardia *et al.* (1979) for further background. In principle, the two modes can be combined, e.g., by applying one to a subvector of $\boldsymbol{\delta}$ and the other to its complement.

There are $\frac{1}{2}p(p+1)$ parameters in $\boldsymbol{\Sigma}$, pr in $\boldsymbol{\Lambda}$, and p in $\boldsymbol{\Theta}$. In exploratory mode the condition that $\boldsymbol{\Lambda}^\top\boldsymbol{\Theta}^{-1}\boldsymbol{\Lambda}$ be diagonal imposes $\frac{1}{2}r(r-1)$ constraints. Therefore, an exploratory factor analysis model is well-defined when

$$\frac{p(p+1)}{2} > p(r+1) - \frac{r(r-1)}{2},$$

or, equivalently, when

$$(p-r)^2 > p+r. \tag{7.3}$$

In confirmatory mode the number of functionally independent constraints on the entries of $\boldsymbol{\Lambda}$ has to be at least $\frac{1}{2}\{(p-r)^2 - (p+r)\}$.

In exploratory mode we seek a set of independent latent variables, or, factors, $\boldsymbol{\delta}$. To facilitate a suitable interpretation for the estimated factor loadings $\hat{\boldsymbol{\Lambda}}$ we consider linear transformations $\hat{\boldsymbol{\Lambda}}\mathbf{A}$ of the original solution which yield an interpretable pattern of its columns. The search is usually restricted to orthonormal matrices $\mathbf{A}$ (rotations of the factors). In confirmatory mode we start with an assumed pattern of the factor loadings and explore the correlation structure of the corresponding latent variables.

Although it is not customary, other constraints on the parameters in the factor analysis model can be imposed, such as equal uniquenesses, or equal elements in $\boldsymbol{\Lambda}$.

7.3 Maximum likelihood estimation

The log-likelihood for the model in (7.1) is

$$l = -\frac{Np}{2}\log(2\pi) - \frac{N}{2}\log(\det \boldsymbol{\Sigma}) - \frac{1}{2}\sum_i \mathbf{e}_i^\top \boldsymbol{\Sigma}^{-1}\mathbf{e}_i, \tag{7.4}$$

where $\mathbf{e}_i = \mathbf{y}_i - \boldsymbol{\mu}$, or $\mathbf{e}_i = \mathbf{y}_i - \mathbf{B}\mathbf{x}_i^\top$ if a regression adjustment is applied. By rearranging the elements of $\mathbf{B}$ and $\mathbf{x}_i$ the linear predictor $\mathbf{B}\mathbf{x}_i^\top$ can be rewritten in the standard form $\mathbf{X}_i\boldsymbol{\beta}$. For example, $\mathbf{x}_i \equiv 1$ corresponds to $\mathbf{X}_i \equiv \mathbf{I}_p$ and $\boldsymbol{\beta} = \boldsymbol{\mu}$. We denote the $Np \times 1$ vector $\mathbf{y} = \{\mathbf{y}_i\} \otimes \mathbf{1}_N$ and its variance matrix $\mathbf{V} = \boldsymbol{\Sigma} \otimes \mathbf{I}_N$ and write the linear predictor as $\mathbf{E}(\mathbf{y}) = \mathbf{X}\boldsymbol{\beta}$, where $\mathbf{X} = \{\mathbf{X}_i\} \otimes \mathbf{I}_N$.

Now the log-likelihood has the standard form

$$l = -\frac{1}{2}\{Np\log(2\pi) + \log(\det \mathbf{V}) + \mathbf{e}^\top \mathbf{V}^{-1}\mathbf{e}\}. \tag{7.5}$$

Differentiation of the log-likelihood (7.5) yields the familiar identities

$$\frac{\partial l}{\partial \boldsymbol{\beta}} = \mathbf{X}^\top \mathbf{V}^{-1}\mathbf{e}, \qquad \frac{\partial^2 l}{\partial \boldsymbol{\beta} \partial \boldsymbol{\beta}^\top} = \mathbf{X}^\top \mathbf{V}^{-1}\mathbf{X},$$

which imply the GLS estimator

$$\hat{\boldsymbol{\beta}} = \left(\mathbf{X}^\top \mathbf{V}^{-1}\mathbf{X}\right)^{-1}\mathbf{X}^\top \mathbf{V}^{-1}\mathbf{y}. \tag{7.6}$$

When no explanatory variables are used, the scoring vector

$$\mathbf{X}^\top \mathbf{V}^{-1} \mathbf{e} = \boldsymbol{\Sigma}^{-1} \sum_i (\mathbf{y}_i - \boldsymbol{\mu})$$

has the unique root $\hat{\boldsymbol{\mu}} = \bar{\mathbf{y}} = n^{-1} \sum_i \mathbf{y}_i$, independently of the variance matrix $\boldsymbol{\Sigma}$.

Differentiation of (7.5) with respect to a parameter involved in $\boldsymbol{\Sigma}$ (a covariance structure parameter) yields

$$\frac{\partial l}{\partial \gamma} = -\frac{N}{2} \text{tr} \left(\boldsymbol{\Sigma}^{-1} \frac{\partial \boldsymbol{\Sigma}}{\partial \gamma} \right) + \frac{1}{2} \sum_i \mathbf{e}_i^\top \boldsymbol{\Sigma}^{-1} \frac{\partial \boldsymbol{\Sigma}}{\partial \gamma} \boldsymbol{\Sigma}^{-1} \mathbf{e}_i, \qquad (7.7)$$

and the matrix derivative $\partial \boldsymbol{\Sigma}/\partial \gamma$ is equal to

$$\frac{\partial \boldsymbol{\Sigma}}{\partial \Lambda_{kh}} = \boldsymbol{\Lambda} \boldsymbol{\Psi} \boldsymbol{\Delta}_{h,r} \boldsymbol{\Delta}_{k,p}^\top + \boldsymbol{\Delta}_{k,p} \boldsymbol{\Delta}_{h,r}^\top \boldsymbol{\Psi} \boldsymbol{\Lambda}^\top,$$

$$\frac{\partial \boldsymbol{\Sigma}}{\partial \Psi_{kh}} = \boldsymbol{\Lambda} \frac{\partial \boldsymbol{\Psi}}{\partial \Psi_{kh}} \boldsymbol{\Lambda}^\top, \qquad (7.8)$$

$$\frac{\partial \boldsymbol{\Sigma}}{\partial \Theta_k} = \boldsymbol{\Delta}_{k,p} \boldsymbol{\Delta}_{k,p}^\top,$$

for respective elements of the matrices $\boldsymbol{\Lambda}$, $\boldsymbol{\Psi}$, and $\boldsymbol{\Theta}$. In (7.8) $\boldsymbol{\Delta}_{k,p}$ is the $p \times 1$ incidence vector, $\boldsymbol{\Delta}_{k,p} = (0, \ldots, 1, \ldots, 0)^\top$; the kth element of the vector is equal to unity and all the other entries are equal to zero. The partial derivative $\partial \boldsymbol{\Psi}/\partial \Psi_{kh}$ is also an incidence matrix,

$$\frac{\partial \boldsymbol{\Psi}}{\partial \Psi_{kh}} = \boldsymbol{\Delta}_{k,p} \boldsymbol{\Delta}_{h,p}^\top + \boldsymbol{\Delta}_{h,p} \boldsymbol{\Delta}_{k,p}^\top \qquad (7.9)$$

for a covariance ($h \neq k$). A variance Ψ_{jj} can always be absorbed in $\boldsymbol{\Lambda}$, and so it is usually set to unity. If it is not, it is advantageous to consider the half-variance parametrization, as in Section 4.4, so that (7.9) holds for all the parameters involved in $\boldsymbol{\Psi}$.

The expected information function for a pair of covariance structure parameters is equal to

$$-\mathbf{E} \left(\frac{\partial^2 l}{\partial \gamma_1 \partial \gamma_2} \right) = \frac{N}{2} \text{tr} \left(\boldsymbol{\Sigma}^{-1} \frac{\partial \boldsymbol{\Sigma}}{\partial \gamma_1} \boldsymbol{\Sigma}^{-1} \frac{\partial \boldsymbol{\Sigma}}{\partial \gamma_2} \right). \qquad (7.10)$$

For example,

$$-\mathbf{E}\left(\frac{\partial^2 l}{\partial \Lambda_{kk'}\, \partial \Psi_{hh'}}\right) \; = \; N\left(\boldsymbol{\Lambda}^\top \boldsymbol{\Sigma}^{-1} \boldsymbol{\Lambda} \boldsymbol{\Psi}\right)_{k'h} \left(\boldsymbol{\Sigma}^{-1}\boldsymbol{\Lambda}\right)_{kh'}$$

$$+ \; N\left(\boldsymbol{\Lambda}^\top \boldsymbol{\Sigma}^{-1} \boldsymbol{\Lambda} \boldsymbol{\Psi}\right)_{k'h'} \left(\boldsymbol{\Sigma}^{-1}\boldsymbol{\Lambda}\right)_{kh}. \tag{7.11}$$

A variety of constraints on the estimated parameters can be implemented by the method of Lagrange multipliers.

7.3.1   EXPLORATORY MODE

For estimation in the exploratory mode there is a substantially simpler algorithm than the one given by (7.7) and (7.10). Let $\mathbf{S} = \sum_i \mathbf{e}_i \mathbf{e}_i^\top$ be the matrix of cross-products of the residuals. The scoring function for an element of $\boldsymbol{\Lambda}$ can be expressed as

$$\frac{\partial l}{\partial \lambda} \; = \; -\frac{N}{2}\,\mathrm{tr}\left(\boldsymbol{\Sigma}^{-1}\frac{\partial \boldsymbol{\Sigma}}{\partial \lambda} - \boldsymbol{\Sigma}^{-1}\mathbf{S}\boldsymbol{\Sigma}^{-1}\frac{\partial \boldsymbol{\Sigma}}{\partial \lambda}\right). \tag{7.12}$$

For a given matrix $\boldsymbol{\Theta}$, finding a root of (7.12) for all the elements of $\boldsymbol{\Lambda}$ corresponds to solving the matrix equation

$$\boldsymbol{\Sigma}^{-1}(\boldsymbol{\Sigma} - \mathbf{S})\boldsymbol{\Sigma}^{-1}\boldsymbol{\Lambda} \; = \; \mathbf{0}, \tag{7.13}$$

which is equivalent to the equation $\mathbf{S}\boldsymbol{\Sigma}^{-1}\boldsymbol{\Lambda} = \boldsymbol{\Lambda}$. The inverse of $\boldsymbol{\Sigma}$ is

$$\boldsymbol{\Sigma}^{-1} \; = \; \boldsymbol{\Theta}^{-1} - \boldsymbol{\Theta}^{-1}\boldsymbol{\Lambda}\mathbf{G}^{-1}\boldsymbol{\Lambda}^\top\boldsymbol{\Theta}^{-1}, \tag{7.14}$$

where $\mathbf{G} = \mathbf{I}_r + \boldsymbol{\Lambda}^\top \boldsymbol{\Theta}^{-1}\boldsymbol{\Lambda}$. Since

$$\boldsymbol{\Sigma}^{-1}\boldsymbol{\Lambda} \; = \; \boldsymbol{\Theta}^{-1}\boldsymbol{\Lambda}\left\{\mathbf{I} - \mathbf{G}^{-1}(\mathbf{G} - \mathbf{I})\right\} \; = \; \boldsymbol{\Theta}^{-1}\boldsymbol{\Lambda}\mathbf{G}^{-1}, \tag{7.15}$$

equation (7.13) can be expressed as

$$\mathbf{S}\boldsymbol{\Theta}^{-1}\boldsymbol{\Lambda} \; = \; \boldsymbol{\Lambda}\mathbf{G}. \tag{7.16}$$

Since $\mathbf{G}$ is diagonal, this can be rewritten in the form of an eigenvector–eigenvalue system of equations:

$$\mathbf{S}^*\boldsymbol{\Lambda}^* \; = \; \boldsymbol{\Lambda}^*\mathbf{G}, \tag{7.17}$$

where $\mathbf{S}^* = \boldsymbol{\Theta}^{-\frac{1}{2}}\mathbf{S}\boldsymbol{\Theta}^{-\frac{1}{2}}$ and $\boldsymbol{\Lambda}^* = \boldsymbol{\Theta}^{-\frac{1}{2}}\boldsymbol{\Lambda}$, so that the diagonal elements of $\mathbf{G}$ are the eigenvalues, and the columns of $\boldsymbol{\Lambda}^*$ the eigenvectors of $\mathbf{S}^*$.

By a similar process we obtain the estimating equation

$$\hat{\boldsymbol{\Theta}} = \text{diag}(\mathbf{S} - \hat{\boldsymbol{\Lambda}}\hat{\boldsymbol{\Lambda}}^{\top}), \tag{7.18}$$

so that the sample variances, the diagonal of $\mathbf{S}$, are matched with the fitted variances, the diagonal of $\hat{\boldsymbol{\Sigma}} = \hat{\boldsymbol{\Lambda}}\hat{\boldsymbol{\Lambda}}^{\top} + \hat{\boldsymbol{\Theta}}$.

Fitting an exploratory factor analysis model consists of iterations of the eigenvalue problem given by (7.17) and variance matching, (7.18). Of course, there are no guarantees that the estimated variances in $\hat{\boldsymbol{\Theta}}$ will be non-negative. The occurrence of a non-positive estimated variance is referred to as a *Heywood case*. In the course of estimation a negative estimated variance is usually set to zero. As an alternative we could abandon the original interpretation of $\boldsymbol{\Theta}$ as a variance matrix, and merely search for a decomposition of the variance matrix $\boldsymbol{\Sigma}$, (7.2), in which case negative estimated elements of $\boldsymbol{\Theta}$ are allowed.

7.4 Two-level factor analysis

There are a number of avenues for generalization of the factor analysis for clustered data but the resulting extensions are essentially identical. We discuss these and then proceed to ML estimation.

First, to extend the analogy with linear regression, we can consider separate factor analysis models for each cluster:

$$\mathbf{y}_{ij} \sim \mathcal{N}_p(\boldsymbol{\mu}_j, \boldsymbol{\Sigma}_{1j}), \quad \text{i.i.d.}, \tag{7.19}$$

$i = 1, \ldots, n_j$, and $j = 1, \ldots, N_2$, with an appropriate parametrization for the variance matrices $\boldsymbol{\Sigma}_{1j}$. Equal variance matrices are the only meaningful choice for $\boldsymbol{\Sigma}_{1j}$, although, in principle, $\boldsymbol{\Sigma}_{1j}$ could be modelled as a function of a small number of parameters. We consider only a common value of the variance matrices $\boldsymbol{\Sigma}_{1j}$, and denote it by $\boldsymbol{\Sigma}_1$.

For the mean vectors $\boldsymbol{\mu}_j$ a common value $\boldsymbol{\mu}$ would be too restrictive; (7.19) would then correspond to the classical factor analysis of $N = \sum_j n_j$ vectors $\mathbf{y}_{ij}$. A realistic, and often interpretable, proposition is to assume another factor analysis model for the vectors of conditional expectations $\boldsymbol{\mu}_j$:

$$\boldsymbol{\mu}_j \sim \mathcal{N}_p(\boldsymbol{\mu}, \boldsymbol{\Sigma}_2) \quad \text{i.i.d.} \tag{7.20}$$

We assume that these random vectors are independent of the within-cluster deviations $\{\mathbf{y}_{ij} - \boldsymbol{\mu}_j\}_i$. The covariance structure of the outcome vectors $\{\mathbf{y}_{ij}\}$ is then described by two factor analysis decompositions,

$$\begin{aligned} \boldsymbol{\Sigma}_1 &= \boldsymbol{\Lambda}_1 \boldsymbol{\Psi}_1 \boldsymbol{\Lambda}_1^{\top} + \boldsymbol{\Theta}_1, \\ \boldsymbol{\Sigma}_2 &= \boldsymbol{\Lambda}_2 \boldsymbol{\Psi}_2 \boldsymbol{\Lambda}_2^{\top} + \boldsymbol{\Theta}_2, \end{aligned} \tag{7.21}$$

for some matrices $\boldsymbol{\Lambda}_m$, $\boldsymbol{\Psi}_m$, and $\boldsymbol{\Theta}_m$, $m = 1, 2$. Usually, $\boldsymbol{\Psi}_m$ and $\boldsymbol{\Theta}_m$ are assumed to be non-negative definite, although for the development presented in this and the following sections it suffices that the matrices $\boldsymbol{\Sigma}_1$ and $\boldsymbol{\Sigma}_2$ be non-negative definite.

Alternatively, a decomposition of the variance matrix $\boldsymbol{\Sigma} = \mathrm{var}(\mathbf{y}_{ij})$ into its within- and between-cluster components can be considered,

$$\boldsymbol{\Sigma} = \boldsymbol{\Sigma}_1 + \boldsymbol{\Sigma}_2, \tag{7.22}$$

where $\boldsymbol{\Sigma}_2 = \mathrm{cov}(\mathbf{y}_{i,j}, \mathbf{y}_{i',j})$, $(i \neq j)$. For the component matrices $\boldsymbol{\Sigma}_m$ $(m = 1, 2)$ we can posit parametric forms such as (7.21). Such decompositions can be represented in terms of random variables as

$$\mathbf{y}_{ij} = \boldsymbol{\mu} + \boldsymbol{\Lambda}_2 \boldsymbol{\delta}_{2j} + \boldsymbol{\varepsilon}_{2j} + \boldsymbol{\Lambda}_1 \boldsymbol{\delta}_{1ij} + \boldsymbol{\varepsilon}_{1ij}. \tag{7.23}$$

If $\boldsymbol{\Psi}_m$ and $\boldsymbol{\Theta}_m$ are non-negative definite this is equivalent to (7.21). In a different approach we assume the factor analysis model in (7.1),

$$\mathbf{y}_{ij} = \boldsymbol{\mu} + \boldsymbol{\Lambda} \boldsymbol{\delta}_{ij} + \boldsymbol{\varepsilon}_{ij},$$

with a decomposition of the random terms

$$\boldsymbol{\delta}_{ij} = \boldsymbol{\delta}_{2j} + \boldsymbol{\delta}_{1ij},$$
$$\boldsymbol{\varepsilon}_{ij} = \boldsymbol{\varepsilon}_{2j} + \boldsymbol{\varepsilon}_{1ij},$$

with the usual assumptions of mutual independence of the random samples $\{\boldsymbol{\delta}_{1ij}\}$, $\{\boldsymbol{\delta}_{2j}\}$, $\{\boldsymbol{\varepsilon}_{1ij}\}$, and $\{\boldsymbol{\varepsilon}_{2j}\}$. This would appear to be a submodel of (7.23), with $\boldsymbol{\Lambda}_2 = \boldsymbol{\Lambda}_1$, but different variance matrices for $\boldsymbol{\delta}_1$ and $\boldsymbol{\delta}_2$ correspond to a representation with unrelated matrices $\boldsymbol{\Lambda}$.

For the model equations (7.23) we will use the following terminology: The matrix $\boldsymbol{\Lambda}_m$, $m = 1, 2$, is the matrix of factor loadings at level m $(m = 1, 2)$, the elements of $\boldsymbol{\Theta}_m = \mathrm{var}(\boldsymbol{\varepsilon}_m)$ are the uniquenesses at level m, and $\boldsymbol{\delta}_m$ and matrix $\boldsymbol{\Psi}_m$ are the factors and their variance matrix (structure) at level m.

It is meaningful to insist that $\boldsymbol{\Sigma}_2$ be non-negative definite, or even that its component $\boldsymbol{\Theta}_2$ be non-negative definite. Then the models in (7.22) and (7.23) coincide. In (7.23) the mean vector $\boldsymbol{\mu}$ can be replaced by a linear regression, $\mathbf{B}\mathbf{x}_{ij}^{\top} = \mathbf{X}_{ij}\boldsymbol{\beta}$, as described in Section 7.2.

As is customary, we denote $\mathbf{y}_j = \{\mathbf{y}_{ij}\}_i \otimes \mathbf{1}_{n_j}$ and $\mathbf{y} = \{\mathbf{y}_j\} \otimes \mathbf{1}_{N_2}$, so that $\mathbf{y}$ is a $Np \times 1$ random vector $(N = \sum_j n_j)$,

$$\mathbf{y} \sim \mathcal{N}_{Np}(\mathbf{X}\boldsymbol{\beta}, \{\mathbf{V}_j\} \otimes \mathbf{I}_{N_2}), \tag{7.24}$$

where $\mathbf{X} = \{\mathbf{X}_j\} \otimes \mathbf{1}_{N_2}$, $\mathbf{X}_j = \{\mathbf{X}_{ij}\} \otimes \mathbf{1}_{n_j}$, and

$$\mathbf{V}_j = \boldsymbol{\Sigma}_1 \otimes \mathbf{I}_{n_j} + \boldsymbol{\Sigma}_2 \otimes \mathbf{J}_{n_j}. \tag{7.25}$$

We assume throughout that the variance matrix $\boldsymbol{\Sigma}_1$ is positive definite, although $\boldsymbol{\Sigma}_2$ may be singular.

7.4.1 MAXIMUM LIKELIHOOD ESTIMATION

The log-likelihood associated with the vector $\mathbf{y}$ is equal to

$$l = -\frac{1}{2}\left\{Np\log(2\pi) + \log(\det \mathbf{V}) + \mathbf{e}^\top\mathbf{V}^{-1}\mathbf{e}\right\}, \qquad (7.26)$$

where $\mathbf{e} = \mathbf{y} - \mathbf{X}\boldsymbol{\beta}$. For the variance matrices $\mathbf{V}_j$ we have the following inversion and determinant formulae:

$$\mathbf{V}_j^{-1} = \boldsymbol{\Sigma}_1^{-1}\otimes\mathbf{I}_{n_j} - \left(\mathbf{H}_{2j}^{-1}\boldsymbol{\Sigma}_2\boldsymbol{\Sigma}_1^{-1}\right)\otimes\mathbf{J}_{n_j}, \qquad (7.27)$$

$$\det \mathbf{V}_j = (\det\boldsymbol{\Sigma}_1)^{n_j-1}\det\mathbf{H}_{2j}, \qquad (7.28)$$

where $\mathbf{H}_{2j} = \boldsymbol{\Sigma}_1 + n_j\boldsymbol{\Sigma}_2$. Note that $\mathbf{H}_{2j}$ is proportional to the variance of the within-cluster mean, $\mathbf{H}_{2j} = n_j^{-1}\mathrm{var}(\sum_i\mathbf{y}_{ij})$. Equation (7.27) can be proved by direct multiplication, and equation (7.28) by block-diagonalization (block-sweeping) of the matrix $\mathbf{V}_j$. Note also that the matrix $\mathbf{H}_{2j}^{-1}\boldsymbol{\Sigma}_2\boldsymbol{\Sigma}_1^{-1}$ is symmetric and that

$$\boldsymbol{\Sigma}_1^{-1} - n_j\mathbf{H}_{2j}^{-1}\boldsymbol{\Sigma}_2\boldsymbol{\Sigma}_1^{-1} = \mathbf{H}_{2j}^{-1}. \qquad (7.29)$$

Equations (7.27) and (7.28) enable evaluation of the log-likelihood given by (7.26). For an arbitrary $n_jp \times 1$ vector $\mathbf{u}$

$$\mathbf{u}^\top\mathbf{V}_j^{-1}\left(\boldsymbol{\Lambda}_2\otimes\mathbf{1}_{n_j}\right) = \mathbf{u}^\top\left(\mathbf{H}_{2j}^{-1}\boldsymbol{\Lambda}_2\right)\otimes\mathbf{1}_{n_j}$$
$$= \sum_i\mathbf{u}_i\mathbf{H}_{2j}^{-1}\boldsymbol{\Lambda}_2, \qquad (7.30)$$

where $\mathbf{u}_i$ is the $p\times 1$ subvector of $\mathbf{u}$ corresponding to observation i in cluster j. The contribution of cluster j to the quadratic form $\mathbf{e}^\top\mathbf{V}^{-1}\mathbf{e}$ in (7.26), equal to $\mathbf{e}_j^\top\mathbf{V}_j^{-1}\mathbf{e}_j$, is

$$\sum_i(\mathbf{e}_{ij} - \bar{\mathbf{e}}_j)^\top\boldsymbol{\Sigma}_1^{-1}(\mathbf{e}_{ij} - \bar{\mathbf{e}}_j) + n_j\bar{\mathbf{e}}_j^\top\left(\boldsymbol{\Sigma}_1^{-1} - n_j\mathbf{H}_{2j}^{-1}\boldsymbol{\Sigma}_2\boldsymbol{\Sigma}_1^{-1}\right)\bar{\mathbf{e}}_j$$

$$= \sum_i(\mathbf{e}_{ij} - \bar{\mathbf{e}}_j)^\top\boldsymbol{\Sigma}_1^{-1}(\mathbf{e}_{ij} - \bar{\mathbf{e}}_j) + n_j\bar{\mathbf{e}}_j^\top\mathbf{H}_{2j}^{-1}\bar{\mathbf{e}}_j, \qquad (7.31)$$

where $\bar{\mathbf{e}} = n_j^{-1}\sum_i\mathbf{e}_{ij}$ is the within-cluster mean of the residuals $\mathbf{e}_{ij} = \mathbf{y}_{ij} - \mathbf{X}_{ij}\boldsymbol{\beta}$. We define the decomposition of the total sum of squares and cross-products of the residual vectors,

$$\mathbf{S} = \sum_j\sum_i\mathbf{e}_{ij}\mathbf{e}_{ij}^\top, \qquad (7.32)$$

into its within- and between-cluster components,

$$\mathbf{S}_1 = \sum_j \sum_i (\mathbf{e}_{ij} - \bar{\mathbf{e}}_j)(\mathbf{e}_{ij} - \bar{\mathbf{e}}_j)^\top \qquad (7.33)$$

and $\mathbf{S}_2 = \sum_j \mathbf{D}_j$ respectively, where

$$\mathbf{D}_j = n_j \bar{\mathbf{e}}_j \bar{\mathbf{e}}_j^\top. \qquad (7.34)$$

Note that $\mathbf{S}_1$ does not depend on the model parameters, since $\mathbf{e}_{ij} - \bar{\mathbf{e}}_j = \mathbf{y}_{ij} - \bar{\mathbf{y}}_{ij}$, whereas $\mathbf{S}_2$ (each $\mathbf{D}_j$) does.

By combining identities (7.27) and (7.31) we obtain a tractable formula for the log-likelihood (7.26),

$$l = -\frac{1}{2}\left\{ Np \log(2\pi) + (N - N_2)\log(\det \boldsymbol{\Sigma}_1) + \sum_j \log(\det \mathbf{H}_{2j}) \right.$$

$$\left. + \ \mathrm{tr}(\boldsymbol{\Sigma}_1^{-1}\mathbf{S}_1) + \sum_j \mathrm{tr}(\mathbf{H}_{2j}^{-1}\mathbf{D}_j) \right\}. \qquad (7.35)$$

In the two-level factor analysis model with trivial regression design ($\mathbf{X}_{ij} \equiv \mathbf{I}_p$) the matrix of sample cross-products $\sum_j \sum_i \mathbf{y}_{ij}\mathbf{y}_{ij}^\top$ and the vectors of within-cluster totals $\{\sum_i \mathbf{y}_{ij}\}$ form a set of sufficient statistics. In the balanced case ($n_j \equiv n$) the matrices $\mathbf{H}_{2j}$ coincide ($\mathbf{H}_{2j} \equiv \mathbf{H}_2$), and (7.35) simplifies to

$$l = -\frac{1}{2}\{N_2 np \log(2\pi) + N_2(n - 1)\log(\det \boldsymbol{\Sigma}_1) + N_2 \log(\det \mathbf{H}_2)$$

$$+ \mathrm{tr}(\boldsymbol{\Sigma}_1^{-1}\mathbf{S}_1) + \mathrm{tr}(\mathbf{H}_2^{-1}\mathbf{S}_2)\}. \qquad (7.36)$$

In this case the matrices $\sum_j \sum_i \mathbf{y}_{ij}\mathbf{y}_{ij}^\top$ and $\sum_j \bar{\mathbf{y}}_j\bar{\mathbf{y}}_j^\top$ are a set of sufficient statistics.

For the regression parameter vector $\boldsymbol{\beta}$ the first- and second-order partial derivatives of the log-likelihood l, given by (7.26), again yield the GLS equation

$$\hat{\boldsymbol{\beta}} = \left(\mathbf{X}^\top\mathbf{V}^{-1}\mathbf{X}\right)^{-1}\mathbf{X}^\top\mathbf{V}^{-1}\mathbf{y}, \qquad (7.37)$$

and the quadratic forms involved in this equation can be evaluated using (7.27) or the decomposition analogous to (7.31). For the trivial regression design, $\mathbf{X}_{ij} \equiv \mathbf{I}_p$, (7.27) and (7.29) imply that

$$\mathbf{X}^\top\mathbf{V}^{-1}\mathbf{X} = \sum_j \mathbf{H}_{2j}^{-1},$$

$$\mathbf{X}^{\top}\mathbf{V}^{-1}\mathbf{y} \;=\; \sum_{j}\mathbf{H}_{2j}^{-1}\sum_{i}\mathbf{y}_{ij}\,.$$

The derivative with respect to a general covariance structure parameter γ is

$$\frac{\partial l}{\partial \gamma} \;=\; -\frac{1}{2}(N - N_2)\,\mathrm{tr}\left(\boldsymbol{\Sigma}_1^{-1}\frac{\partial \boldsymbol{\Sigma}_1}{\partial \gamma}\right) - \frac{1}{2}\sum_{j}\mathrm{tr}\left(\mathbf{H}_{2j}^{-1}\frac{\partial \mathbf{H}_{2j}}{\partial \gamma}\right)$$

$$+\; \frac{1}{2}\,\mathrm{tr}\left(\boldsymbol{\Sigma}_1^{-1}\frac{\partial \boldsymbol{\Sigma}_1}{\partial \gamma}\boldsymbol{\Sigma}_1^{-1}\mathbf{S}_1\right) + \frac{1}{2}\sum_{j}\mathrm{tr}\left(\mathbf{H}_{2j}^{-1}\frac{\partial \mathbf{H}_{2j}}{\partial \gamma}\mathbf{H}_{2j}^{-1}\mathbf{D}_j\right),$$

$$(7.38)$$

and the matrix derivatives with respect to the specific types of parameters are:

$$\frac{\partial \boldsymbol{\Sigma}_m}{\partial \Lambda_{m,kh}} \;=\; \boldsymbol{\Lambda}_m\boldsymbol{\Psi}_m\boldsymbol{\Delta}_{h,r_m}\boldsymbol{\Delta}_{k,p}^{\top} \;+\; \boldsymbol{\Delta}_{k,p}\boldsymbol{\Delta}_{h,r_m}^{\top}\boldsymbol{\Psi}_m\boldsymbol{\Lambda}_m\,,$$

$$\frac{\partial \boldsymbol{\Sigma}_m}{\partial \Psi_{m,kh}} \;=\; \boldsymbol{\Lambda}_m\boldsymbol{\Delta}_{k,r_m}\boldsymbol{\Delta}_{h,r_m}^{\top}\boldsymbol{\Lambda}_m^{\top} \;+\; \boldsymbol{\Lambda}_m\boldsymbol{\Delta}_{h,r_m}\boldsymbol{\Delta}_{k,r_m}^{\top}\boldsymbol{\Lambda}_m^{\top} \qquad (k \neq h),$$

$$\frac{\partial \boldsymbol{\Sigma}_m}{\partial \Theta_{m,kk}} \;=\; \boldsymbol{\Delta}_{k,p}\boldsymbol{\Delta}_{k,p}^{\top}\,, \qquad\qquad\qquad (7.39)$$

for $m = 1, 2$. The derivatives of $\mathbf{H}_{2j}$ are linear combinations of the corresponding derivatives of $\boldsymbol{\Sigma}_1$ and $\boldsymbol{\Sigma}_2$; for any parameter γ involved in $\boldsymbol{\Sigma}_2$, but not in $\boldsymbol{\Sigma}_1$,

$$\frac{\partial \mathbf{H}_{2j}}{\partial \gamma} \;=\; n_j\frac{\partial \boldsymbol{\Sigma}_2}{\partial \gamma}\,.$$

Thus, a general matrix derivative in (7.39) has the form

$$\frac{\partial \boldsymbol{\Sigma}_m}{\partial \gamma} \;=\; \mathbf{A}\boldsymbol{\Delta}_1\boldsymbol{\Delta}_2^{\top}\mathbf{B} \;+\; \mathbf{B}\boldsymbol{\Delta}_2\boldsymbol{\Delta}_1^{\top}\mathbf{A}, \qquad (7.40)$$

where $\mathbf{A}$ and $\mathbf{B}$ are suitable matrices (scalar multiples of $\mathbf{I}$, $\boldsymbol{\Lambda}_m$, or $\boldsymbol{\Lambda}_m\boldsymbol{\Psi}_m$, $m = 1, 2$) and $\boldsymbol{\Delta}_1$ and $\boldsymbol{\Delta}_2$ are some indicator vectors. For a parameter γ involved in $\boldsymbol{\Sigma}_1$ the terms in (7.38) can be expressed as

$$\mathrm{tr}\left(\boldsymbol{\Sigma}_1^{-1}\frac{\partial \boldsymbol{\Sigma}_1}{\partial \gamma}\boldsymbol{\Sigma}_1^{-1}\mathbf{T}\right) \;=\; 2\boldsymbol{\Delta}_1^{\top}\mathbf{A}\boldsymbol{\Sigma}_1^{-1}\mathbf{T}\boldsymbol{\Sigma}_1^{-1}\mathbf{B}\boldsymbol{\Delta}_2 \qquad (7.41)$$

(and similarly for terms involving $\mathbf{H}_{2j}^{-1}$), and for a parameter in $\boldsymbol{\Sigma}_2$ as

$$\operatorname{tr}\left(\mathbf{H}_{2j}^{-1}\frac{\partial\boldsymbol{\Sigma}_2}{\partial\gamma}\mathbf{H}_{2j}^{-1}\mathbf{T}\right) \;=\; 2n_j\boldsymbol{\Delta}_1^{\top}\mathbf{A}\mathbf{H}_{2j}^{-1}\mathbf{T}\mathbf{H}_{2j}^{-1}\mathbf{B}\boldsymbol{\Delta}_2,\qquad(7.42)$$

where $\mathbf{T}$ stands for $\boldsymbol{\Sigma}_1$, $\mathbf{H}_{2j}$, $\mathbf{S}_1$, or $\mathbf{D}_j$. Thus, the scoring function (7.38) requires evaluation of elements and of certain quadratic forms in $\boldsymbol{\Sigma}_1^{-1}$, $\boldsymbol{\Sigma}_1^{-1}\mathbf{S}_1\boldsymbol{\Sigma}_1^{-1}$, $\mathbf{H}_{2j}^{-1}$, and $\mathbf{H}_{2j}^{-1}\mathbf{D}_j\mathbf{H}_{2j}^{-1}$.

It can be shown, as in Section 4.4, that the expected information matrix for all the model parameters is block-diagonal, with the blocks corresponding to the regression parameters $\boldsymbol{\beta}$ and the covariance structure parameters; we have

$$\mathbf{E}\left(\frac{\partial l^2}{\partial\boldsymbol{\beta}\partial\gamma}\right) \;=\; \mathbf{X}^{\top}\frac{\partial\mathbf{V}^{-1}}{\partial\gamma}\mathbf{E}(\mathbf{e}) \;=\; \mathbf{0}.$$

The expected information function for a pair of covariance structure parameters is

$$-\,\mathbf{E}\left(\frac{\partial^2 l}{\partial\gamma_1\,\partial\gamma_2}\right) \;=\; \frac{1}{2}\operatorname{tr}\left(\mathbf{V}^{-1}\frac{\partial\mathbf{V}}{\partial\gamma_1}\mathbf{V}^{-1}\frac{\partial\mathbf{V}}{\partial\gamma_2}\right)$$

$$=\; \frac{1}{2}(N-N_2)\operatorname{tr}\left(\boldsymbol{\Sigma}_1^{-1}\frac{\partial\boldsymbol{\Sigma}_1}{\partial\gamma_1}\boldsymbol{\Sigma}_1^{-1}\frac{\partial\boldsymbol{\Sigma}_1}{\partial\gamma_2}\right)$$

$$+\;\frac{1}{2}\sum_j\operatorname{tr}\left(\mathbf{H}_{2j}^{-1}\frac{\partial\mathbf{H}_{2j}}{\partial\gamma_1}\mathbf{H}_{2j}^{-1}\frac{\partial\mathbf{H}_{2j}}{\partial\gamma_2}\right).\qquad(7.43)$$

Note that the information function for a parameter from $\boldsymbol{\Sigma}_1$ and one from $\boldsymbol{\Sigma}_2$ is in general not equal to zero. For example,

$$-\,\mathbf{E}\left(\frac{\partial^2 l}{\partial\Lambda_{2,kh}\,\partial\Theta_{1k'k'}}\right) \;=\; \sum_j n_j\left(\mathbf{H}_{2j}^{-1}\Lambda_2\boldsymbol{\Psi}_2\right)_{hk'}\left(\mathbf{H}_{2j}^{-1}\right)_{kk'}\qquad(7.44)$$

for suitable subscripts k, h, and k'. It is easy to see that (7.43) can be evaluated as a cluster-wise sum of products of elements of quadratic forms in certain $p\times p$ matrices involving $\boldsymbol{\Sigma}_1^{-1}$ and $\mathbf{H}_{2j}^{-1}$.

The number of parameters in a two-level factor analysis model can be quite large, and it may be advantageous to avoid inversion of the fitted information matrix (or to solve the associated system of linear equations) even at the price of slower convergence. An obvious solution is to replace the information matrix in the Fisher scoring algorithm by a 'similar' matrix that is easier to invert. Some or all of the off-diagonal entries of the information matrix can be replaced by zero, such as the entries corresponding to an element of the matrix Λ_m and a parameter not involved in the same column of Λ_m. More generally, the entire vector of the estimated parameters $\boldsymbol{\Xi}$ can be partitioned into a set of non-overlapping subvectors,

$\boldsymbol{\Xi} = \{\boldsymbol{\Xi}_k\} \otimes \mathbf{1}_K$, and the elements of the substitute for the information matrix corresponding to parameters from two different subvectors can be set to zero.

A different approach is based on direct estimation of the inverse of the information matrix. There are a variety of such methods known as *quasi-Newton* or *conjugate gradient methods*. In general, they do not require evaluation of the matrix of second-order partial derivatives (or of their expectations), but the inverse of this matrix is estimated from the first-order partial derivatives in the previous iteration(s).

When the number of components p is much larger than the numbers of factors r_1 and r_2 it is advantageous to avoid direct inversion of the matrices $\mathbf{H}_{2j}$ and $\boldsymbol{\Sigma}_1$. The following identities express the inverse of $\mathbf{H}_{2j}$ in terms of the inverses of matrices with dimensions $r_1 \times r_1$ and $r_2 \times r_2$. First,

$$\mathbf{H}_{2j}^{-1} = \mathbf{H}_{1j}^{-1} - n_j \mathbf{H}_{1j}^{-1} \boldsymbol{\Lambda}_2 \boldsymbol{\Psi}_2 \mathbf{G}_{2j}^{-1} \boldsymbol{\Lambda}_2^\top \mathbf{H}_{1j}^{-1}, \qquad (7.45)$$

where $\mathbf{G}_{2j} = \mathbf{I}_{r_2} + n_j \boldsymbol{\Lambda}_2^\top \mathbf{H}_{1j}^{-1} \boldsymbol{\Lambda}_2 \boldsymbol{\Psi}_2$ and $\mathbf{H}_{1j} = \boldsymbol{\Sigma}_1 + n_j \boldsymbol{\Theta}_2$. Next,

$$\mathbf{H}_{1j}^{-1} = \boldsymbol{\Theta}^{-1} - \boldsymbol{\Theta}^{-1} \boldsymbol{\Lambda}_1 \boldsymbol{\Psi}_1 \mathbf{G}_{1j}^{-1} \boldsymbol{\Lambda}_1^\top \boldsymbol{\Theta}^{-1}, \qquad (7.46)$$

where $\mathbf{G}_{1j} = \mathbf{I}_{r_1} + \boldsymbol{\Lambda}_1^\top \left(\boldsymbol{\Theta}^{(j)} \right)^{-1} \boldsymbol{\Lambda}_1^\top \boldsymbol{\Psi}_1$, $\boldsymbol{\Theta}^{(j)} = \boldsymbol{\Theta}_1 + n_j \boldsymbol{\Theta}_2$, and $\boldsymbol{\Theta} = \boldsymbol{\Theta}^{(1)}$. Also, we have

$$\det \mathbf{H}_{2j} = \det \mathbf{H}_{1j} \det \mathbf{G}_1 \det \boldsymbol{\Theta}^{(j)}. \qquad (7.47)$$

7.4.2 CONSTRAINED MAXIMIZATION

Fitting factor analysis models by a maximum likelihood method involves constrained optimization. One set of constraints is due to the conditions of non-negative definiteness of estimated variance matrices, another set ensures model identifiability, and yet another set may reflect the specific context of the problem (e.g., in confirmatory mode).

To ensure that an estimated variance matrix is non-negative definite its Cholesky decomposition may be estimated ($\boldsymbol{\Sigma} = \mathbf{LL}^\top$). As an alternative we may estimate the variance parameter without a constraint, and then make appropriate adjustments. If one estimated variance is negative, it is set to zero. If two estimated variances are negative then we should in turn set each estimated variance to zero, reestimate the rest of the parameters, and compare the fitted models. The one with the higher log-likelihood is adopted as the ML. The analogous procedure for a higher number of negative estimated variances is rather complex, though. Various penalty methods can be applied as alternatives.

For the orthogonality constraints the method of Lagrange multipliers is applied. Let $\mathbf{u}(\boldsymbol{\Xi})$ be a vector function of the parameters such that the constraints (of orthogonality, and others) are expressible in the form $\mathbf{u} = \mathbf{0}$.

Suppose $\mathbf{u}$ is continuously differentiable, and $\partial\mathbf{u}/\partial\gamma \neq 0$ for any parameter γ throughout the parameter space. Let

$$l^* = l + \boldsymbol{\lambda}^\top \mathbf{u},$$

where $\boldsymbol{\lambda}$ is a vector of constants. Then l has a maximum, subject to the constraints $\mathbf{u} = \mathbf{0}$, only if the derivatives $\partial l^*/\partial\gamma$ vanish for each parameter γ and arbitrary vector $\boldsymbol{\lambda}$, and $\mathbf{u} = \mathbf{0}$. In practice, the function l^* is maximized simultaneously for the parameters $\boldsymbol{\Xi}$ and $\boldsymbol{\lambda}$. See Fletcher (1981) for further background.

Of particular importance are constraints that involve parameters at both levels since the processes modelled by the two variance matrices $\boldsymbol{\Sigma}_1$ and $\boldsymbol{\Sigma}_2$ may be closely related. Then it is desirable to model them by related factor structures. A natural hypothesis is that the factor loading matrices $\boldsymbol{\Lambda}_1$ and $\boldsymbol{\Lambda}_2$ span the same linear space, that is, they contain the same number of factors ($r_1 = r_2$), and $\boldsymbol{\Lambda}_1 = \boldsymbol{\Lambda}_2\mathbf{A}$ for a non-singular matrix $\mathbf{A}$. Such a constraint is awkward to implement, especially when pr_1 is large. It is advantageous to absorb the matrix $\mathbf{A}$ in $\boldsymbol{\Psi}_2$, so that

$$\mathrm{var}(\mathbf{y}_j) = \{\boldsymbol{\Lambda}\boldsymbol{\Psi}_1\boldsymbol{\Lambda}^\top + \boldsymbol{\Theta}_1\} \otimes \mathbf{I}_{n_j} + \{\boldsymbol{\Lambda}\boldsymbol{\Psi}_2\boldsymbol{\Lambda}^\top + \boldsymbol{\Theta}_2\} \otimes \mathbf{J}_{n_j}, \qquad (7.48)$$

where $\boldsymbol{\Lambda}$ is the matrix of common factor loadings at both levels. In the exploratory mode (for both levels) $\boldsymbol{\Psi}_1 = \mathbf{I}$, but in order to accomodate the absorbed matrix $\mathbf{A}$ it is necessary to estimate all the variances and covariances in $\boldsymbol{\Psi}_2$. Alternatively, $\boldsymbol{\Psi}_2$ can be set to the identity matrix and $\boldsymbol{\Psi}_1$ estimated.

7.5   Restricted maximum likelihood

In most cases the focus of factor analysis is on the covariance (correlation) structure, and the regression parameters (the mean vector) are either of secondary interest or are nuisance parameters. We can therefore invoke the argument about unbiased estimation of covariance structure parameters, discussed in Section 2.6, and consider the factor analysis of the error contrasts.

In unilevel factor analysis the ML and REML analyses essentially coincide because the estimated variance matrices in the two methods differ only by a scalar multiple. This property of invariance does not carry over to the multilevel case. However, the adjustment of the log-likelihood, as in (2.44), can be used again, and so the REML estimator can be obtained by an adaptation of the ML procedure. Differences between ML and REML estimates can be appreciable for the cluster-level parameters, but only when the number of clusters N_2 is small and the sampling variance matrix of the mean vector $\bar{\mathbf{y}}$ is large.

7.6   Saturated model (starting solution)

Methods for assessment of two-level factor analysis model fits can be motivated by procedures for unilevel factor analysis. We define the *saturated model* for two-level factor analysis as the model with arbitrary covariance structures at both levels. The saturated model contains $p(p+1)$ covariance structure parameters, and so the ML algorithm described in Section 7.4.1 is computationally inefficient, even for moderately large p. In this section we discuss a simple non-iterative method based on moment matching.

The saturated model fit is often of interest as a 'benchmark' because all the considered factor analysis models are its submodels. The fit for the saturated model is usually compared with the fits for these more parsimonious models, for instance by the likelihood ratio criterion.

Moment estimators of the variance matrices $\boldsymbol{\Sigma}_1$ and $\boldsymbol{\Sigma}_2$ can be derived from the matrices $\mathbf{S}_1$ and $\mathbf{S}_2$. We have

$$\mathbf{E}\{\mathbf{S}_1; \boldsymbol{\mu}, \boldsymbol{\Theta}\} = (N - N_2)\boldsymbol{\Sigma}_1\,,$$

$$\mathbf{E}\{\mathbf{S}_2; \boldsymbol{\mu}, \boldsymbol{\Theta}\} = N_2\boldsymbol{\Sigma}_1 + \left(N - N^{-1}\sum_j n_j^2\right)\boldsymbol{\Sigma}_2\,. \tag{7.49}$$

We emphasize dependence on the parameter vector $\boldsymbol{\mu}$ because these identities also have the REML version:

$$\mathbf{E}\{\mathbf{S}_1; \boldsymbol{\Theta}\} = (N - N_2)\boldsymbol{\Sigma}_1\,,$$

$$\mathbf{E}\{\mathbf{S}_2; \boldsymbol{\Theta}\} = (N_2 - 1)\boldsymbol{\Sigma}_1 + \left(N - N^{-1}\sum_j n_j^2\right)\boldsymbol{\Sigma}_2\,, \tag{7.50}$$

in which the expectations are calculated with regard to random variation of the sample mean $\hat{\boldsymbol{\mu}} = \bar{\mathbf{y}}$. The REML and ML estimators for $\boldsymbol{\Sigma}_1$ coincide; the only difference between (7.49) and (7.50) is in the factor for $\boldsymbol{\Sigma}_1$ in the equation for $\mathbf{S}_2$. The moment estimators of the variance matrices $\boldsymbol{\Sigma}_1$ and $\boldsymbol{\Sigma}_2$ are

$$\hat{\boldsymbol{\Sigma}}_1 = (N - N_2)^{-1}\mathbf{S}_1\,,$$

$$\hat{\boldsymbol{\Sigma}}_2 = \left(N - N^{-1}\sum_j n_j^2\right)^{-1}(\mathbf{S}_2 - M\hat{\boldsymbol{\Sigma}}_1)\,, \tag{7.51}$$

where $M = N_2$ for ML, (7.49), and $M = N_2 - 1$ for REML, (7.50). Whereas $\hat{\boldsymbol{\Sigma}}_1$ is always non-negative definite, $\hat{\boldsymbol{\Sigma}}_2$ may have negative eigenvalues. If no negative eigenvalues arise, the estimators in (7.51) are maximum likelihood. To prove it, note that the log-likelihood (7.26) belongs to the exponential family of distributions; it has the form

$$a_f(\boldsymbol{\mu}) + a_c(\boldsymbol{\Theta}) + \{\mathbf{u}(\boldsymbol{\mu}, \boldsymbol{\Theta})\}^\top \mathbf{b}(\mathbf{y}), \tag{7.52}$$

where a_f and a_c are real functions, and $\mathbf{u}$ and $\mathbf{b}$ are vector functions of the parameters and the data respectively. In our case $\mathbf{b}$ is a linear function of the matrices $\mathbf{S}_1$ and $\{\mathbf{D}_j\}$. The scoring vector for (7.52) is

$$\frac{\partial a_c(\boldsymbol{\Theta})}{\partial \boldsymbol{\Theta}} + \left\{ \frac{\partial \mathbf{u}(\boldsymbol{\mu}, \boldsymbol{\Theta})}{\partial \boldsymbol{\Theta}} \right\} \mathbf{b}(\mathbf{y}), \tag{7.53}$$

and since the expectation of this vector is $\mathbf{0}$, we have

$$\mathbf{E}\{\mathbf{b}(\mathbf{y}); \boldsymbol{\mu}, \boldsymbol{\Theta}\} = - \left[\left\{ \frac{\partial \mathbf{u}(\boldsymbol{\mu}, \boldsymbol{\Theta})}{\partial \boldsymbol{\Theta}} \right\}^\top \right]^{-1} \frac{\partial a_c(\boldsymbol{\Theta})}{\partial \boldsymbol{\Theta}}. \tag{7.54}$$

Hence, any root of the scoring vector is equal to the moment estimator.

Another important function of the saturated solution is to provide a starting solution for the Fisher scoring algorithm. The estimates $\hat{\boldsymbol{\Sigma}}_1$ and $\hat{\boldsymbol{\Sigma}}_2$ can be subjected to factor analyses and the resulting estimates $\hat{\boldsymbol{\Theta}}_1$, $\hat{\boldsymbol{\Theta}}_2$, $\hat{\boldsymbol{\Lambda}}_1$, $\hat{\boldsymbol{\Lambda}}_2$, $\hat{\boldsymbol{\Psi}}_1$, and $\hat{\boldsymbol{\Psi}}_2$ used as the starting values. Often, this starting solution is very close to the ML estimate, especially when N_2 is much smaller than N.

If $\boldsymbol{\Sigma}_2 = \mathbf{0}$ then the factor analysis for $\boldsymbol{\Sigma}_1$ would be based on $\mathbf{S}$, which has the Wishart distribution with $N - 1$ degrees of freedom, denoted by $\mathcal{W}_p(N - 1, \boldsymbol{\Sigma})$. The solution given by (7.50) is based on $\mathbf{S}_1$, which has the Wishart distribution with $N - N_2$ degrees of freedom, $\mathcal{W}_p(N - N_2, \boldsymbol{\Sigma})$. The MLE of $\boldsymbol{\Sigma}_1$ for unknown $\boldsymbol{\Sigma}_2$ is associated with a statistic with 'degrees of freedom' between $N - N_2$ and $N - 1$. If the number of degrees of freedom lost is small relative to the sample size, that is, N_2 is much smaller than N, then the estimator (7.50) is only slightly less efficient than the MLE.

7.7 Inference about $\bar{\mathbf{y}}$

The trivial estimator for the cluster mean is the arithmetic mean

$$\bar{\mathbf{y}}_j = n_j^{-1} \sum_i \mathbf{y}_{ij}; \tag{7.55}$$

it is conditionally unbiased given the within-cluster mean $\boldsymbol{\mu}_j$. Its conditional variance matrix is

$$\operatorname{var}(\bar{\mathbf{y}}_j \mid \boldsymbol{\mu}_j) = n_j^{-1} \boldsymbol{\Sigma}_1. \tag{7.56}$$

This estimator can be improved upon by incorporating between-cluster information. The joint conditional distribution of the random vectors $\boldsymbol{\delta}_{2j}$ and $\boldsymbol{\varepsilon}_{2j}$, given the data $\mathbf{y}$, is normal with mean

$$\begin{pmatrix} n_j \boldsymbol{\Psi}_2 \boldsymbol{\Lambda}_2^\top \mathbf{H}_{2j}^{-1} \\ n_j \boldsymbol{\Theta}_2 \mathbf{H}_{2j}^{-1} \end{pmatrix} (\bar{\mathbf{y}}_j - \boldsymbol{\mu}) \tag{7.57}$$

and variance matrix

$$\begin{pmatrix} \boldsymbol{\Psi}_2 & \mathbf{0} \\ \mathbf{0} & \boldsymbol{\Theta}_2 \end{pmatrix} - n_j \begin{pmatrix} \boldsymbol{\Psi}_2 \, \boldsymbol{\Lambda}_2^\top \\ \boldsymbol{\Theta}_2 \, \boldsymbol{\Psi}_2 \end{pmatrix}^\top \mathbf{H}_{2j}^{-1} \begin{pmatrix} \boldsymbol{\Psi}_2 \, \boldsymbol{\Lambda}_2^\top \\ \boldsymbol{\Theta}_2 \, \boldsymbol{\Psi}_2 \end{pmatrix}^\top . \tag{7.58}$$

The linear combination $\boldsymbol{\Lambda}_2 \boldsymbol{\delta}_{2j} + \boldsymbol{\varepsilon}_{2j}$ then has the conditional distribution

$$\mathcal{N} \left\{ n_j \boldsymbol{\Sigma}_2 \mathbf{H}_{2j}^{-1} (\bar{\mathbf{y}}_j - \boldsymbol{\mu}), \ \boldsymbol{\Sigma}_2 - n_j \boldsymbol{\Sigma}_2 \mathbf{H}_{2j}^{-1} \boldsymbol{\Sigma}_2 \right\}. \tag{7.59}$$

The conditional expectation in (7.59) is an estimator of the realized deviation $\boldsymbol{\mu}_j - \boldsymbol{\mu}$. The related estimator of the mean $\boldsymbol{\mu}_j$ is

$$\bar{\mathbf{y}} + n_j \boldsymbol{\Sigma}_2 \mathbf{H}_{2j}^{-1} (\bar{\mathbf{y}}_j - \bar{\mathbf{y}}) = \boldsymbol{\Sigma}_1 \mathbf{H}_{2j}^{-1} \bar{\mathbf{y}} + n_j \boldsymbol{\Sigma}_2 \mathbf{H}_{2j}^{-1} \bar{\mathbf{y}}_j;$$

it is a multivariate shrinkage estimator. The sample mean $\bar{\mathbf{y}}$ and the within-cluster mean $\bar{\mathbf{y}}_j$ are combined proportionately to the conditional variance matrix of the former, $n_j^{-1} \boldsymbol{\Sigma}_1$, and the variance matrix $\boldsymbol{\Sigma}_2$.

The equations involve $\boldsymbol{\Sigma}_1$ or its components only through $\mathbf{H}_{2j}$. In practice, the matrices $\boldsymbol{\Sigma}_1$ and $\boldsymbol{\Sigma}_2$ are replaced by their estimators. Then (7.59) holds only approximately, although the errors are insubstantial when $\boldsymbol{\Sigma}_1$ and $\boldsymbol{\Sigma}_2$ are estimated with small sampling variation.

7.8   Measuremement error models

In this section a general class of measurement error models for independent observations is introduced which is then extended for clustered observations in Section 7.9.

A general measurement error model for independent outcomes can be described as a linear regression

$$y_i = \mathbf{x}_i \boldsymbol{\beta} + \varepsilon_i, \tag{7.60}$$

$\varepsilon_i \sim \mathcal{N}(0, \sigma^2)$, i.i.d., in which the regressor vector $\mathbf{x}$ is partitioned into two subvectors, $\mathbf{x} = (\mathbf{x}_f, \mathbf{x}_r)$; the values of the vector $\mathbf{x}_f$ are known, while the vector $\mathbf{x}_r$ is observed indirectly and subject to error. Instead of the *latent* variables $\mathbf{x}_r$, a vector of *manifest* variables, $\mathbf{s}$, is observed, and $\mathbf{s}$ is related to $\mathbf{x}_r$ by a linear regression formula,

$$\mathbf{s}_i = \boldsymbol{\Lambda}\mathbf{x}_{r,i}^\top + \boldsymbol{\xi}_i, \tag{7.61}$$

where $\boldsymbol{\xi}_i \sim \mathcal{N}(\mathbf{0}, \boldsymbol{\Theta})$, i.i.d. Further, we assume that the vectors of latent variables also form a random sample,

$$\mathbf{x}_{r,i} \sim \mathcal{N}(\boldsymbol{\mu}_r, \boldsymbol{\Psi}), \qquad \text{i.i.d.;} \tag{7.62}$$

the elements of $\boldsymbol{\mu}_r$ and $\boldsymbol{\Psi}$ may be known or unknown. We refer to the assumptions (7.60)–(7.62) as the regression, measurement error equations and the distribution of the latent variables respectively.

In a typical case the matrix $\boldsymbol{\Lambda}$ is known; for instance, when a univariate latent variable x_r is observed by three exchangeable measurements with error variance σ_r^2, then $\boldsymbol{\Lambda} = (1, 1, 1)^\top$ and $\boldsymbol{\xi} \sim \mathcal{N}(\mathbf{0}, \sigma_r^2 \mathbf{I}_3)$. On the other hand, (7.61) on its own, with $\boldsymbol{\Lambda}$ unknown, is a factor analysis model. Thus, equations (7.60)–(7.62) combine measurement error models and factor analysis. In the subsequent development, each element of $\boldsymbol{\Lambda}$ may be assumed either known or unknown.

When the regressors $\mathbf{x}$ are measured subject to error, the estimators of the parameters have standard errors greater than the nominal standard errors that result from OLS. Also, the OLS regression parameter estimators are biased. An insight into errors associated with the application of OLS when the regressors are measured subject to error can be gained by the following comparison.

Suppose $\mathbf{X}$ is the 'true' regression design, containing the latent values of the regressors, and $\mathbf{X}^*$ is the matrix of its manifest (observed) values. If $\mathbf{X}$ were observed (in addition to the outcomes y), then, under certain conditions, the OLS estimator,

$$\hat{\boldsymbol{\beta}}_l = \left(\mathbf{X}^\top\mathbf{X}\right)^{-1}\mathbf{X}^\top\mathbf{y},$$

would be the optimal estimator for $\boldsymbol{\beta}$. The OLS estimator using the 'manifest' design matrix $\mathbf{X}^*$ yields the estimator

$$\hat{\boldsymbol{\beta}}_m = \left(\mathbf{X}^{*\top}\mathbf{X}^*\right)^{-1}\mathbf{X}^{*\top}\mathbf{y}.$$

Taking the expectations over the measurement errors we have

$$\mathbf{E}(\mathbf{X}^{*\top}\mathbf{y}) = \mathbf{X}^\top\mathbf{y},$$

and

$$\mathbf{E}\left(\mathbf{X}^{*\top}\mathbf{X}^*\right) = \mathbf{X}^\top\mathbf{X} + \boldsymbol{\Theta}.$$

Since $\boldsymbol{\Theta}$ is non-negative definite the components of $\boldsymbol{\beta}_m$ are likely to be smaller in absolute value than the corresponding components of $\boldsymbol{\beta}_l$. Also, the nominal standard errors based on the 'manifest' matrix of cross-products, $\mathbf{X}^\top \mathbf{X}$, are smaller than those for the 'latent' matrix $\mathbf{E}\left(\mathbf{X}^{*\top}\mathbf{X}^*\right)$.

Figure 7.1 illustrates this phenomenon for simple regression using simulated data. The latent variable x was generated as a random draw from $\mathcal{N}(4, 10)$, and the manifest variable as $x + r$, where r is a random draw from $\mathcal{N}(0, 1)$. The outcome variable is generated as $y = x + \varepsilon$, where ε is a random draw from $\mathcal{N}(0, 0.25^2)$. The three random samples, $\{x_i\}, \{r_i\}$, and $\{\varepsilon_i\}$, are mutually independent.

The left-hand panel contains the plot of the outcome on the latent variable; the linear regression appears to be well-determined; the OLS fit (using the generated values of the latent variable which would not be available in a more realistic setting) is $\hat{\boldsymbol{\beta}}_L = (-0.03, 1.02)^\top$ with standard errors $(0.063, 0.027)^\top$ and estimated residual variance $\hat{\sigma}_L^2 = 0.081$. The right-hand panel contains the plot of the outcome on the manifest variable; the OLS fit using the manifest variable as a regressor is $\hat{\boldsymbol{\beta}}_M = (0.22, 0.90)^\top$, with standard errors $(0.134, 0.057)^\top$ and residual variance $\hat{\sigma}_M^2 = 0.393$. Even though the measurement error variance is only a modest fraction of the variation of the latent variable x, replacement of x by its observed version $x + r$ inflates the standard errors of the estimated regression parameters more than twofold, and the estimate of the residual variance is inflated almost fivefold. Also, the regression parameter estimate is smaller than the true or estimated values for the regression on the latent variable.

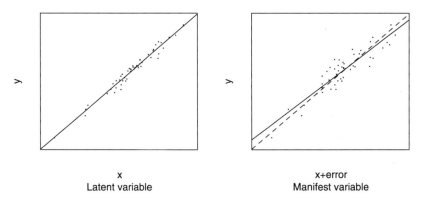

x
Latent variable

x+error
Manifest variable

Fig. 7.1. The effect of the measurement error on the parameter estimates. Simulated example. The left-hand panel contains the plot of the simulated outcome variable against the latent variable and in the right-hand panel the outcome variable is plotted against the manifest variable. The fitted OLS regressions are indicated by solid lines. The OLS regression for the latent variable is reproduced in the right-hand panel by a dashed line.

In the remainder of this section we describe a ML procedure for measurement error models, and in Section 7.9 we define two-level measurement error models and describe an extension of this procedure for two-level models.

Let $\boldsymbol{\beta} = \begin{pmatrix} \boldsymbol{\beta}_f \\ \boldsymbol{\beta}_r \end{pmatrix}$ be the partitioning of $\boldsymbol{\beta}$ compatible with that for $\mathbf{x}$, so that

$$y_i \sim \mathcal{N}(\mathbf{x}_{f,i}\boldsymbol{\beta}_f + \boldsymbol{\mu}_r\boldsymbol{\beta}_r, \; \boldsymbol{\beta}_r^\top \boldsymbol{\Psi}\boldsymbol{\beta}_r + \sigma^2). \tag{7.63}$$

The constant variance σ^2 can be replaced by a function of $\mathbf{x}_f$, or of other variables. For example, sampling weights $\{w_i\}$ can be accomodated by replacing σ^2 with $w_i\sigma^2$ or $\boldsymbol{\beta}_r^\top \boldsymbol{\Psi}\boldsymbol{\beta}_r + \sigma^2$ with $w_i(\boldsymbol{\beta}_r^\top \boldsymbol{\Psi}\boldsymbol{\beta}_r + \sigma^2)$. For simplicity, we consider a constant variance σ^2 throughout the rest of the chapter. The manifest variables have the distribution

$$\mathbf{s}_i \sim \mathcal{N}(\boldsymbol{\Lambda}\boldsymbol{\mu}_r, \; \boldsymbol{\Lambda}\boldsymbol{\Psi}\boldsymbol{\Lambda}^\top + \boldsymbol{\Theta}), \tag{7.64}$$

and the covariance of the outcomes and the latent variables is

$$\mathrm{cov}(y_i, \mathbf{s}_i) = \boldsymbol{\Lambda}\boldsymbol{\Psi}\boldsymbol{\beta}_r. \tag{7.65}$$

We denote $\mathbf{u}_i = \begin{pmatrix} y_i \\ \mathbf{s}_i \end{pmatrix}$, $\mathbf{u} = \{\mathbf{u}_i\} \otimes \mathbf{I}_N$, and $\boldsymbol{\Sigma}_u = \mathrm{var}(\mathbf{u}_i)$.

For illustration we consider the simple regression in which the explanatory variable is latent, that is, $x_{f,i} \equiv 1$, and $x_{r,i} = x_i$ is observed by means of k repeated measurements;

$$\mathbf{s}_i = \mathbf{1}_k x_{r,i} + \boldsymbol{\xi}_i. \tag{7.66}$$

If $k^{-1}\mathrm{var}(\xi)$ were small we could ignore the measurement variance of $k^{-1}\mathbf{s}_i\mathbf{1}_k$, and use this variable as the regressor instead of $x_{r,i}$. Then the assumption in (7.62) is redundant, and (7.60) is reduced to a simple regression model. Such an approximation may be suitable for a preliminary analysis, but is too crude for any other purpose.

The log-likelihood associated with the vector of observations $\{\mathbf{u}_i\}$ is equal to

$$l = -\frac{1}{2}\left\{ Np\log(2\pi) + N\log(\det\boldsymbol{\Sigma}_u) + \sum_i \mathbf{e}_i^\top \boldsymbol{\Sigma}_u^{-1}\mathbf{e}_i \right\}, \tag{7.67}$$

where $\mathbf{e}_i = \mathbf{u}_i - \mathbf{E}(\mathbf{u}_i) = \begin{pmatrix} y_i - \mathbf{x}_{f,i}\boldsymbol{\beta}_f - \boldsymbol{\mu}_r\boldsymbol{\beta}_r \\ \mathbf{s}_i - \boldsymbol{\Lambda}\boldsymbol{\mu}_r \end{pmatrix}$. The parameters $\boldsymbol{\beta}_f$ and $\boldsymbol{\mu}_f$ are involved only in $\mathbf{E}(\mathbf{u}_i)$, and $\boldsymbol{\Psi}$, $\boldsymbol{\Theta}$, and σ^2 only in $\boldsymbol{\Sigma}_u$, but $\boldsymbol{\beta}_r$ is involved in both the mean and variance structures. This does not create insurmountable problems, though. The first-order partial derivative with respect to a general parameter γ has the form

$$-\frac{1}{2}\left\{ N\operatorname{tr}\left(\boldsymbol{\Sigma}_u^{-1}\frac{\partial\boldsymbol{\Sigma}_u}{\partial\gamma}\right) - \sum_i \mathbf{e}_i^\top \boldsymbol{\Sigma}_u^{-1}\frac{\partial\boldsymbol{\Sigma}_u}{\partial\gamma}\boldsymbol{\Sigma}_u^{-1}\mathbf{e}_i - 2\sum_i \mathbf{e}_i^\top \boldsymbol{\Sigma}_u^{-1}\frac{\partial\mathbf{E}(\mathbf{u}_i)}{\partial\gamma}\right\}.$$

Although the equations for the second-order partial derivatives are rather complex their negative expectations simplify to

$$-\mathbf{E}\left(\frac{\partial^2 l}{\partial\gamma_1\partial\gamma_2}\right) =$$

$$\sum_i \left\{\frac{\partial\mathbf{E}(\mathbf{u}_i)}{\partial\gamma_1}\right\}^\top \boldsymbol{\Sigma}_u^{-1}\frac{\partial\mathbf{E}(\mathbf{u}_i)}{\partial\gamma_2} + \frac{1}{2}\sum_i \operatorname{tr}\left(\boldsymbol{\Sigma}_u^{-1}\frac{\partial\boldsymbol{\Sigma}_u}{\partial\gamma_1}\boldsymbol{\Sigma}_u^{-1}\frac{\partial\boldsymbol{\Sigma}_u}{\partial\gamma_2}\right).$$
$$(7.68)$$

These equations further simplify for the parameters not involved in both expectations and variances of the observations, in which case one of the terms in (7.68) vanishes.

7.9   A two-level measurement error model

We consider the general setting of clustered multivariate observations in which there are several sets of random vectors $\{\mathbf{u}_{ij}\}$, $i = 1,\ldots,n_j$, $j = 1,\ldots,N_2$. Let $N = \sum_{j=1}^{N_2} n_j$. Each vector $\mathbf{u}_{ij}$ contains an outcome variable y_{ij} and a vector of manifest variables $\mathbf{s}_{ij}$, $\mathbf{u}_{ij} = (y_{ij}, \mathbf{s}_{ij})^\top$. For the outcomes $\{y_{ij}\}$ we assume a two-level regression model,

$$y_{ij} = \mathbf{x}_{f,ij}\boldsymbol{\beta}_{f,j} + \mathbf{x}_{r,ij}\boldsymbol{\beta}_r + \varepsilon_{ij};\qquad(7.69)$$

the regression coefficients $\boldsymbol{\beta}_{f,ij}$ are assumed to vary across the clusters j but the parameters $\boldsymbol{\beta}_r$ are assumed to be universal (constant). For $\{\boldsymbol{\beta}_{f,j}\}$ and $\{\varepsilon_{ij}\}$ we adopt the usual assumptions

$$\boldsymbol{\beta}_{f,j} \sim \mathcal{N}(\boldsymbol{\beta}_f, \boldsymbol{\Sigma}_\beta), \qquad \text{i.i.d.}$$
$$\varepsilon_{ij} \sim \mathcal{N}(0, \sigma^2), \qquad \text{i.i.d.}$$
$$(7.70)$$

Similarly, the measurement error equations involve a decomposition of the error terms:

$$\mathbf{s}_{ij} = \boldsymbol{\Lambda}\mathbf{x}_{r,ij}^\top + \boldsymbol{\xi}_{2,j} + \boldsymbol{\xi}_{1,ij},\qquad(7.71)$$

where $\boldsymbol{\xi}_{1,ij} \sim \mathcal{N}(\mathbf{0}, \boldsymbol{\Theta}_1)$ and $\boldsymbol{\xi}_{2,j} \sim \mathcal{N}(\mathbf{0}, \boldsymbol{\Theta}_2)$ are two mutually independent random samples. The distributional assumption for the latent vectors can also be extended to accomodate within-cluster correlations:

$$\mathbf{x}_{r,ij} = \boldsymbol{\mu}_r + \boldsymbol{\zeta}_{2,j} + \boldsymbol{\zeta}_{1,ij}, \tag{7.72}$$

where $\boldsymbol{\zeta}_{1,ij} \sim \mathcal{N}(\mathbf{0}, \boldsymbol{\Psi}_1)$, i.i.d., and $\boldsymbol{\zeta}_{2,j} \sim \mathcal{N}(\mathbf{0}, \boldsymbol{\Psi}_2)$, i.i.d. The random samples $\{\boldsymbol{\beta}_{f,j}\}$, $\{\varepsilon_{ij}\}$, $\{\boldsymbol{\xi}_{1,ij}\}$, $\{\boldsymbol{\xi}_{2,j}\}$, $\{\boldsymbol{\zeta}_{1,ij}\}$, and $\{\boldsymbol{\zeta}_{2,j}\}$ are mutually independent.

The model given by (7.69)–(7.72) has a formidable number (and variety) of parameters. For a realistic application, constraints on the parameters have to be considered, both to ensure model identifiability and to promote model parsimony. The rules for model identifiability, based on independent replication, carry over from the models for independent observations to two-level models, but they are far from sufficient. There is no simple set of rules to supplement them, and intuition has to be relied on when setting up models with complex structures of random terms (errors).

We proceed now with a description of the Fisher scoring algorithm. The random vectors $\mathbf{u}_{ij}$ have the distribution

$$\mathcal{N}\left\{\begin{pmatrix} \mathbf{x}_{f,ij}\boldsymbol{\beta}_f + \boldsymbol{\mu}_f\boldsymbol{\beta}_r \\ \boldsymbol{\Lambda}\boldsymbol{\mu}_r^\top \end{pmatrix}, \begin{pmatrix} \sigma^2 + \mathbf{x}_{f,ij}\boldsymbol{\Sigma}_\beta\mathbf{x}_{f,ij}^\top + \boldsymbol{\beta}_r^\top\boldsymbol{\Psi}\boldsymbol{\beta}_r & \boldsymbol{\beta}_r^\top\boldsymbol{\Psi}\boldsymbol{\Lambda}^\top \\ \boldsymbol{\Lambda}\boldsymbol{\Psi}\boldsymbol{\beta}_r & \boldsymbol{\Lambda}\boldsymbol{\Psi}\boldsymbol{\Lambda}^\top + \boldsymbol{\Theta} \end{pmatrix}\right\} \tag{7.73}$$

where $\boldsymbol{\Psi} = \boldsymbol{\Psi}_1 + \boldsymbol{\Psi}_2$ and $\boldsymbol{\Theta} = \boldsymbol{\Theta}_1 + \boldsymbol{\Theta}_2$.

The covariance of two vectors $\mathbf{u}_{ij}$ from the same cluster j is

$$\operatorname{cov}(\mathbf{u}_{ij}, \mathbf{u}_{i'j}) = \begin{pmatrix} \mathbf{x}_{f,ij}\boldsymbol{\Sigma}_\beta\mathbf{x}_{f,i'j}^\top + \boldsymbol{\beta}_r^\top\boldsymbol{\Psi}_2\boldsymbol{\beta}_r & \boldsymbol{\beta}_r^\top\boldsymbol{\Psi}_2\boldsymbol{\Lambda}^\top \\ \boldsymbol{\Lambda}\boldsymbol{\Psi}_2\boldsymbol{\beta}_r & \boldsymbol{\Lambda}\boldsymbol{\Psi}_2\boldsymbol{\Lambda}^\top + \boldsymbol{\Theta}_2 \end{pmatrix} \tag{7.74}$$

$(i \neq i')$, so that the variance matrix for the vector $\mathbf{u}_j = \mathbf{u}_{ij} \otimes \mathbf{1}_{n_j}$ is

$$\mathbf{V}_j = \mathbf{A} \otimes \mathbf{I}_{n_j} + \mathbf{B} \otimes \mathbf{J}_{n_j} + \mathbf{Z}_j\boldsymbol{\Sigma}_\beta\mathbf{Z}_j^\top, \tag{7.75}$$

where

$$\mathbf{A} = \begin{pmatrix} \sigma^2 + \boldsymbol{\beta}_r^\top\boldsymbol{\Psi}_1\boldsymbol{\beta}_r & \boldsymbol{\beta}_r^\top\boldsymbol{\Psi}_1\boldsymbol{\Lambda}^\top \\ \boldsymbol{\Lambda}\boldsymbol{\Psi}_1\boldsymbol{\beta}_r & \boldsymbol{\Lambda}\boldsymbol{\Psi}_1\boldsymbol{\Lambda}^\top + \boldsymbol{\Theta}_1 \end{pmatrix},$$

$$\mathbf{B} = \begin{pmatrix} \boldsymbol{\beta}_r^\top\boldsymbol{\Psi}_2\boldsymbol{\beta}_r & \boldsymbol{\beta}_r^\top\boldsymbol{\Psi}_2\boldsymbol{\Lambda}^\top \\ \boldsymbol{\Lambda}\boldsymbol{\Psi}_2\boldsymbol{\beta}_r & \boldsymbol{\Lambda}\boldsymbol{\Psi}_2\boldsymbol{\Lambda}^\top + \boldsymbol{\Theta}_2 \end{pmatrix},$$

and the (variation) design matrix $\mathbf{Z}_j$ is formed from the rows $\mathbf{x}_{f,ij}$ supplemented by $p - 1$ rows of zeros, so that $\mathbf{Z}_j = \mathbf{Z}_{ij} \otimes \mathbf{1}_{n_j}$ and

$$\mathbf{Z}_{ij}\boldsymbol{\Sigma}_\beta\mathbf{Z}_{ij}^\top = \begin{pmatrix} \mathbf{x}_{f,ij}\boldsymbol{\Sigma}_\beta\mathbf{x}_{f,ij}^\top & \mathbf{0}^\top \\ \mathbf{0} & \mathbf{0}\mathbf{0}^\top \end{pmatrix}. \tag{7.76}$$

The variance matrix for $\mathbf{u} = \mathbf{u}_j \otimes \mathbf{1}_{N_2}$ is $\{\mathbf{V}_j\} \otimes \mathbf{I}_{N_2}$.

As before, we seek analytical expressions for the inverse and determinant of the variance matrix $\mathbf{V}$ $(\mathbf{V}_j)$, so that computation of the log-likelihood and of its partial derivatives would be feasible even for large clusters and for several latent and manifest variables.

Let $\mathbf{W}_j = \mathbf{A} \otimes \mathbf{I}_{n_j} + \mathbf{B} \otimes \mathbf{J}_{n_j}$, so that $\mathbf{V}_j = \mathbf{W}_j + \mathbf{Z}_j \boldsymbol{\Sigma}_\beta \mathbf{Z}_j^\top$. For the inverse we have

$$\mathbf{V}_j^{-1} = \mathbf{W}_j^{-1} - \mathbf{W}_j^{-1} \mathbf{Z}_j \boldsymbol{\Sigma}_\beta \mathbf{G}_{2j}^{-1} \mathbf{Z}_j^\top \mathbf{W}_j^{-1}, \qquad (7.77)$$

where $\mathbf{G}_{2j} = \mathbf{I}_p + \mathbf{Z}_j^\top \mathbf{W}_j^{-1} \mathbf{Z}_j \boldsymbol{\Sigma}_\beta$, and from equation (7.27) for block-patterned matrices,

$$\mathbf{W}_j^{-1} = \{\mathbf{A}^{-1}\} \otimes \mathbf{I}_{n_j} - \{\mathbf{H}_{2j}^{-1} \mathbf{B} \mathbf{A}^{-1}\} \otimes \mathbf{J}_{n_j}, \qquad (7.78)$$

where $\mathbf{H}_{2j} = \mathbf{A} + n_j \mathbf{B}$. Equation (7.77) is proved by direct multiplication, see (6.18).

For the determinants we have the identity

$$\det \mathbf{V}_j = (\det \mathbf{A})^{n_j - 1} \det \mathbf{H}_{2j} \det \mathbf{G}_{2j}. \qquad (7.79)$$

Equations (7.77)–(7.79) require inversion of N_2 matrices of size $p \times p$ and N_2 matrices of size $p_r \times p_r$. Although rather complex, the equations avoid the inversion of large matrices. Description of an efficient algorithm for the general case is not possible, but the computations are usually simplified if instead of the matrices in (7.77) and (7.78) all the quadratic forms involving them are evaluated. When $n_j p$ is small it may be advantageous to invert the matrices $\mathbf{V}_j$ directly.

The log-likelihood for the model in (7.69)–(7.72) is

$$l = -\frac{1}{2} \sum_j \left\{ n_j \log(2\pi) + \log(\det \mathbf{V}_j) + \mathbf{e}_j^\top \mathbf{V}_j^{-1} \mathbf{e}_j \right\}, \qquad (7.80)$$

with $\mathbf{V}_j$ given by (7.75). The quadratic form $\mathbf{e}^\top \mathbf{V}^{-1} \mathbf{e} = \sum_j \mathbf{e}_j^\top \mathbf{V}_j^{-1} \mathbf{e}_j$ can be expressed in terms of within- and between-cluster totals of cross-products of the residuals:

$$\mathbf{e}^\top \mathbf{V}^{-1} \mathbf{e} = \mathrm{tr}(\mathbf{A}^{-1} \mathbf{S}_1) + \sum_j \mathrm{tr}\left(\mathbf{H}_{2j}^{-1} \mathbf{D}_j\right) - \sum_j \mathrm{tr}\left(\boldsymbol{\Sigma}_\beta \mathbf{G}_{2j}^{-1} \mathbf{f}_j^\top \mathbf{f}_j\right), \quad (7.81)$$

where $\mathbf{S}_1 = \sum_j \sum_i (\mathbf{e}_{ij} - \bar{\mathbf{e}}_j)(\mathbf{e}_{ij} - \bar{\mathbf{e}}_j)^\top$, $\mathbf{D}_j = n_j(\bar{\mathbf{e}}_j - \bar{\mathbf{e}})(\bar{\mathbf{e}}_j - \bar{\mathbf{e}})^\top$, $\bar{\mathbf{e}}_j = n_j^{-1} \sum_j \mathbf{e}_{ij}$, $\bar{\mathbf{e}} = N^{-1} \sum_j n_j \bar{\mathbf{e}}_j$, and $\mathbf{f}_j = \mathbf{Z}_j^\top \mathbf{W}_j^{-1} \mathbf{e}_j$. The latter expression can be evaluated by using (7.78) and taking advantage of the structure of $\mathbf{Z}_j$:

$$\mathbf{f}_j = \sum_i (\mathbf{e}_{ij} - \bar{\mathbf{e}}_j)^\top \mathbf{A}^{-1} \mathbf{Z}_{ij} + \bar{\mathbf{e}}_j^\top \left(\mathbf{A}^{-1} - n_j \mathbf{H}_{2j}^{-1} \mathbf{B} \mathbf{A}^{-1} \right) \boldsymbol{\Sigma}_\beta \mathbf{Z}_{ij}$$

$$= \sum_i (\mathbf{e}_{ij} - \bar{\mathbf{e}}_j)^\top \mathbf{a}^{(1)} \mathbf{x}_{f,ij} + n_j \bar{\mathbf{e}}_j^\top \mathbf{h}_{2j}^{(1)} \bar{\mathbf{x}}_{f,j} , \qquad (7.82)$$

where $\mathbf{a}^{(1)}$ and $\mathbf{h}^{(1)}$ denote the respective first columns of $\mathbf{A}^{-1}$ and $\mathbf{H}_{2j}^{-1}$, and $\bar{\mathbf{x}}_{f,j}$ is the within-cluster mean of $\mathbf{x}_f$ for cluster j, $\bar{\mathbf{x}}_{f,j} = n_j^{-1} \sum_i \mathbf{x}_{f,ij}$.

7.10   General covariance structures

When a complex network of relationships is considered, model specification by explicit transcription of the hypothetical (additive) influences among the variables may not be realistic. A model constructed in this way may be in conflict with the description of the covariance structure of the observations. It is often more expedient to specify models by a regression function and a variance matrix decomposition. Let $\{\mathbf{y}_{ij}\}$ be the observed vectors of outcomes. We specify their expectation by a linear function of some explanatory variables $\mathbf{x}$,

$$\mathbf{E}(\mathbf{y}_{ij}) = \mathbf{x}_{ij} \boldsymbol{\beta}, \qquad (7.83)$$

and their variance matrix by a decomposition into its elementary- and cluster-level components,

$$\mathrm{var}(\mathbf{y}_{ij}) = \boldsymbol{\Sigma}_{1ij} + \boldsymbol{\Sigma}_{2iij} , \qquad (7.84)$$

where $\boldsymbol{\Sigma}_{2ii'j} = \mathrm{cov}(\mathbf{y}_{ij}, \mathbf{y}_{i'j})$, $(i \neq i')$, and the matrices $\boldsymbol{\Sigma}_{1ij}$ and $\boldsymbol{\Sigma}_{2ii'j}$ are quadratic functions of certain variables and parameters.

   In order that any ML procedure be feasible when the cluster sample sizes n_j are large, it is expedient to choose a covariance structure parametrization for which the variance matrix

$$\mathrm{var}(\mathbf{y}_j) = \{\boldsymbol{\Sigma}_{1ij}\} \otimes \mathbf{I}_{n_j} + \{\boldsymbol{\Sigma}_{2ii'j}\} \otimes \mathbf{J}_{n_j}$$

can be inverted by computationally efficient methods (by taking advantage of the pattern of its blocks and their elements). Then the general outline of the Fisher scoring algorithm (Section 1.9) can be followed.

7.11   Example. Second International Mathematics Study

As an illustration of two-level factor analysis we use the mathematics achievement data on US eighth-grade students from the Second International Mathematics Study (SIMS) carried out in 1982. The dataset contains the test results of 819 students from 179 classrooms in 112 schools.

Since the sampling of classrooms within schools is very thin, we consider the classrooms as aggregate units. Representation of the sampled classrooms in the data is sparse and very uneven. Ten classrooms have only one student each in the sample and 18 have two students each; the classroom with the largest representation has ten students in the sample.

The mathematics test consisted of 70 items, each scored right or wrong. The items are classified by subject matter into eight categories: ratio, proportion, and percent (eleven items), common and decimal fractions (twelve items), equations and expressions (ten items), integers and numbers (nine items); standard units and estimation (six items), area and volume (five items), coordinates and visualization (nine items), and plane figures (eight items). We consider the number of correct responses to the items within these categories as an 8-variate outcome for students. Note that some of the categories have small numbers of items which renders validity of the factor analysis contentious.

There are several causes of student-level variation: natural variation in ability, variation due to the nature of the test instrument, temporal variation in human performance, and others. Of these causes, variation due to the test instrument can be assessed by various measures of test reliability, and temporal variation could in principle be assessed by replication. While other causes are nuisance factors, natural variation is of considerable interest in educational research.

Sources of variation of the theoretical within-classroom means fall into two distinct categories: background and instruction. A classroom may be achieving better results because of the favourable (economic, social, intellectual) climate of the families in the school's recruitment area. The potential effect of instruction is well appreciated by all parties with a stake in the educational process. Better instruction results in students having better developed skills, abilities, and deeper knowledge and understanding of the subject of instruction.

The data enable differentiation of two sources of variation: students and schools (or classrooms). It appears that the two sources are unrelated; student-level variation refers to imperfection of the test instrument, natural variation among students, and variation within-students (over time or in replication), whereas classroom variation refers to the school climate, neighbourhood characteristics, and instruction. This provides a rationale for considering two unrelated factor structures underlying the student- and classroom/school-level variation. On the other hand, other, more complex and more difficult to describe, influences are at work which may operate at both levels of clustering.

The (total) sample variance matrix $\mathbf{S}$ has the decomposition $\mathbf{S} = \mathbf{S}_1 + \mathbf{S}_2$, with

$$S_1 = \begin{pmatrix} 5.034 & 2.630 & 2.000 & 2.194 & 1.376 & 1.246 & 1.451 & 1.700 \\ 2.630 & 5.125 & 2.019 & 1.951 & 1.327 & 1.207 & 1.322 & 1.594 \\ 2.000 & 2.019 & 2.966 & 1.397 & 0.873 & 0.783 & 0.937 & 1.000 \\ 2.194 & 1.951 & 1.397 & 3.154 & 0.939 & 0.930 & 1.098 & 1.208 \\ 1.376 & 1.327 & 0.873 & 0.939 & 1.603 & 0.507 & 0.765 & 0.741 \\ 1.246 & 1.207 & 0.783 & 0.930 & 0.507 & 1.418 & 0.671 & 0.741 \\ 1.451 & 1.322 & 0.937 & 1.098 & 0.765 & 0.671 & 2.266 & 0.890 \\ 1.700 & 1.594 & 1.000 & 1.208 & 0.741 & 0.741 & 0.890 & 2.623 \end{pmatrix}$$

(7.85)

and

$$S_2 = \begin{pmatrix} 2.949 & 3.315 & 2.786 & 2.476 & 1.484 & 1.132 & 1.437 & 1.897 \\ 3.315 & 3.824 & 3.161 & 2.847 & 1.683 & 1.282 & 1.527 & 2.090 \\ 2.786 & 3.161 & 3.255 & 2.651 & 1.505 & 1.196 & 1.525 & 1.908 \\ 2.476 & 2.847 & 2.651 & 2.330 & 1.301 & 1.039 & 1.301 & 1.549 \\ 1.484 & 1.683 & 1.505 & 1.301 & 0.853 & 0.569 & 0.739 & 0.963 \\ 1.132 & 1.282 & 1.196 & 1.039 & 0.569 & 0.571 & 0.596 & 0.771 \\ 1.437 & 1.527 & 1.525 & 1.301 & 0.739 & 0.596 & 0.836 & 0.933 \\ 1.897 & 2.090 & 1.908 & 1.549 & 0.963 & 0.771 & 0.933 & 1.377 \end{pmatrix}.$$

(7.86)

This decomposition corresponds to the model fit with unconstrained patterns of variation within and between classrooms (the saturated solution). Its deviance is 23 776.81. We consider first the exploratory factor analysis model with a single factor at each level. As the starting solution we take the factor analysis decompositions of the matrices S_1 and S_2; they are

$$\hat{\Lambda}_{1,0} = (1.716\ \ 1.632\ \ 1.136\ \ 1.240\ \ 0.791\ \ 0.723\ \ 0.867\ \ 0.979)^\top,$$
$$\hat{\Lambda}_{2,0} = (1.708\ \ 1.944\ \ 1.756\ \ 1.537\ \ 0.897\ \ 0.703\ \ 0.873\ \ 1.121)^\top,$$

with $\hat{\Theta}_{h,0} = \text{diag}\left(S_h - \hat{\Lambda}_{h,0}\hat{\Lambda}_{h,0}^\top\right)$. The deviance associated with this model fit is 23 837.18. Successive iterations of the Fisher scoring algorithm yield model fits with decrements of deviance 2.37, 0.136, 0.0175, 0.0026, 0.0004, and 5×10^{-5}, so that after seven cycles the iterations can be terminated. The resulting ML estimates are given in Table 7.1; the associated deviance is equal to 23 834.65. This is much higher than the deviance for the saturated model, 23 776.81, given by (7.85) and (7.86). The difference in the number of free parameters is $7 \times 8 - 4 \times 8 = 24$. Note that in terms of the deviance the starting solution has been improved only by about 2.5 points, and most of the improvement occurred in the first iteration. The 'observed–fitted' matrices $S_m - \hat{\Lambda}_m\hat{\Lambda}_m^\top - \hat{\Theta}_m$ have several large entries (0.2 or larger in absolute value) for both levels. We therefore proceed by

Table 7.1. Two-level factor analysis with single factors at each level. SIMS data. Maximum likelihood solution. Seven iterations were used. The deviance of the fitted model is 23 834.65

				Estimates				
Θ_1	2.102	2.502	1.668	1.625	1.002	0.885	1.502	1.666
Λ_1	1.725	1.641	1.133	1.236	0.795	0.722	0.865	0.984
Θ_2	0.095	0.232	0.297	0.078	0.074	0.108	0.113	0.179
Λ_2	1.727	1.942	1.770	1.551	0.898	0.707	0.879	1.121

fitting the factor analysis model with two factors at each level. The starting and the ML solutions are given in Table 7.2. The ML solution is only a modest improvement on the starting solution, even though a large number of iterations (16) were required to achieve convergence. The reduction of deviance due to the second factors is 26.27, for an addition of 14 parameters (a column of factor loadings at each level, minus one parameter each for the constraint of orthogonality). The elements of both 'observed–fitted' matrices are in the range –0.020 to 0.020, greatly reduced in comparison with the single factors model fit. Note that two of the uniquenesses at classroom level are equal to zero. They were constrained to zero after the third iteration when its estimate became negative and a matrix $\mathbf{G}_j$ had a negative eigenvalue (the deviance contains the term $\log(\det \mathbf{G}_j)$).

The first factors at both levels can be interpreted as measures of ability; their loadings are all positive, and roughly proportional to the square roots of the numbers of items within the sections of the test. The second factors are much less important. It is difficult to make an informal judgement about the hypothesis of common factor structures $\Lambda_1 = \Lambda_2$, because rotation of the factors can change the values of the matrices Λ_m substantially. The likelihood ratio test statistic for the hypothesis $\Lambda_1 = \Lambda_2$ is equal to 19.07 (null distribution χ^2 with 11 degrees of freedom); its p-value is equal to 6 per cent. Rotation of the factors at the two levels would facilitate a search for their suitable interpretation.

7.12 Hearing loss data

In this section we continue the analysis of the hearing loss data from Section 6.8 by exploring the covariance structure of hearing loss across the eight frequencies. We consider first the saturated factor analysis model with no constraints on the ear- and subject-level variance matrices, and then seek

Table 7.2.   Two-level factor analysis with two unrelated factors at each level. SIMS data. Moment (starting) and maximum likelihood solutions. The latter required 16 iterations. The asterisk * signifies that the parameter estimate was constrained to zero

	Starting solution							
Θ_1	2.108	2.377	1.474	1.603	0.989	0.880	1.438	1.637
Λ_1	1.709	1.628	1.156	1.240	0.783	0.726	0.877	0.972
	0.073	−0.309	−0.394	0.115	0.019	0.103	0.243	0.202
Θ_2	0.038	0.061	0.146	0.071	0.088	0.099	0.076	0.174
Λ_2	1.699	1.925	1.697	1.487	0.875	0.678	0.844	1.096
	0.158	0.236	−0.480	−0.217	−0.011	−0.109	−0.214	0.022
Deviance	23 811.18							

	Maximum likelihood solution							
Θ_1	2.044	−0.137	1.803	1.637	1.013	0.931	1.507	1.699
Λ_1	1.909	2.257	1.504	1.528	0.942	0.799	0.971	1.140
	−0.509	1.163	−0.198	−0.367	−0.140	−0.150	−0.314	−0.209
Θ_2	0.000*	0.206	0.000*	0.042	0.079	0.072	0.066	0.149
Λ_2	1.469	1.630	1.460	1.221	0.724	0.585	0.734	0.932
	0.303	0.235	−0.496	−0.093	−0.002	0.021	−0.088	0.069
Deviance	23 808.38							

a more parsimonious description in terms of some two-level factor analysis models with similar factor structures at the ear and subject levels.

The deviance for the saturated solution is equal to 192.99. The single factor decomposition for the saturated solution yields the vectors of factor loadings and uniquenesses

$$\Lambda_1^\top = (0.121, 0.101, 0.165, 0.170, 0.175, 0.127, 0.097, 0.118),$$
$$\Lambda_2^\top = (0.264, 0.324, 0.311, 0.344, 0.362, 0.342, 0.285, 0.279).$$

$$\Theta_1 = \mathrm{diag}(0.025, 0.056, 0.035, 0.022, 0.011, 0.020, 0.024, 0.036),$$
$$\Theta_2 = \mathrm{diag}(0.090, 0.148, 0.061, 0.037, 0.015, 0.022, 0.027, 0.048).$$

The deviance associated with this model fit is equal to 390.75. The ML solution has a deviance somewhat smaller, but it is still substantially higher than the 'benchmark' of 192.99. The deviance for the model fit with two factors at each level (obtained by decomposition of the saturated solution) is equal to 297.46. Again, even without iterative fitting we can conclude that the description of the underlying structure of the hearing loss by two factors at each level is not sufficient. The deviance for the model with three factors at each level is equal to 224.16, and the deviance for the ML solution (after iterative fitting) is equal to 217.73. The likelihood ratio test statistic for this model fit is equal to $217.73 - 192.99 = 24.74$, and has χ^2 null-distribution with $72 - 58 = 14$ degrees of freedom. It seems that only very complex factor analysis models can even remotely approach the 'benchmark' value of the saturated model deviance.

Next we consider the hypothesis that the factor loadings at ear- and subject-levels form the same linear space, that is, the respective variance (matrix) components are

$$\Sigma_1 = \Theta_1 + \Lambda\Lambda^\top,$$
$$\Sigma_2 = \Theta_2 + \Lambda\Psi_2\Lambda^\top.$$

This three-factor model contains only 43 free parameters, 15 fewer than the model with unrelated three-factor structures. The estimates of the 'common' factor loadings are:

$$\hat{\Lambda} = \begin{pmatrix} 0.150 & 0.293 & 0.239 & 0.139 & 0.082 & 0.063 & 0.053 & 0.040 \\ 0.051 & 0.149 & -0.157 & -0.006 & 0.117 & 0.116 & 0.117 & 0.107 \\ -0.004 & -0.242 & 0.152 & 0.151 & 0.135 & 0.122 & 0.079 & 0.127 \end{pmatrix}^\top,$$

with the estimated subject-level covariance matrix

$$\hat{\Psi}_2 = \begin{pmatrix} 1.72 & 1.05 & 1.17 \\ 1.05 & 1.25 & 1.10 \\ 1.17 & 1.10 & 1.12 \end{pmatrix}$$

and the uniquenesses

$$\hat{\Theta}_1 = \mathrm{diag}\,(0.025, 0., 0.028, 0.019, 0.014, 0.017, 0.024, 0.031)\,,$$
$$\hat{\Theta}_2 = \mathrm{diag}\,(0.056, 0., 0.028, 0.019, 0.017, 0.018, 0.024, 0.043)\,.$$

The model fit has several undesirable features. First, the uniquenesses for the second component at both levels had to be constrained to zero (Heywood cases). As a consequence the original identification constraint ($\Lambda^\top\Theta_2^{-1}\Lambda = 0$) had to be abandoned. Instead, we constrained three arbitrarily chosen elements of Λ ($\Lambda_{8,2}$, $\Lambda_{7,3}$, and $\Lambda_{8,3}$) to their starting values. After finding a ML solution the $\hat{\Lambda}$ obtained can be transformed to be orthogonal (transforming $\hat{\Psi}_2$ in the process).

The fitted covariance matrix $\hat{\boldsymbol{\Psi}}_2$ is almost singular; in particular the second and third factors are highly correlated. The deviance of the ML fit is 242.31, 49.32 higher than the deviance of the saturated model. This appears to be a suitable compromise between parsimony and adequacy of the fitted model. Rotation of the factor loadings at each level can be explored to find an appealing interpretation.

7.13   Bibliographical notes

The problems of analysis of clustered multivariate data with a complex description of dependence, such as structural equation models, were first addressed only recently. Muthén and Satorra (1989) provide an insight into the problem, and Goldstein and McDonald (1988) and McDonald and Goldstein (1989) discuss ML estimation for a general class of models.

Lee (1990) presents a general framework and estimation method for structural equation models. Model specification is in terms of the vector of expectations (regression) and the variance matrix of the observations. This approach is very general, but leaves the considerable task of defining the covariance structure to the analyst. A similar approach in a somewhat different context is pursued by Browne and du Toit (1992), who even delegate the task of providing routines for partial derivatives of the log-likelihood to the party directly involved in the data analysis. Computations are based on the Fisher scoring algorithm. These methods are computationally very intensive; developments in computational hardware and statistical software are likely to provide a strong impetus for this field in the near future.

8
GLM with random coefficients

8.1 Introduction

Generalized linear models (GLM) are an extension of the linear regression models beyond the realm of the normal distribution. Their unified formulation, as opposed to a set of distinct methods for different distributional assumptions, is due to Nelder and Wedderburn (1972) and Wedderburn (1974).

This chapter defines GLM with random coefficients and discusses procedures for their estimation. Section 8.2 summarizes GLM and prepares the ground for extensions for dependent observations in Section 8.4. Section 8.3 introduces an alternative way of defining a GLM using quasilikelihood. Section 8.5 discusses methods for approximating the log-likelihood for GLM and methods for ML estimation based on these approximations. Section 8.6 analyses the dependence of the information about variation on the nesting design. The chapter is concluded with two examples of clustered binary data analysis.

8.2 Models for independent observations

A GLM is specified by the outcome variable and three model components:

- distributional assumption;
- linear regression (predictor);
- relationship of the expectation of the outcome to the predictor (link function).

We assume that the observations $\{y_i\}_{i=1,N}$ are independent and have discrete or absolutely continuous distributions given by the density

$$f(y_i; \theta_i, \phi) \;=\; \exp\left\{\frac{y_i\theta_i - b(\theta_i)}{a(\phi)} + c(y_i; \phi)\right\}, \tag{8.1}$$

where a, b, and c are some real differentiable functions and ϕ is a parameter (called the *scale* or the *dispersion parameter*). The scale function a is positive and monotone; usually $a(\phi) = \phi$. In fact, no generality is lost by assuming that $a(\phi) = \phi$ ($\phi > 0$), because this can be arranged by reparametrization. The function $b(\theta)$ is twice differentiable, and its second-order derivative is positive. Note that this implies convexity of b. The role of the quantities $\{\theta_i\}$ is elaborated below. When the parameter ϕ is known the densities (8.1) belong to the exponential family.

The mean and variance of an observation are:

$$\mathbf{E}(y) = b'(\theta),$$

$$\text{var}(y) = b''(\theta)a(\phi),$$

(8.2)

where b' and b'' denote the first- and second-order derivatives of b with respect to θ. For proof, see Section 1.10. Note that the expectation is a strictly monotone differentiable function of θ.

If ϕ, or $a(\phi)$, is given, the expectation and the variance of y depend on the density (8.1) only through the function b. In some settings it may be advantageous to specify the distribution of the observations by the relationship of the expectation and the variance. The variance as a function of the expectation of an observation is denoted by V, so that $\text{var}(y) = V\{\mathbf{E}(y)\}$, or

$$V\{b'(\theta)\} = b''(\theta)a(\phi). \tag{8.3}$$

We assume that the expectation $b'(\theta)$ is a function of the linear predictor $\mathbf{x}\boldsymbol{\beta}$:

$$\eta\{\mathbf{E}(y)\} = \mathbf{x}\boldsymbol{\beta}, \tag{8.4}$$

or, equivalently,

$$b'(\theta) = \eta^{-1}(\mathbf{x}\boldsymbol{\beta}).$$

The function η is referred to as the *link* (or link function). For example, the ordinary regression model corresponds to the identity link, $\eta(\mu) = \mu$, with scale equal to the residual variance, $a(\phi) = \sigma^2$, $b(\theta) = \frac{1}{2}\theta^2$, and $c(y; \phi) = -\frac{1}{2}\{\log(2\pi\sigma^2) + y^2/\sigma^2\}$. In general, η is a strictly monotone function of the linear predictor $\nu = \mathbf{x}\boldsymbol{\beta}$.

GLMs are commonly referred to by hyphenating the link function and the distribution, e.g., log-Poisson, logit-binomial model, and so on, although some of the link functions are used almost exclusively with certain distributions (e.g., logit with binary or binomial, and reciprocal with gamma distributions).

Since any non-singular linear transformation of a link function can be compensated by the same transformation of the linear predictor $\mathbf{x}\boldsymbol{\beta}$, the link functions form classes of equivalence. By convention, one member of this class of equivalence is identified, such as the identity, the logarithm, or the logit function.

Frequently used examples include the logit model for binary outcomes and the log-Poisson model. In the logit model the outcomes have a binary distribution,

$$P(y = 1) = p,$$
$$P(y = 1) = 1 - p,$$

where $0 < p < 1$, so that $a(\phi) = 1$, $b(\theta) = -\log(1 - p) = \log\{1 + \exp(\theta)\}$, and $c(y, \phi) = 1$. The probability of outcome 1 is related to the linear predictor by the logit function

$$\mathbf{x}\boldsymbol{\beta} = \log\left(\frac{p}{1 - p}\right). \tag{8.5}$$

The log-Poisson model assumes outcomes with Poisson distribution,

$$P(y = k) = \frac{e^{-\lambda}\lambda^k}{k!}$$

($k = 0, 1, \dots$), and the parameter λ, which is both the expectation and the variance of the distribution, is related to the linear predictor by the logarithmic link

$$\mathbf{x}\boldsymbol{\beta} = \log(\lambda).$$

Note that for both logit-binary and log-Poisson models the linear predictor $\mathbf{x}\boldsymbol{\beta}$ coincides with the coefficient θ.

Throughout the chapter we assume that the scale ϕ is known. For many one-parameter distributions there is an obvious value for the scale, such as $a(\phi) = \phi = 1$ for the binary and Poisson distributions. For gamma distribution, though, there is no obvious choice for the scale parameter.

Many distributions have a limited range of expectations; for example the binary expectations are in the range $[0, 1]$, the Poisson expectations are always positive, and so on. The link function is usually chosen so as to map the range of possible expectations to the real axis. Then any value of the linear predictor can be realized.

The link function corresponding to linear dependence of θ on $\mathbf{x}\boldsymbol{\beta}$, that is, $\theta = \mathbf{x}\boldsymbol{\beta}$, is called the *canonical* link. For each distribution in the exponential family there is an essentially unique canonical link (a unique class of equivalence of functions). For example, identity is the canonical link for the normal distribution, logit for the binary (and binomial) distribution, and logarithm for the Poisson distribution.

In principle, any combination of the three model elements, the linear predictor, the distributional assumption, and the link, can be specified. In practice, each distributional assumption is associated with a small number of link functions that are usually considered (canonical link being one of them). The choice of the linear predictor is a problem analogous to its counterpart in ordinary regression.

ML estimation of the regression parameters can be carried out by standard methods, such as Newton–Raphson or Fisher scoring. The scoring vector for $\boldsymbol{\beta}$ is

$$\frac{\partial l}{\partial \boldsymbol{\beta}} = \frac{1}{a(\phi)} \sum_i \frac{\partial \theta_i}{\partial \nu_i} \{y_i - b'(\theta_i)\} \mathbf{x}_i , \tag{8.6}$$

where $\nu_i = \mathbf{x}_i \boldsymbol{\beta}$ is the linear predictor. For the canonical link we have $\partial \theta / \partial \nu \equiv 1$, and the scoring equations simplify to

$$\sum_i \{y_i - b'(\theta_i)\} \mathbf{x}_i = \mathbf{0}. \tag{8.7}$$

Canonical links are generally preferred because they lead to computationally simpler procedures (although sometimes that is the only reason for preferring them). There are distributions, however, for which the canonical link does not map the linear predictor on to the entire real axis $(-\infty, +\infty)$. A case in point is the gamma distribution; its canonical link is the reciprocal function. A negative value of the linear predictor is not admissible in this model.

The second-order partial derivatives of the log-likelihood are

$$\frac{\partial^2 l}{\partial \boldsymbol{\beta} \, \partial \boldsymbol{\beta}^\top} = \frac{-1}{a(\phi)} \sum_i \left[b''(\theta_i) \left(\frac{\partial \theta_i}{\partial \nu_i} \right)^2 - \{y_i - b'(\theta_i)\} \frac{\partial^2 \theta_i}{\partial \nu_i^2} \right] \mathbf{x}_i \mathbf{x}_i^\top. \tag{8.8}$$

The expected information matrix is

$$-\mathbf{E} \left(\frac{\partial^2 l}{\partial \boldsymbol{\beta} \partial \boldsymbol{\beta}^\top} \right) = \frac{1}{a(\phi)} \sum_i b''(\theta_i) \left(\frac{\partial \theta_i}{\partial \nu_i} \right)^2 \mathbf{x}_i \mathbf{x}_i^\top. \tag{8.9}$$

For the canonical link $\partial^2 \theta_i / \partial \nu_i^2 = 0$, and so (8.8) and (8.9) coincide and are equal to

$$-\mathbf{E} \left(\frac{\partial^2 l}{\partial \boldsymbol{\beta} \partial \boldsymbol{\beta}^\top} \right) = \frac{1}{a(\phi)} \sum_i b''(\theta_i) \mathbf{x}_i \mathbf{x}_i^\top . \tag{8.10}$$

Since (8.9) is a somewhat simpler expression than (8.8) the Fisher scoring algorithm is in general preferred to the Newton–Raphson method (for canonical links the algorithms coincide, however). A brief outline of these algorithms is given in Section 1.9.1, and their application to GLM, using (8.7) and (8.9), is, in principle, straightforward. Details are given in the next section.

Note that the parameter ϕ is not relevant for the estimation of the regression parameters $\boldsymbol{\beta}$, since it appears as a multiplicative factor in both the scoring and information functions, (8.6) and (8.9).

8.2.1   GENERALIZED LEAST SQUARES

We denote

$$w_i = b''(\theta_i) \left(\frac{\partial \theta_i}{\partial \nu_i} \right)^2,$$

$$e_i = w_i^{-1} \{y_i - b'(\theta_i)\} \frac{\partial \theta_i}{\partial \nu_i},$$

(8.11)

and $\mathbf{W} = \{w_i\} \otimes \mathbf{I}_N$ and $\mathbf{e} = \{e_i\} \otimes \mathbf{1}_N$. Using this notation, the Fisher scoring iteration has the form of weighted least squares:

$$\hat{\beta}_{new} = \hat{\beta}_{old} + \left(\mathbf{X}^\top \mathbf{W} \mathbf{X} \right)^{-1} \mathbf{X}^\top \mathbf{W} \mathbf{e},$$

(8.12)

see (8.6) and (8.9).

The quantities w_i and e_i are called the *generalized weights* and the *generalized residuals* respectively. The generalized weights $\mathbf{W}$ and residuals $\mathbf{e}$ depend on the linear predictor, and therefore also on the regression parameters, and so they have to be recalculated at every application of (8.12). The algorithm is referred to as the generalized least squares (GLS), or as the iteratively reweighted least squares.

Fitting a GLM involves iterations (except in the identity-normal case), and they require a starting solution. Instead of the starting values for the regression parameters β it suffices to provide starting values for the linear predictor vector $\nu = \mathbf{X}\beta$. Often it is practical to choose the linear predictor that matches the outcomes, even if it cannot be represented as a linear function of the regressor $\mathbf{x}$. The advantage of this choice is that it represents the best possible fit to the data, and the associated value of deviance, $-2l$, can be compared with the deviance of the ML fit for the adopted model. Of course, in the binary case it is not possible to match the outcomes with values of the linear predictor. In general, the constant linear predictor that corresponds to the sample mean of the observations can be used.

8.3   Quasilikelihood

For ML estimation it is not necessary to specify the underlying distribution; it suffices to provide all the functions that appear on the right-hand side of the identity for the scoring vector, (8.6), except for the scale parameter ϕ which is not relevant for the estimation of β. Instead of the function $b(\theta)$ it suffices to specify its derivative $b'(\theta)$. The linear predictor and the link are indispensible, of course, but they determine b' (as a function of $\mathbf{x}\beta$). To find the root of the scoring vector (8.6) it suffices to know b'', or, equivalently, the variance function V. Thus, an economic way of defining a GLM is by the linear predictor and the link and variance functions.

In many settings it is difficult to choose an appropriate distribution but information may be available about some of its properties or features. An alternative definition of the GLM which dispenses with the need to specify the distribution of the observations is in terms of the quasilikelihood. The quasilikelihood is essentially a representation of a density or of the contribution of an observation to the log-likelihood. It is defined by its partial derivative

$$\frac{\partial Q}{\partial \mu} = \frac{y - \mu}{V(\mu)}, \tag{8.13}$$

where y is the observed and μ the expected value of an observation. The definition of the quasilikelihood makes no reference to a distribution, although each distribution in the exponential family, combined with a link, has a quasilikelihood representation. However, there are forms of the quasilikelihood (8.13) which do not correspond to a distribution.

Being a representation (an approximation) of the log-likelihood, the quasilikelihood can be used for (approximate) likelihood ratio testing. Although the quasilikelihood is defined only by its derivative, and, therefore subject to an unknown additive constant, the value of this constant is not relevant because it is cancelled in any approximation to the likelihood ratio test statistic by the difference of two quasilikelihoods.

It is customary to set the value of the quasilikelihood to zero for a designated model. In many settings there is a *saturated* model which fits the data exactly (expectations equal to observed values), and the quasilikelihood for this model fit is declared equal to zero. For example, for the identity-normal model the saturated model corresponds to $\mathbf{X}\boldsymbol{\beta} = \mathbf{y}$, and the associated value of log-likelihood (assuming known residual variance σ^2) is $-\frac{1}{2}N\log(2\pi\sigma^2)$. Thus the quasilikelihood is obtained from the log-likelihood by subtracting this constant, and the value of -2 quasilikelihood for a model is equal to the likelihood ratio test statistic for the test of this model against the saturated model.

8.3.1 EXTENDED QUASILIKELIHOOD

The quasilikelihood (8.13) can be used for comparison of models with different linear predictors and/or links, but it does not allow comparisons of variance functions. Nelder and Pregibon (1987) defined the *extended quasilikelihood* with which variance functions (as well as links and linear predictors) can be compared using an approximation to the likelihood ratio test statistic (the quasilikelihood ratio). For a single observation the extended quasilikelihood is defined by the normal look-alike formula

$$Q^*(y; \mu) = -\frac{1}{2}\left[\log\{2\pi\phi V(y)\} + \frac{D(y; \mu)}{\phi}\right], \tag{8.14}$$

where $V(y)$ is the variance function, $\mu = \mathbf{E}(y)$, and $D(y; \mu)$ is the *deviance function*, defined as

$$D(y; \mu) = -2 \int_y^\mu \frac{y - \mu}{V(\mu)} du. \qquad (8.15)$$

In the original formulation the variance function may depend on a parameter, but this generalization is not important for our development. The deviance function is non-negative, and $D(\mu; \mu) = 0$; D is decreasing in $(-\infty, \mu)$ and increasing in $(\mu; +\infty)$. The extended quasilikelihood for a set of independent observations $\{y_i\}$ is defined as the sum of the quasilikelihoods (8.14):

$$Q^*(\mathbf{y}; \boldsymbol{\mu}) = \sum_i Q^*(y_i; \mu_i). \qquad (8.16)$$

The contribution (8.15) to the quasilikelihood (8.16) can be interpreted as a measure of discrepancy between the observation and its expected value.

For most familiar combinations of distribution and link the extended quasilikelihood is a very good approximation to the exact log-likelihood. For example, for the binomial, Poisson, and gamma distributions with their respective canonical links (logit, logarithm, and reciprocal) the exact log-likelihoods differ from the extended quasilikelihoods only in that the terms involving factorials or gamma functions are replaced by their Stirling approximations $(\Gamma(k+1) \approx (2\pi k)^{\frac{1}{2}} k^k e^{-k})$. For binary outcomes, or binomial outcomes with small denominators, the Stirling approximation is not satisfactory. This problem is resolved by a simple modification of the variance function. Unlike the original approximation, the modified Stirling approximation,

$$\Gamma(k+1) = \{2\pi(k+c)\}^{\frac{1}{2}} k^k e^{-k},$$

e.g. for $c = \frac{1}{6}$, is suitable for small k. Its application to the binomial log-likelihood results in the extended quasilikelihood (8.14) with the variance function

$$V(y; K) = \frac{(y+c)(K-y+c)}{N+c},$$

where K is the binomial denominator.

See Nelder and Pregibon (1987) for further details.

8.4 Models for clustered observations

Generalized linear models lend themselves to natural extensions to models with random coefficients. The extension described in this section generalizes the models for normally distributed outcomes.

We consider a random sample of clusters, indexed $j = 1, 2, \ldots, N_2$, with elementary units $i = 1, 2, \ldots, n_j$. Within each cluster the observations satisfy generalized linear regressions with a common distribution and link function. For the predictor in these models we consider linear functions $\mathbf{x}_{ij}\boldsymbol{\beta}_j$, where some components of $\boldsymbol{\beta}_j$ are identical across the clusters, and others vary:

$$(y_{ij} \mid \boldsymbol{\beta}_j) \ \sim \ \mathcal{F}\left\{\eta^{-1}(\mathbf{x}_{ij}\boldsymbol{\beta}_j)\right\}, \tag{8.17}$$

where $\mathcal{F}$ is the distribution given by the density (8.1). Instead of this density the quasilikelihood or the extended quasilikelihood representation may be used.

The set of regressions $\mathbf{x}_{ij}\boldsymbol{\beta}_j$ can be described by a single ANCOVA model. As in Chapters 4 and 6 we assume that the regression coefficients $\boldsymbol{\beta}_j$ are a random sample from a multivariate normal distribution,

$$\boldsymbol{\beta}_j \ \sim \ \mathcal{N}(\boldsymbol{\beta}, \ \boldsymbol{\Sigma}^*).$$

Regression coefficients constant across clusters correspond to zero variances in $\boldsymbol{\Sigma}^*$. Interpretation of the variance matrix $\boldsymbol{\Sigma}^*$ is similar to its counterpart in normal models – it describes the between-cluster variation of the regression coefficients. Some difficulties may arise due to non-linearity or the link function; the variance matrix $\boldsymbol{\Sigma}^*$ refers to the linear scale of the predictor (under the link).

To illustrate these models we consider clustered binary data $\{y_{ij}\}$ with no explanatory variables ($\mathbf{x}_{ij} \equiv 1$), and assume that the within-cluster (conditional) logits, given the random terms $\{\delta_j\}$, are

$$\text{logit}\{\mathbf{E}(y_{ij} \mid \delta_j)\} \ = \ \mu + \delta_j.$$

Marginally, we assume that $\delta_j \sim \mathcal{N}(0, \tau^2)$. If $\mu = 0$, then, owing to symmetry of the normal distribution and of the logit function, $\mathbf{E}(y_{ij}) = \text{logit}^{-1}(0) = \frac{1}{2}$. Otherwise, the expectation $\mathbf{E}(y_{ij})$, i.e., the marginal probability $P(y = 1)$, is not equal to the transformed value of the expectation $\text{logit}^{-1}(\mu)$. Since we have no elementary-level explanatory variables, the within-cluster totals $y_j = \sum_i y_{ij}$ have binomial conditional distributions,

$$\mathcal{B}\{n_j, \text{logit}^{-1}(\mu + \delta_j)\},$$

given δ_j. When $\tau^2 > 0$ their unconditional distributions are different from $\mathcal{B}\{n_j, \text{logit}^{-1}(\mu)\}$. In particular, the proportions y_j/n_j display more variation than would be expected if the elementary observations y_{ij} were mutually independent. This phenomenon motivated several researchers, e.g. Williams (1982), to define the variation in excess of that expected if the observations were independent as a measure of within-cluster homogeneity, or, equivalently, of between-cluster variation.

The terminology for the normal models extends to GLMs. We have the regression part of the model, given by the design matrix $\mathbf{X}$ formed by stacking the row-vectors of regressors, $\mathbf{X} = \mathbf{X}_j \otimes \mathbf{1}_{N_2}$ and $\mathbf{X}_j = \mathbf{x}_{ij} \otimes \mathbf{1}_{n_j}$, and the variation part of the model given by the matrix of regressors $\mathbf{Z}$, defined in analogy with $\mathbf{X}$, and the cluster-level variance matrix $\mathbf{\Sigma}^*$.

We will use the representation for the within-cluster regressions in terms of the variables included in the regression and variation parts:

$$\mathbf{x}_{ij}\boldsymbol{\beta} + \mathbf{z}_{ij}\mathbf{\Sigma}^{\frac{1}{2}}\boldsymbol{\delta}_j \,,$$

with random vectors $\boldsymbol{\delta}_j \sim \mathcal{N}(\mathbf{0}, \mathbf{I})$, i.i.d.; $\mathbf{\Sigma}$ is the submatrix of $\mathbf{\Sigma}^*$ corresponding to $\mathbf{z}$. We adopt the same conventions as for the minimal variation design in the normal case, and assume first that $\mathbf{\Sigma}$ is positive definite. Later we extend all the results to singular $\mathbf{\Sigma}$. We use the abbreviation RGLM for generalized linear models with random coefficients.

8.5   Maximum likelihood estimation

The observations within clusters are conditionally independent, given the random terms $\boldsymbol{\delta}_j$, and so the log-likelihood associated with $\mathbf{y}$ is

$$l(\boldsymbol{\beta}, \mathbf{\Sigma}; \mathbf{y}) = \sum_j \log \int \cdots \int P_j(\boldsymbol{\delta}_j)\Phi(\boldsymbol{\delta}_j)d\boldsymbol{\delta}_j \quad \left(= \sum_j l_j\right), \quad (8.18)$$

where $\Phi(\boldsymbol{\delta})$ is the density of the multivariate standard normal distribution, and $P_j(\boldsymbol{\delta}_j)$ is the conditional likelihood for cluster j, given the vector $\boldsymbol{\delta}_j$,

$$P_j(\boldsymbol{\delta}_j) = \prod_i f\left\{y_{ij}; \theta_i\left(\mathbf{x}_{ij}\boldsymbol{\beta} + \mathbf{z}_{ij}\mathbf{\Sigma}^{\frac{1}{2}}\boldsymbol{\delta}_j\right), a(\phi)\right\}; \quad (8.19)$$

we write θ_i as a function of the conditional linear predictor $\mathbf{x}_{ij}\boldsymbol{\beta}+\mathbf{z}_{ij}\mathbf{\Sigma}^{\frac{1}{2}}\boldsymbol{\delta}_j$. The integral in (8.18) is known to have a closed form only for the identity-normal model. Otherwise, we have to face the somewhat unpleasant task of numerical evaluation of these integrals. Note that the normal density $\Phi(\boldsymbol{\delta})$ in (8.18) can, in principle, be replaced by a different distribution. Particularly attractive, from the computational viewpoint, would be distributions for which the integral (8.18) can be expressed in closed form.

As an alternative to the model implied by the conditional likelihood (8.19) for the binary case with no elementary-level explanatory variables, we can posit a distribution for the conditional (within-cluster) probabilities. The beta-binomial model is a familiar example in which the integrated log-likelihood has a closed form. The outcomes are binary with constant (unknown) unconditional probability p, and the conditional within-cluster probabilities have a beta distribution, $\mathcal{B}(a, b)$, given by the density

$$f(x) = C(a,b)x^{a-1}(1-x)^{b-1} \qquad (0 < x < 1)$$
$$ = \qquad 0 \qquad\qquad \text{otherwise,} \tag{8.20}$$

where $C(a,b)$ is the reciprocal of the beta function,

$$C(a,b) = \frac{\Gamma(a+b)}{\Gamma(a)\Gamma(b)}.$$

Let $u_j = \sum_i y_{ij}$ be the number of 'successful' outcomes in cluster j. Then the marginal likelihood for cluster j is equal to

$$C(a,b) \int p^{u_j+a-1}(1-p)^{n_j-u_j+b-1} dp = \frac{C(a,b)}{C(u_j+a, n_j-u_j+b)}, \tag{8.21}$$

which, as a function of u_j, is the (probability) density of the hypergeometric distribution. This approach can be extended to accomodate regressors defined for clusters, but no elementary-level variables can be incorporated.

We describe two algorithms for ML estimation with RGLM.

8.5.1 DIRECT MAXIMIZATION

For the moment we ignore the computational complexity of equation (8.18), and proceed with its differentiation;

$$\frac{\partial l}{\partial \boldsymbol{\beta}} = \sum_j \exp(-l_j) \int \cdots \int P_j(\boldsymbol{\delta}) \mathbf{s}_{x,j}(\boldsymbol{\delta}) \boldsymbol{\Phi}(\boldsymbol{\delta}) d\boldsymbol{\delta}, \tag{8.22}$$

where l_j is the summand of the log-likelihood in (8.18) corresponding to cluster j, and

$$\mathbf{s}_{x,j}(\boldsymbol{\delta}) = \sum_i \left\{ y_{ij} - b'\left(\mathbf{x}_{ij}\boldsymbol{\beta} + \mathbf{z}_{ij}\boldsymbol{\Sigma}^{\frac{1}{2}}\boldsymbol{\delta} \right) \right\} \mathbf{x}_{ij}.$$

For a parameter involved in $\boldsymbol{\Sigma}^{\frac{1}{2}}$ (a covariance structure parameter) we have

$$\frac{\partial l}{\partial \tau} = \sum_j \exp(-l_j) \int \cdots \int P_j(\boldsymbol{\delta}) \mathbf{s}_{z,j}(\boldsymbol{\delta}) \frac{\partial \boldsymbol{\Sigma}^{\frac{1}{2}}}{\partial \tau} \boldsymbol{\delta} \boldsymbol{\Phi}(\boldsymbol{\delta}) d\boldsymbol{\delta}, \tag{8.23}$$

where

$$\mathbf{s}_{z,j}(\boldsymbol{\delta}) = \sum_i \left\{ y_{ij} - b'\left(\mathbf{x}_{ij}\boldsymbol{\beta} + \mathbf{z}_{ij}\boldsymbol{\Sigma}^{\frac{1}{2}}\boldsymbol{\delta} \right) \right\} \mathbf{z}_{ij}.$$

The negative second-order partial derivatives are

$$-\frac{\partial^2 l}{\partial\boldsymbol{\beta}\,\partial\boldsymbol{\beta}^\top} = -\sum_j \frac{\partial l_j}{\partial\boldsymbol{\beta}}\frac{\partial l_j}{\partial\boldsymbol{\beta}^\top}$$

$$+ \sum_j \exp(-l_j)\int\cdots\int P_j(\boldsymbol{\delta})\mathbf{S}_{xx,j}(\boldsymbol{\delta})\Phi(\boldsymbol{\delta})d\boldsymbol{\delta}, \qquad (8.24)$$

$$-\frac{\partial^2 l}{\partial\boldsymbol{\beta}\,\partial\boldsymbol{\tau}} = -\sum_j \frac{\partial l_j}{\partial\boldsymbol{\beta}}\frac{\partial l_j}{\partial\boldsymbol{\tau}}$$

$$+ \sum_j \exp(-l_j)\int\cdots\int P_j(\boldsymbol{\delta})\mathbf{S}_{xz,j}(\boldsymbol{\delta})\frac{\partial\boldsymbol{\Sigma}^{\frac{1}{2}}}{\partial\boldsymbol{\tau}}\boldsymbol{\delta}\,\Phi(\boldsymbol{\delta})d\boldsymbol{\delta},$$

$$(8.25)$$

and

$$-\frac{\partial^2 l}{\partial\tau_1\,\partial\tau_2} = -\sum_j \frac{\partial l_j}{\partial\tau_1}\frac{\partial l_j}{\partial\tau_2} + \sum_j \exp(-l_j)$$

$$\times \int\cdots\int P_j(\boldsymbol{\delta})\left\{\boldsymbol{\delta}^\top\frac{\partial\boldsymbol{\Sigma}^{\frac{1}{2}}}{\partial\tau_1}\mathbf{S}_{zz,j}(\boldsymbol{\delta})\frac{\partial\boldsymbol{\Sigma}^{\frac{1}{2}}}{\partial\tau_2} - \mathbf{s}_{z,j}(\boldsymbol{\delta})\frac{\partial^2\boldsymbol{\Sigma}^{\frac{1}{2}}}{\partial\tau_1\partial\tau_2}\right\}\boldsymbol{\delta}\,\Phi(\boldsymbol{\delta})d\boldsymbol{\delta},$$

$$(8.26)$$

where

$$\begin{aligned}
\mathbf{S}_{xx,j} &= \mathbf{X}_j^\top\mathbf{W}_j(\boldsymbol{\delta})\mathbf{X}_j - \mathbf{s}_{x,j}(\boldsymbol{\delta})\mathbf{s}_{x,j}^\top(\boldsymbol{\delta}), \\
\mathbf{S}_{xz,j} &= \mathbf{X}_j^\top\mathbf{W}_j(\boldsymbol{\delta})\mathbf{Z}_j - \mathbf{s}_{x,j}(\boldsymbol{\delta})\mathbf{s}_{z,j}^\top(\boldsymbol{\delta}), \\
\mathbf{S}_{zz,j} &= \mathbf{Z}_j^\top\mathbf{W}_j(\boldsymbol{\delta})\mathbf{Z}_j - \mathbf{s}_{z,j}(\boldsymbol{\delta})\mathbf{s}_{z,j}^\top(\boldsymbol{\delta}).
\end{aligned}$$

The general outline of the Newton–Raphson algorithm (Section 1.9.1) can now be followed. The GLS provides a natural starting solution for the regression parameters $\boldsymbol{\beta}$; for the elements of $\boldsymbol{\Sigma}^{\frac{1}{2}}$ any non-singular matrix can be used (e.g., the identity matrix). There is some advantage in choosing a diagonal starting solution for $\boldsymbol{\Sigma}^{\frac{1}{2}}$.

The integrals in the scoring vector and the sample information matrix can be approximated by Gaussian quadrature. For most purposes 5-point quadrature suffices, especially when the number of clusters and the cluster sizes are moderate. Nevertheless, the computational load increases very rapidly with the number of random coefficients, and multiple evaluation of third or higher order integrals may not be feasible.

The convergence of the Newton–Raphson algorithm is very rapid for well-identified models. Usually three to nine iterations are sufficient to achieve very high accuracy.

8.5.2 APPROXIMATION OF THE INTEGRAND

The basic idea of this approach is to approximate the logarithms of the integrands in (8.18) by quadratic functions. Then the integrals can be evaluated analytically and the log-likelihood approximated by an expression in closed form.

The conditional log-likelihood function $\log P_j(\boldsymbol{\delta})$ has the Taylor expansion around $\boldsymbol{\delta} = \mathbf{0}$

$$\log P_j(\boldsymbol{\delta}) \;=\; \log P_j(\mathbf{0}) \;+\; \frac{1}{\phi}\mathbf{s}_{z,j}(\mathbf{0})$$

$$-\sum_{k=2}\sum_i \frac{\left(\mathbf{z}_{ij}\boldsymbol{\Sigma}^{\frac{1}{2}}\boldsymbol{\delta}\right)^k}{k!\phi}\; \frac{\partial^{k-2}w_{ij}}{\left\{\partial\left(\mathbf{z}_{ij}\boldsymbol{\Sigma}^{\frac{1}{2}}\boldsymbol{\delta}\right)\right\}^{k-2}}\Bigg|_{\boldsymbol{\delta}=\mathbf{0}}, \quad (8.27)$$

and so each summand in (8.18) can be approximated as

$$l_j(\boldsymbol{\beta}, \boldsymbol{\Sigma}; \mathbf{y}_j) \approx \log\{P_j(\mathbf{0})\} - \frac{p}{2}\log(2\pi) - \int \cdots \int \exp\Bigg[\frac{1}{\phi^2}\mathbf{s}_{z,j}(\mathbf{0})\boldsymbol{\Sigma}^{\frac{1}{2}}\boldsymbol{\delta}$$

$$-\frac{1}{2}\boldsymbol{\delta}^\top\boldsymbol{\Sigma}^{\frac{1}{2}}\left(\boldsymbol{\Sigma}^{-1}+\frac{1}{\phi}\mathbf{Z}_j^\top\mathbf{W}_j\mathbf{Z}_j\right)\boldsymbol{\Sigma}^{\frac{1}{2}}\boldsymbol{\delta}\Bigg]d\boldsymbol{\delta}. \quad (8.28)$$

The approximation is exact when $\log P_j(\boldsymbol{\delta})$ is a quadratic function, as in the identity-normal case. The approximation (8.28) is likely to be most imprecise when $\boldsymbol{\Sigma}$ contains large variances.

For brevity we will omit the arguments of the functions P_j, $\mathbf{W}_j$, $\mathbf{s}_{z,j}$, and $\mathbf{s}_{x,j}$ whenever $\boldsymbol{\delta}_j = \mathbf{0}$. Thus $\mathbf{s}_{z,j} = \mathbf{e}_j^\top\mathbf{W}_j\mathbf{Z}_j$ and $\mathbf{s}_{x,j} = \mathbf{e}_j^\top\mathbf{W}_j\mathbf{X}_j$, where $\mathbf{e}_j = \{e_{ij}\}\otimes\mathbf{1}_{n_j}$ is the vector of generalized residuals defined by (8.11). The right-hand side of (8.28) can be expressed in closed form; by completing the square in the exponential of the integrand we obtain

$$\frac{1}{\phi^2}\mathbf{s}_{z,j}\boldsymbol{\Sigma}^{\frac{1}{2}}\boldsymbol{\delta} - \frac{1}{2}\boldsymbol{\delta}^\top\boldsymbol{\Sigma}^{\frac{1}{2}}\left(\boldsymbol{\Sigma}^{-1}+\frac{1}{\phi}\mathbf{Z}_j^\top\mathbf{W}_j\mathbf{Z}_j\right)\boldsymbol{\Sigma}^{\frac{1}{2}}\boldsymbol{\delta}$$

$$= \frac{1}{\phi^2}\mathbf{s}_{z,j}\mathbf{H}_j^{-1}\mathbf{s}_{z,j}^\top - \frac{1}{2}\left(\mathbf{H}_j^{\frac{1}{2}}\boldsymbol{\Sigma}^{\frac{1}{2}}\boldsymbol{\delta} - \mathbf{s}_{z,j}\mathbf{H}_j^{-\frac{1}{2}}\right)^\top\left(\mathbf{H}_j^{\frac{1}{2}}\boldsymbol{\Sigma}^{\frac{1}{2}}\boldsymbol{\delta} - \mathbf{s}_{z,j}\mathbf{H}_j^{-\frac{1}{2}}\right),$$

$$(8.29)$$

where $\mathbf{H}_j = \boldsymbol{\Sigma}^{-1}+\phi^{-1}\mathbf{Z}_j^\top\mathbf{W}_j\mathbf{Z}_j$. The density of the distribution

$$\mathcal{N}_p\left(\mathbf{s}_{z,j}\mathbf{H}_j^{-1}\boldsymbol{\Sigma}^{-\frac{1}{2}}, \mathbf{H}_j^{\frac{1}{2}}\boldsymbol{\Sigma}\mathbf{H}_j^{\frac{1}{2}}\right)$$

integrates to unity, that is,

$$\int \cdots \int \exp\left\{-\frac{1}{2}\left(\mathbf{H}_j^{\frac{1}{2}}\boldsymbol{\Sigma}^{\frac{1}{2}}\boldsymbol{\delta} - \mathbf{H}_j^{-\frac{1}{2}}\mathbf{s}_{z,j}^\top\right)^\top \left(\mathbf{H}_j^{\frac{1}{2}}\boldsymbol{\Sigma}^{\frac{1}{2}}\boldsymbol{\delta} - \mathbf{H}_j^{-\frac{1}{2}}\mathbf{s}_{z,j}^\top\right)\right\} d\boldsymbol{\delta}$$

$$= \{(2\pi)^p \det(\boldsymbol{\Sigma}\mathbf{H})\}^{\frac{1}{2}},$$

and so

$$l(\boldsymbol{\beta}, \boldsymbol{\Sigma}; \mathbf{y}) \approx \sum_j \log P_j - \frac{N}{2}\log(2\pi\phi)$$

$$- \frac{1}{2}\sum_j \left[\log\{\det(\boldsymbol{\Sigma}\mathbf{H}_j)\} + \phi^{-1}\mathbf{s}_{z,j}\mathbf{H}_j^{-1}\mathbf{s}_{z,j}^\top\right]. \qquad (8.30)$$

In order to extend this formula for singular variance matrices $\boldsymbol{\Sigma}$ we denote $\mathbf{G}_j = \boldsymbol{\Sigma}\mathbf{H}_j$ ($= \mathbf{I} + \phi^{-1}\mathbf{Z}_j^\top\mathbf{W}_j\mathbf{Z}_j\boldsymbol{\Sigma}$) and express (8.30) in terms of $\mathbf{G}_j^{-1}$ and $\det\mathbf{G}_j$:

$$l(\boldsymbol{\beta}, \boldsymbol{\Sigma}; \mathbf{y}) \approx \sum_j \log P_j - \frac{N}{2}\log(2\pi\phi)$$

$$- \frac{1}{2}\sum_j \left\{\log(\det\mathbf{G}_j) + \phi^{-1}\mathbf{s}_{z,j}\boldsymbol{\Sigma}\mathbf{G}_j^{-1}\mathbf{s}_{z,j}^\top\right\}. \qquad (8.31)$$

For the identity-normal case, $\mathbf{G}_j$ coincides with the $\mathbf{G}_j$ defined in Section 4.4. Since all the eigenvalues of $\mathbf{G}_j$ are positive, this expression is well defined even for singular $\boldsymbol{\Sigma}$; in fact we can consider this approximation for any matrix $\boldsymbol{\Sigma}$ for which all the matrices $\mathbf{G}_j$ are positive definite. Of course, for the normal case (8.31) holds exactly. The first component of (8.30), $\sum_j \log P_j$, can be replaced by the quasilikelihood or the extended quasilikelihood.

8.5.3 APPROXIMATE FISHER SCORING ALGORITHM

An approximate MLE can be defined as the exact or an approximate maximizer of the approximate log-likelihood (8.31). Just as in the normal case, it is advantageous to use the scaled variance (matrix) parametrization $(\phi, \boldsymbol{\Omega}) = (\phi, \phi^{-1}\boldsymbol{\Sigma})$. We denote the approximate log-likelihood, the right-hand side of (8.31), by l^*. Its first-order partial derivative with respect to a parameter involved in $\boldsymbol{\Sigma}$ is

$$\frac{\partial l^*}{\partial\omega} = -\frac{1}{2}\sum_j \mathrm{tr}\left(\mathbf{G}_j^{-1}\frac{\partial\mathbf{G}_j}{\partial\omega}\right) + \frac{1}{2\phi}\sum_j \mathbf{s}_{z,j}\mathbf{G}_j^{-1}\frac{\partial\mathbf{G}_j}{\partial\omega}\left(\mathbf{G}_j^{-1}\right)^\top\mathbf{s}_{z,j}^\top,$$

$$(8.32)$$

and $\partial \mathbf{G}_j / \partial \omega = \mathbf{Z}_j^\top \mathbf{W}_j \mathbf{Z}_j \, \partial \mathbf{\Omega} / \partial \omega$. No generality is lost by assuming that all the covariance structure parameters (those involved in $\mathbf{\Sigma}$ or $\mathbf{\Omega}$) are linear in $\mathbf{\Omega}$, so that $\partial \mathbf{\Sigma} / \partial \omega$ and $\partial \mathbf{\Omega} / \partial \omega$ are constant. Partial derivatives for a different parametrization can be obtained by applying the chain rule. If ω is a scaled half-variance or a scaled covariance, then $\partial \mathbf{\Omega} / \partial \omega$ is an incidence matrix;

$$\frac{\partial \mathbf{\Omega}}{\partial \omega} = \mathbf{\Delta}_i \mathbf{\Delta}_j^\top + \mathbf{\Delta}_j \mathbf{\Delta}_i^\top .$$

Leaving out the term containing $\partial^2 \mathbf{\Omega} / (\partial \omega_1 \partial \omega_2)$, for a pair of covariance structure parameters we have

$$\begin{aligned}
\frac{\partial^2 l^*}{\partial \omega_1 \, \partial \omega_2} &= \frac{1}{2\phi} \sum_j \mathrm{tr} \left\{ \mathbf{G}_j^{-1} \frac{\partial \mathbf{\Omega}}{\partial \omega_1} \left(\mathbf{G}_j^{-1} \right)^\top \frac{\partial \mathbf{\Omega}}{\partial \omega_2} \right\} \\
&\quad - \frac{1}{2} \sum_j \mathbf{s}_{z,j} \mathbf{G}_j^{-1} \frac{\partial \mathbf{\Omega}}{\partial \omega_1} \left\{ \mathbf{G}_j^{-1} + \left(\mathbf{G}_j^{-1} \right)^\top \right\} \frac{\partial \mathbf{\Omega}}{\partial \omega_2} \left(\mathbf{G}_j^{-1} \right)^\top \mathbf{s}_{z,j}^\top .
\end{aligned}$$

$$(8.33)$$

The regression parameters $\boldsymbol{\beta}$ are involved in both $\mathbf{s}_{z,j}$ and $\mathbf{G}_j$, although the latter depends on $\boldsymbol{\beta}$ only through the matrix of weights $\mathbf{W}$. If we disregard this dependence (set $\partial \mathbf{G}_j / \partial \boldsymbol{\beta} = \mathbf{0}$), we obtain the following approximations for the scoring vector and the information matrix for $\boldsymbol{\beta}$:

$$\frac{\partial l^*}{\partial \boldsymbol{\beta}} \approx \phi^{-1} \mathbf{X}^\top \mathbf{W} \mathbf{e} - \phi^{-1} \sum_j \mathbf{X}_j^\top \mathbf{W}_j \mathbf{Z}_j \mathbf{\Omega} \mathbf{G}_j^{-1} \mathbf{s}_{z,j} \qquad (8.34)$$

and

$$-\frac{\partial l^{*\,2}}{\partial \boldsymbol{\beta} \partial \boldsymbol{\beta}^\top} \approx \phi^{-1} \mathbf{X}^\top \mathbf{W} \mathbf{X} - \phi^{-1} \sum_j \mathbf{X}_j^\top \mathbf{W}_j \mathbf{Z}_j \mathbf{\Omega} \mathbf{G}_j^{-1} \mathbf{Z}_j^\top \mathbf{W}_j \mathbf{X}_j .$$

$$(8.35)$$

We define the generalized variance matrix for cluster j:

$$\mathbf{V}_j = \phi \left(\mathbf{W}_j^{-1} + \mathbf{Z}_j \mathbf{\Omega} \mathbf{Z}_j^\top \right), \qquad (8.36)$$

and $\mathbf{V} = \mathbf{V}_j \otimes \mathbf{I}_{N_2}$. For the normal case $\mathbf{V}$ coincides with the variance matrix for the observations. The inverse and the determinant of $\mathbf{V}_j$ are

$$\mathbf{V}_j^{-1} = \phi^{-1} \mathbf{W}_j - \phi^{-1} \mathbf{W}_j \mathbf{Z}_j \mathbf{\Omega} \mathbf{G}_j^{-1} \mathbf{Z}_j^\top \mathbf{W}_j , \qquad (8.37)$$

$$\det \mathbf{V}_j \;=\; \phi^{n_j} \det \mathbf{W}_j^{-1} \det \mathbf{G}_j \,, \tag{8.38}$$

and so the approximate Fisher scoring algorithm based on (8.34) and (8.35) can be expressed in a concise form of GLS as

$$\hat{\boldsymbol{\beta}}_{new} \;=\; \hat{\boldsymbol{\beta}}_{old} + \left(\mathbf{X}^{\top}\mathbf{V}^{-1}\mathbf{X}\right)^{-1}\mathbf{X}^{\top}\mathbf{V}^{-1}\mathbf{e}. \tag{8.39}$$

Note that the approximate sample information matrix for $\boldsymbol{\beta}$, (8.35), is equal to $\mathbf{X}^{\top}\mathbf{V}^{-1}\mathbf{X}$. The form of the corrections (8.39) is identical to the Fisher scoring corrections for the normal random coefficient model, with the generalized variance matrix $\mathbf{V}$ and the generalized residuals $\mathbf{e}$ in place of their counterparts for the normal case.

The scoring function for a covariance structure parameter can be expressed in terms of $\mathbf{V}$ as

$$\frac{\partial l^*}{\partial \omega} \;=\; -\frac{\phi}{2}\sum_j \operatorname{tr}\left(\mathbf{V}_j^{-1}\frac{\partial \boldsymbol{\Omega}}{\partial \omega}\right) + \frac{\phi}{2}\sum_j \mathbf{e}_j^{\top}\mathbf{V}_j^{-1}\frac{\partial \boldsymbol{\Omega}}{\partial \omega}\mathbf{V}_j^{-1}\mathbf{e}_j \,, \tag{8.40}$$

and the negative second-order partial derivatives as

$$\begin{aligned}
-\frac{\partial^2 l^*}{\partial \omega_1 \partial \omega_2} \;=\; &-\frac{\phi^2}{2}\sum_j \operatorname{tr}\left(\mathbf{V}_j^{-1}\frac{\partial \boldsymbol{\Omega}}{\partial \omega_1}\mathbf{V}_j^{-1}\frac{\partial \boldsymbol{\Omega}}{\partial \omega_2}\right) \\
&+\; \phi^2 \sum_j \mathbf{e}_j^{\top}\mathbf{V}_j^{-1}\frac{\partial \boldsymbol{\Omega}}{\partial \omega_1}\mathbf{V}_j^{-1}\frac{\partial \boldsymbol{\Omega}}{\partial \omega_2}\mathbf{V}_j^{-1}\mathbf{e}_j \,,
\end{aligned} \tag{8.41}$$

both of which coincide with their respective normal counterparts (2.18) and (2.19).

To extend the similarity of the normal procedure with that for the exponential family (or quasilikelihood) we consider the information matrix associated with $(\boldsymbol{\beta}, \boldsymbol{\Omega})$. Unlike the normal case, owing to the non-linearity of the link, and non-symmetry of η^{-1} as a function of $\boldsymbol{\delta}_j$, the generalized residuals do not have zero expectations:

$$\mathbf{E}(y_{ij}) \;=\; \int_{-\infty}^{+\infty}\int \cdots \int y f\left\{y; \eta^{-1}(\mathbf{x}_{ij}\boldsymbol{\beta} + \mathbf{z}_{ij}\boldsymbol{\Sigma}^{\frac{1}{2}}\boldsymbol{\delta})\right\} d\boldsymbol{\delta}dy \;\neq\; \eta^{-1}(\mathbf{x}_{ij}\boldsymbol{\beta}). \tag{8.42}$$

However, the delta method can be used to show that equality in (8.42) holds approximately.

The expectation of the approximate scoring function (8.34) is close to zero, and therefore

$$\operatorname{var}(\mathbf{e}_j) \;\approx\; \mathbf{V}_j \,, \tag{8.43}$$

and so the expected information matrix is

$$-\mathbf{E}\left(\frac{\partial l^{*\,2}}{\partial \omega_1 \partial \omega_2}\right) \approx \frac{\phi^2}{2}\sum_j \text{tr}\left(\mathbf{V}_j^{-1}\frac{\partial \mathbf{\Omega}}{\partial \omega_1}\mathbf{V}_j^{-1}\frac{\partial \mathbf{\Omega}}{\partial \omega_2}\right). \tag{8.44}$$

Similarity of the Fisher scoring equations for GLMs with random coefficients to their normal counterparts enables us to adapt any implementation of a ML algorithm for normal random coefficient models for GLMs. The adaptation can be summarized by the following steps:

1. Use GLS in place of OLS to obtain a starting solution.
2. Apply the Fisher scoring iterations (Section 4.4), with the residuals replaced by the generalized residuals, and all the cross-products $(\mathbf{e}_j, \mathbf{X}_j)^{\top}(\mathbf{e}_j, \mathbf{X}_j)$ replaced by the weighted cross-products

$$(\mathbf{e}_j, \mathbf{X}_j)^{\top}\mathbf{W}_j(\mathbf{e}_j, \mathbf{X}_j).$$

In most cases comparison of the GLS fit (corresponding to $\mathbf{\Omega} = \mathbf{0}$) with the random coefficient model fit is of interest, and so GLS as a starting solution is also useful for model comparison with the RGLM solution to assess the importance (significance) of between-cluster variation. Solely for the purpose of defining a suitable starting solution for the Fisher scoring algorithm, one or two iterations of GLS would suffice. Note that the role of the normal elementary-level variance is played by the scale. Since the scale is fixed, it is not estimated.

Of course, the approximation in (8.31) can be used for the estimation of the scale ϕ. Using the $(\phi, \mathbf{\Omega})$ parametrization the scale ϕ can be extracted from the approximate scoring function for ϕ. Given all the other model parameters, it has the unique root

$$\hat{\phi} = \frac{1}{N}\left(\mathbf{e}^{\top}\mathbf{e} - \sum_j \mathbf{e}_j^{\top}\hat{\mathbf{\Omega}}\mathbf{G}_j^{-1}\mathbf{e}_j\right). \tag{8.45}$$

The complexity of the approximate Fisher scoring algorithm can be compared to that of the (normal) Fisher scoring algorithm for the same dataset. In the former the starting solution requires iterations, but for the sole purpose of fitting random coefficient models, one iteration may be sufficient. In the (approximate) Fisher scoring iterations the weighted cross-products have to be recalculated each time the linear predictor (or, equivalently, $\boldsymbol{\beta}$) is altered in the previous iteration, whereas in the normal Fisher scoring procedure all the required cross-products are calculated only once (e.g., in the process of evaluating the items for OLS). Also, in the normal case the residuals $\{e_{ij}\}$ are not required, because their cross-products with $\mathbf{X}$ and $\mathbf{Z}_j$ can be obtained directly from the other cross-products. For GLM the generalized residuals have to be calculated at each iteration, unless the linear predictor was not altered.

In the Fisher scoring algorithm any reasonable parametrization for the covariance structure parameters can be used. If the variance matrix Ω is estimated and its ML estimate is positive definite, intermediate solutions $\hat{\Omega}$ are less likely to have a negative eigenvalue when the starting solution is a positive definite diagonal matrix. If negative eigenvalues do occur, they can be removed by step-halving.

8.6   Information about variation

The approximate information matrix given by (8.44) is essentially in a closed form which can be used to discuss a number of issues related to the estimation of covariance structure parameters in RGLM models. Since its form is similar to that for normal random coefficient models, precision and sample size issues for RGLM can be related to those for normal random coefficient models.

For the moment, supose that all linear predictors $\eta_{ij} = \mathbf{x}_{ij}\boldsymbol{\beta}$ are identical (equal to η), or that their variation can be ignored. Let w be the generalized weight common to all the observations. Note that for the normal case with $\sigma^2 = 1$ the weight is $w = 1$. The approximate information matrix in (8.44) is a function of the 'generalized' crossproducts $w(\mathbf{X}_j, y_j)^\top \mathbf{Z}_j$. Thus, information contained in a given RGLM (dataset) is approximately equivalent to that in a normal model with the same number of clusters in which each linear predictor $\mathbf{x}_{ij}\boldsymbol{\beta}$ is equal to η, and the cluster sizes are w multiples of the cluster sizes in the RGLM model.

Such a comparison is of particular importance for binary/binomial data. Suppose $w = 1/6$ (corresponding to binary data with $p = 0.79$ or $p = 0.21$). Then a particular dataset with normal outcomes and $\eta = \text{logit}(0.79)$ corresponds to its binary counterpart dataset with within-cluster designs replicated six times: $\mathbf{X}_j \otimes \mathbf{1}_6$ and $\mathbf{Z}_j \otimes \mathbf{1}_6$. This comparison is approximate, and is meaningful only for large cluster sizes; however, it illustrates the paucity of information about variation in binary data.

This illustration can be extended to data with a non-constant linear predictor by matching the design for normal data with one for binary data in which each normal observation y_{ij} is replaced by $1/w_{ij}$ binary observations in the same cluster with the same vector of explanatory variables $\mathbf{x}_{ij}$. This implies that in order to make an inference about variation in clustered binary data, much more extensive data are required than in the normal case, and that modelling of complex patterns of variation (random slopes) is usually not feasible. Since a typical clustered binary dataset (e.g., with cluster sizes 10–40) corresponds to a normal dataset with very thin clustered sampling, the GLS fit is often a good approximation to the ML (as OLS is in the normal case). A similar approach can be used to assess information about regression parameters in RGLM.

8.7 Restricted maximum likelihood

The issue of unbiased estimation of the covariance structure parameters carries over from normal random coefficient models to RGLM. The motivation by estimation based on a set of error contrasts does not extend to the exponential family of distributions because, unlike the normal distribution, these distributions are not closed with respect to arithmetic operations. Integrating out the regression parameters from the joint likelihood is problematic just when it would be most useful to account for the 'degrees of freedom'. The advantages of REML are most important when there are a relatively large number of regression parameters, and then numerical integration of the likelihood is most computer intensive (and subject to numerical inaccuracy). In this section we describe a method for approximate REML estimation with RGLM models which involves no numerical integration other than that required for ML estimation.

For an RGLM dataset we define the restricted maximum likelihood by the expression

$$l_R = l_F + R, \tag{8.46}$$

where l_F is the log-likelihood defined by (8.18) and

$$R = -\frac{1}{2} \log \left\{ \det \left(\mathbf{X}^\top \mathbf{V}^{-1} \mathbf{X} \right) \right\}.$$

The log-likelihood l_F can be replaced by the quasilikelihood, or by one of its approximations derived in Section 8.5. The definition in (8.46) is motivated by the analogy of the models and methods of estimation between normal random coefficient models and RGLMs. Its properties were explored in a modest simulation study reported in Longford (1993), and it appears that addition of the term R to the log-likelihood reduces the bias of the estimator of the between-cluster variance in logit-binary regression, with no perceptible loss of efficiency vis-à-vis the full MLE.

The implementation of REML for RGLM is straightforward. The first-order partial derivatives with respect to the covariance structure parameters have to be adjusted by the derivatives of the correction term R in (8.46). These can be evaluated in complete analogy with the normal case (see Sections 4.4 and 4.7). As in the derivation for the approximate ML method we ignore the dependence of the generalized variance matrix $\mathbf{V}$ on the linear predictor. Thus, REML estimation of the regression parameters involves the same GLS equation as does ML estimation. In practice it is not necessary to adjust the information matrix by the second-order partial derivatives of R.

The rules for (approximate) likelihood ratio testing carry over from REML in the normal case. In particular, REML cannot be used for comparison of two models with different regression designs.

8.8   Example. Interviewer variability

Eliciting information from human subjects is fraught with a number of sources of imprecision. Subjects usually regard the interviewer, a letter requesting completion of the enclosed questionnaire, a phone call, or the like, as an intrusion, and often lack motivation and integrity to respond as precisely as the format of the inquiry would otherwise permit. The subjects are under no legal or ethical obligation to respond, and when they do respond there are no repercussions for deliberate or unintentional inaccuracies. Such inaccuracies are especially prevalent when responding to items with confidential content or items which require the respondent's memory recall or abstraction from records (e.g., income, property value, and date of the last medical treatment). The questionnaire items may appear ambiguous to some of the respondents, they may invoke different connotations depending on the setting of the interview (time of day, mode and circumstances of contact with the respondent), the psychological, medical or (socio-)economic state of the respondent, and a variety of everyday influences. Respondents in surveys have been known to give conflicting responses even to clearly stated questions on ubiquitous topics.

One crucial element of the interviewing process is the interviewer. In a typical survey a number of interviewers are appointed and each interviewer is assigned to contact a list of selected subjects and to confront them with the protocol of the survey. This process is usually standardized; the interviewers are instructed on how to approach the subject, how to conduct themselves, how to deal with various contingencies, and so on. Despite all these attempts at uniformity it has been observed in numerous surveys that interviewers appear to influence the responses of the contacted subjects. It is believed that some interviewers, for reasons explainable or inextricable, tend to elicit responses of a certain kind more often than other interviewers, had they been assigned to the same respondents. To save costs, interviewers are usually assigned to geographical regions. Then differences among the interviewers are confounded with differences among the regions.

For illustration, we analyse a dataset collected in a survey of public awareness of political issues in Los Angeles. The survey, directed by Professor Shure of the Psychology Department, UCLA, was part of a large scale study aimed at understanding interviewer variability. To simplify the description we present only a skeletal version of the problem. The respondents were given a list of questions related to their perception of the role of the US Federal Government in their everyday lives, and their responses were coded on a single ordinal scale 1 to 5 (from least to most significant

Table 8.1. Interviewer loads. Interviewer variability data. Forty interviewers contacted a total of 1008 respondents. The numbers of male and female respondents are given in parentheses for each interviewer

Respondents by interviewer (men + women)				
24 (13+11)	85 (38+47)	13 (5+ 8)	36 (14+22)	47 (22+25)
24 (5+19)	42 (22+20)	28 (6+22)	45 (24+21)	60 (28+32)
14 (7+ 7)	24 (13+11)	24 (11+13)	17 (11+ 6)	12 (4+ 8)
20 (9+11)	21 (9+12)	23 (13+10)	21 (8+13)	30 (19+11)
17 (5+12)	20 (9+11)	26 (17+ 9)	15 (9+ 6)	31 (16+15)
18 (8+10)	23 (15+ 8)	22 (15+ 7)	17 (5+12)	15 (5+10)
26 (16+10)	20 (6+14)	19 (10+ 9)	16 (5+11)	12 (5+ 7)
18 (7+11)	23 (5+18)	19 (6+13)	23 (17+ 6)	18 (7+11)

'role'). Since very few respondents were rated by scores 2 or 4, we recode the outcomes to three ordinal categories, 1, 2, and 3, including the rare outcomes 2 and 4 on the original five-point scale in the extreme categories 1 and 3 of the constructed scale. The numbers of respondents of each sex contacted by each interviewer are given in Table 8.1. Forty interviewers were engaged and they interviewed a total of 1008 subjects. The assignment of the respondents to interviewers was conditionally random, given the agreed interviewer load. All the interviewers received the same training and instructions.

Each respondent's gender ($RSEX$) was recorded. For interviewers their gender, $ISEX$, political opinion, $IPOL$, and a self-rated measure of concern for others, $ICON$, were recorded. The variables $IPOL$ and $ICON$ were defined on respective integer scales 1–4 (liberal to conservative) and 1–3 (least concern to a lot of concern). Although these variables involve ordered categories, we regard them throughout as quantitative variables, so as to reduce the number of estimated regression parameters and to simplify the illustration.

For the analysis of the ordered trinomial outcomes y we consider the following method. We construct the binary outcomes $y^{(1)}$ and $y^{(2)}$:

$$
\begin{aligned}
y^{(1)} &= 1 \quad \text{if } y > 1, \\
y^{(1)} &= 0 \quad \text{otherwise,} \\
y^{(2)} &= 1 \quad \text{if } y > 2, \\
y^{(2)} &= 0 \quad \text{otherwise,}
\end{aligned}
$$

so that $y^{(1)}$ indicates responses 3–5 (mid or high role), and $y^{(2)}$ indicates responses 4 and 5 (high role) on the original scale. For each constructed outcome $y^{(h)}$, $h = 1, 2$, we consider a random-effects logistic regression

$$\text{logit}^{-1}\left\{P\left(y_{ij}^{(h)} = 1 \,|\, \delta_j^{(h)}\right)\right\} = \mathbf{x}_{ij}\boldsymbol{\beta}^{(h)} + \delta_j^{(h)}, \qquad (8.47)$$

where $\boldsymbol{\beta}^{(h)}$ is a vector of parameters and the interviewer 'effect' δ_j is drawn at random from $\mathcal{N}(0, \tau_h^2)$. The explanatory variables $\mathbf{x}_{ij}$ contain the intercept 1, and consist of respondent-level variable ($RSEX$) and the three interviewer-level variables;

$$\mathbf{x}_{ij} = (1, RSEX_{ij}, ISEX_j, IPOL_j, ICON_j).$$

The latter three variables represent the adjustment for systematic interviewer differences; they are of considerable interest as descriptors of interviewer variability. The variances τ_h^2 describe the variation unaccounted for by the interviewer-level explanatory variables. The regression parameter corresponding to $RSEX$ has a continuum of interpretations. At one extreme, assuming identical distributions of the studied phenomenon within the two sexes, it is a measure of response bias; at the other extreme, assuming no response bias, it is a measure of the difference between the two sexes. And, of course, it may reflect a mixture of these two causes.

The ML and REML estimates for the model in (8.47) for outcome $y^{(1)}$ are given in Table 8.2. Table 8.3 contains the corresponding results for $y^{(2)}$. For the direct maximum likelihood (MLE), 9-point Gaussian quadrature was used. For each outcome and each estimator type (full or restricted ML) results for the direct and approximate method are given. For comparison, the GLS fit is also included.

The corresponding approximate and exact estimates and their standard errors differ only marginally. The ML and REML estimates are also very similar with the exception of the estimates of the variances τ_1^2.

The fitted logit for $y^{(1)}$, using MLE, is

$$0.593 + 0.269\,RSEX + 0.074\,ISEX + 0.183\,IPOL + 0.113\,ICON.$$

The interpretation of such an equation is analogous to the normal linear regression fit. For example, a male respondent ($RSEX=0$) interviewed by a female interviewer ($ISEX=1$) whose $IPOL=2$ and $ICON=3$ has the fitted logit for $y^{(1)}$ of 1.41, which corresponds to probability 0.804 (of scoring $y > 1$). A change from $ICON=3$ to $ICON=1$ is associated with a decrease in the fitted logit of 0.226, and the resulting probability is 0.765, a drop of 4 per cent. The fitted probabilities vary a fair amount across the various configurations of the explanatory variables. For example, $RSEX=ISEX=0$ and $IPOL=ICON=1$ corresponds to probability 0.71,

Table 8.2. Random-effects model fits for threshold 2. ML and REML estimation. Interviewer variability data. The outcome variable, $y^{(1)}$, is equal to 1 if $y > 1$ and equal to 0 otherwise. The logit link function is used. The estimation methods are: GLS – generalized least squares ($\tau_1^2 = 0$), AML – approximate maximum likelihood, MLE – direct maximum likelihood using 9-point Gaussian quadrature, RAML – approximate restricted maximum likelihood, and REML – direct restricted maximum likelihood using 9-point Gaussian quadrature

	GLM		AML		MLE	
Parameter	Estimate	St.error	Estimate	St.error	Estimate	St.error
Intercept	0.597	(0.294)	0.590	(0.304)	0.593	(0.305)
RSEX	0.271	(0.152)	0.269	(0.152)	0.269	(0.152)
ISEX	0.071	(0.071)	0.074	(0.073)	0.074	(0.073)
IPOL	0.179	(0.137)	0.182	(0.142)	0.183	(0.142)
ICON	0.114	(0.167)	0.113	(0.174)	0.113	(0.174)
τ_1^2			0.014	(0.049)	0.014	
τ_1			0.117	(0.208)	0.117	(0.203)
Deviance	1071.15		1071.05		1071.05	

	RAML		REML	
Intercept	0.582	(0.294)	0.590	(0.325)
RSEX	0.266	(0.152)	0.267	(0.153)
ISEX	0.078	(0.071)	0.079	(0.077)
IPOL	0.187	(0.137)	0.188	(0.151)
ICON	0.110	(0.167)	0.110	(0.186)
τ_1^2	0.039	(0.052)	0.040	
τ_1	0.198	(0.133)	0.199	(0.136)
Deviance (1093.11)	1092.46		1092.45	

but $RSEX = ISEX = 1$, $IPOL = 4$, and $ICON = 3$ corresponds to probability 0.90. Note, however, that all the estimated regression parameters are not significant (at 5 per cent level). Also the interviewer-level variances (standard deviations) τ_h^2 (τ_h) are not significant, although the sizes of their estimates are substantial. For illustration, suppose the fitted logit for a respondent is equal to unity. Then the fitted distribution of the logits for this respondent, if interviewed by different interviewers, is $\mathcal{N}(1, \hat{\tau}_h^2)$, and so conditional logits, given the interviewer's effect, in the range 0.8–1.2 are quite feasible. On the probability scale this corresponds to the range

Table 8.3. Random-effects model fits for threshold 3. ML and REML estimation. Interviewer variability data. The outcome variable, $y^{(2)}$, is equal to 1 if $y > 2$ and to 0 otherwise. The logit link function is used. The layout and notation are the same as in Table 8.2

Parameter	GLM Estimate	St.error	AML Estimate	St.error	MLE Estimate	St.error
Intercept	−1.254	(0.298)	−1.267	(0.331)	−1.280	(0.333)
RSEX	−0.511	(0.151)	−0.527	(0.152)	−0.532	(0.154)
ISEX	−0.017	(0.069)	−0.024	(0.076)	−0.024	(0.054)
IPOL	0.133	(0.137)	0.154	(0.152)	0.156	(0.154)
ICON	0.212	(0.170)	0.223	(0.192)	0.224	(0.194)
τ_2^2			0.050	(0.059)	0.050	
τ_2			0.223	(0.134)	0.225	(0.142)
Deviance	1078.36		1077.56		1077.51	

Parameter			RAML Estimate	St.error	REML Estimate	St.error
Intercept			−1.274	(0.322)	−1.294	(0.355)
RSEX			−0.534	(0.151)	−0.543	(0.155)
ISEX			−0.018	(0.069)	−0.019	(0.068)
IPOL			0.164	(0.137)	0.165	(0.143)
ICON			0.220	(0.170)	0.231	(0.179)
τ_2^2			0.082	(0.063)	0.086	
τ_2			0.287	(0.111)	0.294	(0.120)
Deviance	1100.38		1098.45		1098.29	

0.69–0.77. In the context of the fitted unconditional probabilities (0.71–0.90) such variation is sizeable. The importance of the random effect of the interviewer (the variation of outcomes associated with the hypothetical choice of the interviewer) is comparable to that of the observed attributes of the respondent and interviewer. This is the case when the estimated parameters are regarded as true parameters. On the other hand, for $y^{(1)}$ even the hypothesis that all the regression and variance parameters vanish, that is, that $\{y_{ij}^{(1)}\}$ is a random sample from a binary distribution, is accepted. The corresponding (approximate) likelihood ratio test statistic is equal to $1077.82 - 1071.51 = 6.31$; for $y^{(2)}$ this statistic is equal to 14.86 ($1092.37 - 1077.51$). Both test statistics have χ_5^2 distribution assuming the respective null hypotheses $\tau_1^2 = 0$ and $\tau_2^2 = 0$.

Striking features of the analyses are the discrepancies between the respective ML and REML estimates of the variances τ_1^2 and τ_2^2. For τ_1^2 the respective estimates are 0.014 and 0.040. The difference of these estimates vastly exceeds what would be expected from conventional considerations of degrees of freedom associated with the clusters and the estimated regression parameters. An explanation of the additional discrepancy between ML and REML rests on the limited information associated with binary data, as discussed in Section 8.6. Naturally, one would be inclined to choose the unbiased (REML) estimate. However, in the presence of substantial sampling variation the property of unbiasedness is of limited virtue.

Uncertainty about the variances τ_h^2 is illustrated by the approximations to the profile log-likelihood, or, more accurately, the 'profile deviance', plotted in Figure 8.1. The deviance, evaluated at $\hat{\boldsymbol{\beta}}_{MLE}^{(1)}$, is plotted for a range of values of τ_1^2 using Gaussian quadrature with three, five, and nine points and the approximation given by (8.31). The deviance is very flat in the neighbourhood of zero. The plotted curves also indicate that there is little to choose between the methods for approximation while $\tau_1^2 < 0.4$. For larger values of τ_1^2, first the 3-point Gaussian quadrature, and then the AML approximation break down. It seems that 5-point Gaussian quadrature is sufficient even for extremely large variances.

The non-quadratic nature of dependence of the log-likelihood on two of the most common transformations of the variance is highlighted in Figure 8.2. Here the profile log-likelihood for $y^{(1)}$ is plotted as a function of the

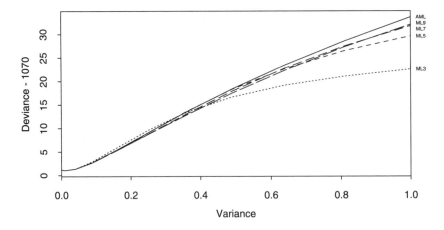

Fig. 8.1. Approximations to the profile log-likelihood for $y^{(1)}$ as a function of the cluster-level variance τ_1^2. Interviewer variability data. The methods of approximation are: ML3 – 3-point Gaussian quadrature, ML5 – 5-point Gaussian quadrature, ML9 – 9-point Gaussian quadrature, and AML – approximate maximum likelihood based on (8.32).

standard deviation τ_1 and of the log-variance $\log(\tau_1^2)$. The profile log-likelihood for the log-variance, in the plotted range, resembles a quadratic curve much more closely than do the plots for the standard deviation or variance. However, the profile log-likelihood is very flat in the interval $(-\infty, -4)$, and so a quadratic approximation to the profile log-likelihood for log-variance in the region of interest is not adequate either.

The analysis for $y^{(1)}$ yields non-significant associations with all the explanatory variables. However, for $y^{(2)}$ the estimated adjusted difference due to the sex of the respondent is substantial. The probability of $y = 1$ is not associated with *RSEX*, but the probability of $y = 3$ is positively associated with women, *RSEX* $= 1$ (the estimate of the *RSEX* regression parameter for $y^{(2)}$ is negative). For other regression parameters the differences between the corresponding estimates in the two analyses are trivial, with the exception of the intercept, which is of no real importance.

The outcome of the analysis is somewhat pessimistic. Even though the data have what would normally be regarded as a large elementary-level sample size, and even though the cluster-level sample size is quite large, all the estimated parameters are subject to substantial sampling variation. All that we can infer with confidence is that certain extreme differences (very large values of the parameters) are unlikely. Inference about interviewer variation is particularly unsatisfactory. On the one hand, the interviewer-level variance may be equal to zero; on the other hand, not even the variance $\tau_1^2 = 0.16$ can be ruled out.

Since estimation of the simplest pattern of between-cluster variation is subject to considerable uncertainty, it is not meaningful to pursue more detailed analysis of the pattern of interviewer-level variation. Conceivably, it would be of interest to know whether the male–female differences are

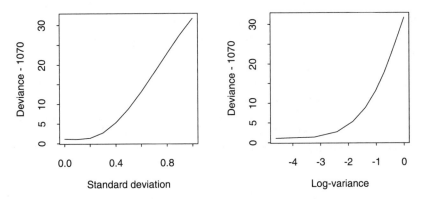

Fig. 8.2. Profile log-likelihoods for $y^{(1)}$ as a function of the standard deviation τ_1, and of the log-variance $\log(\tau_1^2)$. The 9-point Gaussian quadrature is used.

constant across the interviewers. Reliable inference about such differences would require an unrealistically large sample.

8.9 Death rates of Medicare patients

The Health Care Finance Administration (HCFA) monitors the quality of medical care provided for patients insured by the Medicare system in the USA. HCFA publishes annual reports giving, for each of about 5500 acute care hospitals in the USA, the number of cases admitted (a person may be admitted more than once a year), and the number of deaths among the Medicare patients (aged 65 or over) in each of 14 diagnostic categories. For each diagnostic category the within-hospital observed death rates are compared with the national death rate, and (nominally) significant differences are pointed out. Most hospitals have small numbers of patients for each diagnostic category, and so only marked differences in observed death rates are detected as significant. On the other hand, relatively small differences in death rates (e.g., 5 per cent) are generally thought to be important.

The allocation of patients to hospitals does not conform with the simple random sampling design. In most cases a prospective patient is admitted to the nearest acute care hospital. The hospitals differ substantially in characteristics such as socio-economic and demographic composition of their patients, climate, air and water quality, predominant diet and lifestyle in the region, and the like. As a consequence the circumstances associated with admitted patients may vary systematically among the hospitals. Comparisons of observed death rates thus reflect other phenomena than merely the quality of medical care. For more equitable comparisons of the hospitals, adjustment for severity of the case is essential.

In order to assess the need for adjustment for severity a stratified probability sample of 296 hospitals was selected from the list of all US hospitals in each of the accounting years 1981–2 and 1985–6. A set of standardized variables measuring patients' severity of condition at admission was extracted from a small number of randomly selected patient records (three or four cases from each hospital per year). Derivation of the values of these variables is time consuming and costly because it involves detailed abstraction from medical records. A group of internists studied these records and assessed the quality of the procedures employed to treat each case. These assessments were coded in a quantitative variable called *PROCESS*.

For the complete national census for the year 1986 we consider the logistic RANOVA model

$$\text{logit}\{P(y_{ij} = 1 \,|\, \delta_j)\} = \mu + \delta_j, \qquad (8.48)$$

$\delta_j \sim \mathcal{N}(0, \tau^2)$. The estimates and standard errors for the parameters μ and τ^2 for four diagnostic categories with high mortality rates are given in Table 8.4. The raw death rates for these categories are between 15.3–25.5

Table 8.4. Analysis of the census data for the four most frequent conditions. Standard errors are given in parentheses ( ) and the ML9 estimates in brackets [ ]. The standard errors for the ML9 estimates are omitted (to conserve space); they differ from their AML counterparts by less than 0.0002 for the mean logit and less than 0.0006 for the variance τ^2. The likelihood ratio test statistic is for the hypothesis $\tau^2 = 0$; the REML version of the statistic is in brackets [ ]

DEATH RATES AND THEIR VARIATION ACROSS HOSPITALS
CENSUS DATA, 1986

Condition Hospitals	Patients Died	Death rt (%)	GLM	AML [ML9]	τ^2	Likelihood ratio
Pneumonia	415 179		−1.465	−1.506	0.283	1689.03
5628	77 961	18.78		[−1.481]	[0.285]	[1710.12]
			(0.0040)	(0.0059)	(0.0061)	
Heart	465 229		−1.711	−1.718	0.211	559.45
failure	71 223	15.31		[−1.704]	[0.207]	[553.63]
5541			(0.0041)	(0.0054)	(0.0068)	
Stroke	298 306		−1.366	−0.135	0.260	813.81
5310	60 617	20.32		[−1.337]	[0.252]	[805.81]
			(0.0046)	(0.0063)	(0.0075)	
Heart	278 114		−1.072	−1.060	0.208	507.21
attack	70 942	25.51		[−1.041]	[0.203]	[499.11]
5285			(0.0044)	(0.0057)	(0.0071)	

per cent. The column labelled 'GLM' gives the corresponding values on the logit scale (GLS estimators of μ), with the associated nominal standard errors. The ML estimates of the parameters in (8.48) are given in the next two columns. For completeness both the approximate and direct ML estimates are given (the latter, using 9-point Gaussian quadrature, are in brackets, [ ]). The standard errors for $\hat{\mu}$ are essentially identical for the two methods, and so only their common value is given (in parentheses). Note that these standard errors are substantially higher than their counterparts in GLM. The estimates of the between-hospital variances are in the range 0.21–0.28, and the standard errors are smaller than 0.01; for each diagnostic category the death rates have substantial between-hospital variance and its estimate has small sampling variance. The two versions of the likelihood ratio statistics, one based on ML and the other on REML, differ by several points but they confirm the overall conclusion about between-hospital variation.

To illustrate the importance of between-hospital variation of death rates, consider two hospitals with 1000 cases of pneumonia each, and suppose one hospital is 'average', with logit of death rate -1.51 (18.1 per cent), and the other with logit of death rate $\hat{\tau}$ below average (logit -2.03, or 11.6 per cent). The expected numbers of deaths in these hospitals are 181 and 116, a substantial difference. Observed between-hospital differences of 10 per cent or more in death rates of pneumonia cases are not unusual.

As we pointed out there are numerous causes of between-hospital differences. It is of considerable interest to establish whether the hospitals have much more similar within-hospital *adjusted* death rates after adjustment for some of the relevant explanatory variables, in particular for severity of the case.

Our illustration focuses on the analysis for pneumonia using the sample of cases from 296 hospitals in years 1981–2 and 1985–6.

For adjustment by logistic regression we consider the following sets of variables:

- stratification of hospitals (urban/rural, number of beds, and intern to bed ratio);
- severity of the case (11 quantitative variables);
- variable *PROCESS*, a quantitative assessment of the quality of medical care provided to the case.

The logistic regressions using each combination of these sets of variables,

$$\text{logit}\{P(y_{ij} = 1 \,|\, \delta_j) = \mathbf{x}_{ij}\boldsymbol{\beta} + \delta_j\,, \tag{8.49}$$

are summarized in Tables 8.5 and 8.6 for the respective years 1981–2 and 1985–6. The regression parameter estimates are of lesser importance, and are therefore omitted with the exception of *PROCESS* and *APACHE II*, a constructed variable, which is the most important of the severity measures. *PROCESS* is nominally significant in the 1985–6 analysis but not in 1981–2. *APACHE II* is highly significant in both years, justifying the expense in obtaining its values.

With one exception in the 1985–6 data, the fitted between-hospital variance decreases as more variables are used for adjustment. These fitted variances are much smaller than the unadjusted between-hospital variance (0.283). However, the standard errors associated with these estimates are so large that even values higher than the unadjusted variance cannot be ruled out. For all fitted models the likelihood ratio test statistic for the hypothesis of $\tau^2 = 0$ is smaller than unity (null distribution χ_1^2). The negative estimated variances for some of the models in the analysis of 1985–6 data are unrealistic because the corresponding covariances cannot be realized in hospitals with large numbers of cases.

Table 8.5. Logistic regression for death rates (pneumonia), with adjustment for severity, stratification and *PROCESS* (as indicated in the rows), year 1981–2. HOSP denotes the between-hospital differences (variation), STRAT the stratifying (hospital-level) variables, PROC the process variable, and SEVER the measures of severity (they include APACHE as a variable). The standard errors for $\hat{\tau}^2$ are given in parentheses (), and the deviance corresponding to $\tau^2 = 0$ (GLS deviance) in brackets []. The number of regression parameters is given in the column 'd.f.'

Adjustment for	APACHE	PROCESS	$\hat{\tau}^2$	Deviance [GLS deviance]	d.f.
HOSP			0.096 (0.161)	1019.27 [1019.66]	1
HOSP, STRAT			0.071 (0.158)	1014.63 [1014.85]	6
HOSP, PROC		−0.142 (0.084)	0.110 (0.162)	1016.31 [1016.82]	2
HOSP, SEVER	0.956 (0.107)		0.004 (0.212)	692.82 [692.82]	12
HOSP, SEVER PROC	0.988 (0.110)	−0.154 (0.101)	0.009 (0.213)	690.50 [690.51]	13
HOSP, SEVER PROC, STRAT	0.984 (0.111)	−0.163 (0.103)	−0.009 (0.210)	687.57 [687.57]	18

The sample of cases contains very sparse information about between-hospital variation. On the one hand, selecting a larger number of hospitals would increase information about variation, but the associated cost of abstracting the severity measures may become prohibitive. On the other hand, sampling fewer hospitals with a higher number of cases from each sampled hospital would yield much more information about τ^2. For simplicity, suppose that the same number of patients, n, is sampled from each of N_2 selected hospitals. The approximate information for τ^2 obtained from (8.44) is equal to

$$H = \frac{N_2 n^2 w^2}{2(1 + nw\tau^2)^2},\tag{8.50}$$

where $w = p(1 - p)$ is the common conditional variance (given $\delta_j = 0$). For fixed $N = N_2 n$, w, and τ^2, (8.50) attains its unique maximum for $n^* = 1/w\tau^2$. For pneumonia we have $w \doteq 0.15$ and almost certainly $\tau^2 < 0.3$, and so it is very likely that $n^* > 20$. This suggests that a design

Table 8.6. Logistic regression for death rates (pneumonia), with adjustment for severity, stratification and PROCESS, 1985–6. The same notation is used as in Table 8.5

Adjustment for	APACHE	PROCESS	$\hat{\tau}^2$	Deviance [GLS deviance]	d.f.
HOSP			−0.003 (0.124)	1208.60 [1208.60]	1
HOSP, STRAT			−0.029 (0.121)	1204.81 [1204.86]	6
HOSP, PROC		−0.272 (0.076)	0.017 (0.126)	1195.78 [1195.80]	2
HOSP, SEVER	0.803 (0.094)		−0.105 (0.146)	915.31 [915.77]	12
HOSP, SEVER PROC	0.846 (0.096)	−0.199 (0.088)	−0.117 (0.146)	910.29 [910.84]	13
HOSP, SEVER PROC, STRAT	0.852 (0.096)	−0.188 (0.089)	−0.154 (0.150)	906.61 [907.53]	18

with fewer hospitals (e.g., 50) and more cases from each selected hospital (e.g., 24) is most informative about τ^2. For example, if $\tau^2 = 0.25$ and $w = 0.15$, then, for $N = 1200$ and $n = 4$, $H = 40.83$, but for $n = 24$ $H = 89.75$. Adopting the balanced design with $n = 24$ would yield around 2.2 times more information about τ^2 than the design with $n = 4$. The exact choice of the within-hospital sample sizes is not critical; balanced designs with n in the range 20–36 yield $H > 88$.

If the design with $n = 24$ and $N = 1200$ were adopted, little loss of information about the regression parameters would be incurred. However, the improvement in estimation of τ^2 would not be sufficient for the desired comparisons because the standard errors for $\hat{\tau}^2$ would be reduced only about 1.5 times. Since estimation of even the simplest pattern of between-hospital variation is problematic it is meaningless to pursue inference about varying between-hospital regression coefficients.

8.10 Bibliographical notes

Williams (1982) is generally regarded as a milestone in analysis for binomial data with overdispersion. See also Prentice (1986) and Moore (1987). A similar approach is adopted by Breslow (1984) for log-Poisson models.

The EM algorithm as a method for regression analysis of clustered dichotomous observations was developed by Stiratelli, Laird and Ware (1984).

An important feature of the adapted EM algorithm is an approximation for the conditional expectations required for the E-step of the algorithm.

Anderson and Aitkin (1985) and Stokes (1988) use binary variance component models for inference about interviewer variation in surveys. The former paper discusses the limitations of the EM algorithm when applied using the GLIM software.

Bonney (1987) and Connolly and Liang (1988) use conditional binary distributions to define a multivariate binary distribution which can be used for the analysis of clustered binary data. The approach can be applied to other non-normal distributions.

Im and Gianola (1988) take the 'hard-nosed' approach of direct maximization of the log-likelihood for clustered binary data (with three levels of nesting). The log-likelihood is evaluated using Gaussian quadrature, and is maximized by a simplex method.

Other important references on binary clustered data analysis include Stanek and Diehl (1988) and Rosner (1989).

Research on random coefficient methods for non-normally distributed outcomes, and for binary outcomes in particular, exploded after the definition of the GEE (generalized estimation equations) method by Zeger and Liang (1986) and Liang and Zeger (1986); see Zeger et al. (1988), Zeger and Qadish (1988), Prentice (1988), Zhao and Prentice (1990), Liang et al. (1992) for examples and refinements of the method.

The GEE method is a general computational algorithm, motivated by the GLS, which is applicable for outcomes from any distribution in the exponential family, or, more generally, for the distributions specified by the quasilikelihood.

Morton (1987) and Firth and Harris (1991) present methods for variance component analysis with multiplicative random terms. Log-Poisson regression analysis with random coefficients is the most important application of these methods.

Although ML methods for clustered binary data are now firmly established among the computational methods in statistics, there is no consensus as to the relevance of the methods for analysis of clustered data with other distributional assumptions. Transformation of such data to promote the normal assumptions is an alternative method which avoids both computational problems and simplifies the interpretation of the results of the analysis.

9
Appendix. Asymptotic theory

We have derived the standard errors for the estimated parameters in a random coefficient model from the inverse of the information matrix. This approach is justified in a variety of methods by reference to results from asymptotic theory. A typical asymptotic result for maximum likelihood estimators (MLE) in models with independent observations states that under certain *regularity conditions*, as the sample size n increases beyond all bounds, a MLE $\hat{\boldsymbol{\theta}}$ of a parameter vector $\boldsymbol{\theta}$ exists with probability converging to one, and the sampling distribution of

$$\sqrt{n}(\hat{\boldsymbol{\theta}}_n - \boldsymbol{\theta}_0)$$

converges to the limiting distribution of the sequence $\{\mathcal{N}(\mathbf{0}, n\mathbf{H}_n^{-1})\}$, where $\mathbf{H}_n$ is the expected information matrix for $\boldsymbol{\theta}$, evaluated at its true value $\boldsymbol{\theta}_0$, based on n observations. The limiting distribution and its properties are referred to as the *asymptotic* distribution (properties). Extension of the well-known asymptotic results for independent observations to models with clustered observations raises several issues discussed in this appendix.

Maximum likelihood estimators have a closed form and their finite sample distributions can be determined exactly only in some simple cases. For example, given the residual variance σ^2 the asymptotic and finite-sample distributions of the regression parameter estimators in ordinary regression coincide:

$$\hat{\boldsymbol{\beta}} \sim \mathcal{N}\left\{\boldsymbol{\beta}, \sigma^2\left(\mathbf{X}^\top\mathbf{X}\right)^{-1}\right\}.$$

Maximum likelihood estimators in random coefficient models rarely have a closed form and their distributions are in general difficult to explore. Nevertheless, the MLE for a parameter is often looked upon as a standard, or benchmark, for alternative estimators because of the anticipated asymptotic result and because its asymptotic variance is the lower bound for the variance of any unbiased estimator of the parameter.

The meaning of 'large samples' in models with independent observations is unambiguous. For clustered observations it is not immediately clear whether asymptotic results apply as the elementary-level sample size N increases (e.g., holding the number of clusters constant), or if it is necessary for the number of clusters to increase beyond all bounds. Also, whereas for independent observations the same normalization ($\sqrt{n}$) applies to each parameter, for random coefficient models it is not always the case, as we illustrate below. Therefore, it is meaningful to set our goal to establish for a range of realistic conditions that, as the elementary- and/or cluster-level sample sizes increase, the MLE $\hat{\boldsymbol{\theta}}$ of the parameter vector $\boldsymbol{\theta}$ in a random coefficient model satisfies the asymptotic distributional identity

$$(\hat{\boldsymbol{\theta}} - \boldsymbol{\theta})\mathbf{M} \sim \mathcal{N}(\mathbf{0}, \mathbf{M}\mathbf{H}^{-1}\mathbf{M}), \tag{9.1}$$

where $\mathbf{M}$ is a diagonal matrix (e.g., containing entries $N^{\frac{1}{2}}$ and $N_2^{\frac{1}{2}}$ on its diagonal), and $\mathbf{H}$ is the expected information matrix for $\boldsymbol{\theta}$.

To illustrate the problems associated with asymptotic normality we consider the random coefficient model

$$y_{ij} = \mathbf{x}_{ij,r}\boldsymbol{\beta}_{j,r} + \mathbf{x}_{ij,f}\boldsymbol{\beta}_f + \mathbf{x}_{j,c}\boldsymbol{\beta}_c + \varepsilon_{ij}, \tag{9.2}$$

$i = 1, \ldots, n_j$, $j = 1, \ldots, N_2$, $\boldsymbol{\beta}_{j,r} \sim \mathcal{N}(\mathbf{0}, \boldsymbol{\Sigma})$, i.i.d., $\varepsilon_{ij} \sim \mathcal{N}(\mathbf{0}, \sigma^2)$, i.i.d., in which the elementary-level regressors $\mathbf{x}_r$ are associated with variation, and elementary- and cluster-level regressors, $\mathbf{x}_f$ and $\mathbf{x}_c$, are not (i.e., the regression slopes for $\mathbf{x}_f$ and $\mathbf{x}_c$ are constant). Intercept 1 is subsumed in $\mathbf{x}_r$. For simplicity, we assume that $\mathbf{x}_f$ is orthogonal to $\mathbf{x}_r$ and $\mathbf{x}_c$, and the eigenvalues of $N^{-1}\mathbf{X}_f^{\top}\mathbf{X}_f$ converge to finite positive values ($\mathbf{X}_f$ is the design matrix corresponding to the regressors $\mathbf{x}_f$). Then the asymptotic result for the regression parameters $\boldsymbol{\beta}_f$ holds as the elementary-level sample size grows to infinity, even when the data comprise a single cluster ($N_2 = 1$) or clusters with single elementary-level units ($n_j \equiv 1$) since in these cases the ordinary regression theory applies. Presumably, the result extends to any other sequence of clustering designs. Obviously, asymptotic results do not hold for the regression parameters $\boldsymbol{\beta}_r = \mathbf{E}(\boldsymbol{\beta}_{j,r})$ unless the cluster-level sample size N_2 increases to infinity. For estimation of the covariance structure parameters it is also necessary that $N_2 \to \infty$, and that the cluster sizes n_j be greater than 1. These conditions are not sufficient, however.

Proofs of asymptotic results usually require an extensive technical and analytic apparatus. We outline such a proof for two-level random coefficient models. We will only sketch some of the details, concentrating instead on the generality of the assumptions. The method of proof is motivated by standard approaches applied to models for independent observations, see Fahrmeir and Kaufmann (1985) and Moore (1986). Hartley and Rao (1967) and Miller (1977) derived asymptotic properties for MLEs in a general context overlapping random coefficient models. A typical such proof consists of the following steps:

1. limited variance of the scoring function $\mathbf{s}(\boldsymbol{\theta}_0)$ as $n \to \infty$;

2. asymptotic normality of $n^{-1}\mathbf{s}(\boldsymbol{\theta}_0)$;

3. existence and consistency of the estimator;

4. asymptotic distribution of the MLE.

Here, and in the following, the subscript 0 denotes the true parameter vector. Before discussing each of these steps we set some of the assumptions. For simplicity, we assume that the sequence of datasets is formed by adding new clusters. First, we assume that the number of clusters grows to infinity and that the cluster sizes n_j are bounded, $n_j < c_2$. This is a realistic assumption and it dispenses with the problem of differential convergence to the asymptotic distribution. In essence, the problem with allowing for increasing cluster sizes (as $N_2 \to \infty$) is that a small number of clusters may contain a disproportionately large amount of information about certain parameters. Further, we assume that all the regressors are bounded; $|X_{ijk}| < X_{max}$ (and similarly for the variation regressors Z_{ijk}, $|Z_{ijk}| < Z_{max}$). Next, to ensure that the data contain increasing information about the estimated parameters we assume that the smallest eigenvalues of $\mathbf{X}^\top\mathbf{X}$ diverge to $+\infty$ as $N_2 \to \infty$, and that the minimum and maximum eigenvalues of the matrices $\mathbf{Z}_j^\top\mathbf{Z}_j$ (denoted by $\lambda_{N_2,min}$ and $\lambda_{N_2,max}$ respectively) have upper and lower essential limits:

$$0 < \lambda_{min} = \liminf_{N_2 \to \infty} \lambda_{N_2,min} \leq \limsup_{N_2 \to \infty} \lambda_{N_2,max} = \lambda_{max} < +\infty.$$

Note that this condition implies that, apart from a finite number of clusters j, each matrix $\mathbf{Z}_j^\top\mathbf{Z}_j$ is non-singular. Further, we assume that the true parameter vector lies in the interior of the parameter space. If there are no constraints other than those of non-negative definiteness of the variance matrix $\boldsymbol{\Sigma}$ and positiveness of the variance σ^2, then this amounts to assuming that $\boldsymbol{\Sigma}$ is non-singular.

Consistency of the estimator of a parameter (vector) is defined as convergence to the true parameter (vector) as the sample size increases. For each definition of convergence (weak, strong, etc.), there is a corresponding definition of consistency. Weak consistency (of the estimator $\hat{\boldsymbol{\theta}}$), most frequently considered, is formally defined as follows: for any $\epsilon > 0$

$$P(\| \hat{\boldsymbol{\theta}} - \boldsymbol{\theta}_0 \| > \epsilon) \to 0, \tag{9.3}$$

as $N_2 \to \infty$. Note that consistency implies the existence of a MLE for sufficiently large datasets with high probability.

Let $\boldsymbol{\theta} = \left(\boldsymbol{\beta}^{\top}, \sigma^2, \boldsymbol{\omega}^{\top}\right)^{\top}$ be the vector of all the model parameters in a random coefficient model. Since the clusters are mutually independent,

$$\mathbf{s}(\boldsymbol{\theta}) = \sum_{j} \mathbf{s}_j(\boldsymbol{\theta}),$$

where the subvectors of $\mathbf{s}_j$ corresponding to $\boldsymbol{\beta}$, σ^2, and $\boldsymbol{\omega}$ are

$$\mathbf{X}_j \mathbf{V}_j^{-1} \mathbf{e}_j \, ,$$

$$-\frac{1}{2} \left\{ \operatorname{tr}\left(\mathbf{V}_j^{-1}\right) - \mathbf{e}_j^{\top} \mathbf{V}_j^{-2} \mathbf{e}_j \right\} , \tag{9.4}$$

$$-\frac{1}{2} \left\{ \operatorname{tr}\left(\mathbf{Z}_j^{\top} \mathbf{V}_j^{-1} \mathbf{Z}_j \frac{\partial \boldsymbol{\Sigma}}{\partial \theta}\right) - \mathbf{e}_j^{\top} \mathbf{V}_j^{-1} \mathbf{Z}_j \frac{\partial \boldsymbol{\Sigma}}{\partial \theta} \mathbf{Z}_j^{\top} \mathbf{V}_j^{-1} \mathbf{e}_j \right\} ,$$

respectively ($\mathbf{e}_j = \mathbf{y}_j - \mathbf{X}_j \boldsymbol{\beta}$).

We now discuss steps 1–4 of the proof of asymptotic distribution of the MLE.

9.1   Limited variance of the scoring function

The variance matrices of the components of the scoring function are

$$\operatorname{var}\left(\mathbf{X}_j^{\top} \mathbf{V}_j^{-1} \mathbf{e}_j\right) = \mathbf{X}_j^{\top} \mathbf{V}_j^{-1} \mathbf{V}_{j,0} \mathbf{V}_j^{-1} \mathbf{X}_j$$

$$+ \mathbf{X}_j^{\top} \mathbf{V}_j^{-1} \mathbf{X}_j (\boldsymbol{\beta} - \boldsymbol{\beta}_0)(\boldsymbol{\beta} - \boldsymbol{\beta}_0)^{\top} \mathbf{X}_j^{\top} \mathbf{V}_j^{-1} \mathbf{X}_j \, ,$$

$$\frac{1}{4} \operatorname{var}\left(\mathbf{e}_j^{\top} \mathbf{V}_j^{-2} \mathbf{e}_j\right) = \frac{1}{2} \operatorname{tr}\left(\mathbf{V}_{0,j}^2 \mathbf{V}_j^{-4}\right) + \frac{1}{4}\{\operatorname{tr}(\mathbf{V}_{0,j}^2 \mathbf{V}_j^{-4})$$

$$+ (\boldsymbol{\beta} - \boldsymbol{\beta}_0)^{\top} \mathbf{X}_j^{\top} \mathbf{V}_j^{-2} \mathbf{X}_j (\boldsymbol{\beta} - \boldsymbol{\beta}_0)\}^2,$$

$$\tag{9.5}$$

$$\tfrac{1}{4} \operatorname{var}\left(\mathbf{e}_j^{\top} \mathbf{V}_j^{-1} \mathbf{Z}_j \tfrac{\partial \boldsymbol{\Sigma}}{\partial \theta} \mathbf{Z}_j^{\top} \mathbf{V}_j^{-1} \mathbf{e}_j\right) = \tfrac{1}{2} \operatorname{tr}\left\{ \left(\mathbf{Z}_j^{\top} \mathbf{V}_j^{-1} \mathbf{V}_{j,0} \mathbf{V}_j^{-1} \mathbf{Z}_j \tfrac{\partial \boldsymbol{\Sigma}}{\partial \theta}\right)^2 \right\}$$

$$+ \tfrac{1}{4} \left\{ \operatorname{tr}\left(\mathbf{Z}_j^{\top} \mathbf{V}_j^{-1} \mathbf{V}_{j,0} \mathbf{V}_j^{-1} \mathbf{Z}_j \tfrac{\partial \boldsymbol{\Sigma}}{\partial \theta}\right) \right.$$

$$\left. + (\boldsymbol{\beta} - \boldsymbol{\beta}_0)^{\top} \mathbf{X}_j^{\top} \mathbf{V}_j^{-1} \mathbf{Z}_j \tfrac{\partial \boldsymbol{\Sigma}}{\partial \theta} \mathbf{Z}_j^{\top} \mathbf{V}_j^{-1} \mathbf{X}_j (\boldsymbol{\beta} - \boldsymbol{\beta}_0) \right\}^2 .$$

Since the regressors X_{ijk} and Z_{ijk} are bounded and $\sigma_0^2 > 0$, these variance matrices are continuous and bounded in a neighbourhood of the true parameter vector $\boldsymbol{\theta}_0$.

For each component of $\mathbf{s}$ we use the strong law of large numbers which states that if independent centred random variables s_j have variances such that

$$\sum_j^\infty \frac{\mathrm{var}(s_j)}{j^2} < +\infty, \tag{9.6}$$

then the sequence of partial means converges to 0 almost surely:

$$\frac{1}{J} \sum_j^J s_j \to 0 \qquad \text{(a.s.)}$$

as $J \to \infty$. The condition (9.6) is known as the *Kolmogorov* condition. When the variances $\mathrm{var}(s_j)$ are bounded, this condition is trivially satisfied.

9.2 Asymptotic normality of the scoring vector

Asymptotic normality of the scoring function can be proved by use of the Lyapunov theorem. Applied in our context the theorem states that when the random variables $\{s_j\}$, with respective variances $\{v_j\}$, are independent, centred, have finite third moments, and

$$\frac{1}{\sum_j v_j^{\frac{3}{2}}} \sum_j \mathbf{E}\,|s_j|^3 \to 0, \tag{9.7}$$

then the normalized partial totals of $\{s_j\}$ converge to the standard normal distribution;

$$\sum_{j=1}^J \frac{s_j}{\sqrt{v_j}} \to \mathcal{N}(0,1). \tag{9.8}$$

Since (9.7) refers to univariate convergence, we apply it to an arbitrary linear combination of the parameters, $\mathbf{b}^\top \boldsymbol{\theta}$. Checking the condition in (9.7) involves computing the third moments of the scoring function; although tedious, this is elementary. All the equations are based on the properties of the multivariate normal distribution. Boundedness of the means and variances then implies (9.7).

9.3 Consistency of MLE

Consistency of the regression parameter estimates is not a very strong result. In Section 2.8 we showed that for random coefficient models even the OLS estimator of $\boldsymbol{\beta}$ is unbiased, and since its variance matrix converges to zero it is consistent. The MLE of $\boldsymbol{\beta}$ is consistent by the same token when the covariance structure parameters are given, or are consistently estimated.

Next we prove consistency of the root of the scoring function (implying the existence of a root for large N_2). A key analytical tool for this purpose is the inverse function theorem. The scoring vector maps the parameter space into an open set of the same dimension. The partial derivatives of $\mathbf{s}$ exist and are continuous in the parameter space, and the matrix of partial derivatives, $\partial \mathbf{s}/\partial \boldsymbol{\theta}$, has an inverse at the true value $\boldsymbol{\theta}_0$. Then there is a neighbourhood of $\boldsymbol{\theta}_0$ in which, for any $\boldsymbol{\Theta}_1$ and $\boldsymbol{\Theta}_2$,

$$|\mathbf{s}(\boldsymbol{\theta}_1) - \mathbf{s}(\boldsymbol{\theta}_2)| \geq 2\lambda \|\boldsymbol{\theta}_1 - \boldsymbol{\theta}_2\|, \tag{9.9}$$

for some positive λ, and the image $\mathbf{s}(\boldsymbol{\theta}_0)$ contains an open neighbourhood of $\mathbf{s}(\boldsymbol{\theta}_0)$.

In the neighbourhood of $\boldsymbol{\theta}_0$ the matrix of second-order partial derivatives converges in probability to the information matrix. The information matrix at $\boldsymbol{\theta}_0$ is non-singular, and so the matrix of second-order partial derivatives is also non-singular in a neighbourhood of $\boldsymbol{\theta}_0$. Also, for large N_2 the scoring function is close to $\mathbf{0}$, and so, according to the inverse function theorem, it has a root in the neighbourhood (with high probability). Now a continuity argument implies that this sequence of roots converges to $\boldsymbol{\theta}_0$.

9.4 Asymptotic normality of MLE

We expand the scoring vector evaluated at its root,

$$(\mathbf{s}(\hat{\boldsymbol{\theta}}) =)\quad \mathbf{0} = \mathbf{s}(\boldsymbol{\theta}_0) + (\hat{\boldsymbol{\theta}} - \boldsymbol{\theta}_0)^\top \mathbf{H}(\boldsymbol{\theta}_0) + \ldots . \tag{9.10}$$

Asymptotic normality of $\hat{\boldsymbol{\theta}}$ is obtained by application of the central limit theorem for $N_2^{-\frac{1}{2}} \mathbf{H}^{-\frac{1}{2}} (\hat{\boldsymbol{\theta}} - \boldsymbol{\theta}_0)$.

Owing to the independence of clusters the information matrix for $\boldsymbol{\theta}$ is a sum of matrix contributions from the clusters, $\mathbf{H} = \sum_j \mathbf{H}_j$. First we need to relate the eigenvalues of $\mathbf{H}$ to those of $\mathbf{H}_j$. Let λ_1 be the smallest eigenvalue of $\mathbf{H}$, and $\lambda_{1,j}$ be the smallest eigenvalue of $\mathbf{H}_j$. Then

$$\lambda_1 \geq \sum_j \lambda_{1,j} .$$

To prove this let $\mathbf{v}$ be an eigenvector associated with λ_1 ($\mathbf{H}\mathbf{v} = \lambda_1 \mathbf{v}$). Then

$$\lambda_1 = \mathbf{v}^\top \mathbf{H} \mathbf{v} = \sum_j \mathbf{v}^\top \mathbf{H}_j \mathbf{v}.$$

If $\mathbf{v}^\top \mathbf{H}_j \mathbf{v} < \lambda_{1,j}$, then premultiplying this inequality by $\mathbf{v}$ implies an eigenvalue of $\mathbf{H}_j$ smaller than $\lambda_{1,j}$, in contradiction with the assumptions. Similarly, it can be shown that the largest eigenvalue of $\mathbf{H}$ is not greater than the sum of the largest eigenvalues of $\mathbf{H}_j$.

We now proceed to find suitable sufficient conditions for boundedness of the eigenvalues of $\mathbf{H}_j$. Since $\mathbf{H}_j$ is block-diagonal, we can deal separately with the contributions to the information about regression parameters,

$$\mathbf{H}_{\beta,j} = \mathbf{X}_j^\top \mathbf{V}_j^{-1} \mathbf{X}_j ,$$

and the information about covariance structure parameters,

$$\mathbf{H}_{\Omega,j} = \frac{1}{2} \left(\mathbf{V}_j^{-1} \frac{\partial \mathbf{V}_j}{\partial \omega_1} \mathbf{V}_j^{-1} \frac{\partial \mathbf{V}_j}{\partial \omega_2} \right) .$$

For the former, assuming that $\mathbf{X}_j^\top \mathbf{X}_j$ is non-singular, we have

$$\mathbf{H}_{\beta,j} = \left\{ \left(\mathbf{X}_j^\top \mathbf{X}_j \right)^{-1} + \mathbf{\Omega}^* \right\}^{-1} ,$$

where $\mathbf{\Omega}^*$ is the scaled variance matrix of the regression coefficients $\boldsymbol{\beta}_j$. Let λ_ω be the largest eigenvalue of $\mathbf{\Omega}^*$. Then the eigenvalues of $\mathbf{H}_\beta$ are bounded by $\lambda_{1,j}$ and $(\lambda_{2,j}^{-1} + \lambda_\omega)^{-1}$, where $\lambda_{2,j}$ is the largest eigenvalue of $\mathbf{H}_j$.

Bounds for the eigenvalues of $\mathbf{H}_{\Omega,j}$ are much more difficult to establish because they cannot be related to the eigenvalues of the cross-products $\mathbf{Z}_j^\top \mathbf{Z}_j$. The element of $\mathbf{H}_\Omega$ corresponding to a pair of elements of the scaled variance matrix $\mathbf{\Omega}$ is the sum of two products of two elements of the matrix $\mathbf{S}_j = \mathbf{Z}_j^\top \mathbf{W}_j^{-1} \mathbf{Z}_j$. We conjecture that the eigenvalues of $\mathbf{H}_{\Omega,j}$ are bounded from 0 and $+\infty$ if the eigenvalues of $\mathbf{S}_j$ are. Assuming non-singularity of $\mathbf{Z}_j^\top \mathbf{Z}_j$ we have

$$\mathbf{S}_j = \left\{ \left(\mathbf{Z}_j^\top \mathbf{Z}_j \right)^{-1} + \mathbf{\Omega} \right\}^{-1} ,$$

where $\mathbf{\Omega}$ is the scaled variance matrix corresponding to the variation design given by $\mathbf{Z}$. Now bounds on the eigenvalues of $\mathbf{S}_j$ can be related to those for $\mathbf{Z}_j^\top \mathbf{Z}_j$ and $\mathbf{\Omega}$ in complete analogy with the bounds for $\mathbf{H}_{\beta,j}$. Bounds on the eigenvalues of $\mathbf{Z}_j^\top \mathbf{Z}_j$ imply that the extreme eigenvalues of $N_2^{-1} \mathbf{H}_\Omega$ are bounded from 0 and $+\infty$. Then the central limit theorem implies that, asymptotically,

$$\mathbf{H}^{-\frac{1}{2}} (\hat{\boldsymbol{\Theta}} - \boldsymbol{\Theta}_0) \sim \mathcal{N}(\mathbf{0}, \mathbf{I}).$$

Technical details of these asymptotic results are beyond the scope of this book. The interested reader is referred to Hartley and Rao (1967), Miller (1977), Kass and Steffey (1989), and the references in these papers.

References

Ahrens, H. (1965). Standardfehler geschätzter Varianzkomponenten eines unbalanzier-ten Versuchsplanes in r-stufiger hierarchischer Klassifikation. *Monatsbeschrifte der Deu-tschen Akademie für Wissenschaften*, **7**, 89–94.

Airy, G.B. (1861). *On the algebraical and numerical theory of errors of observations and the combination of observations*. McMillan, London.

Aitkin, M. (1987). Modelling variance heterogeneity in normal regression using GLIM. *Applied Statistics*, **36**, 332–9.

Aitkin, M., Anderson, D., and Hinde, J. (1981). Statistical modelling of data on teaching styles. *Journal of the Royal Statistical Society*, Ser. A, **144**, 148–61.

Aitkin, M., Anderson, D., Francis, B., and Hinde, J. (1989). *Statistical modelling in GLIM*. Oxford University Press.

Aitkin, M. and Longford, N.T. (1986). Statistical modelling issues in school effectiveness studies. *Journal of the Royal Statistical Society*, Ser. A, **149**, 1–43.

Anderson, D. and Aitkin, M. (1985). Variance component models with binary response: Interviewer variability. *Journal of the Royal Statistical Society*, Ser. B, **47**, 203–10.

Bartholomew, D.J. (1987). *Latent variable models and factor analysis*. C. Griffin & Co., London and Oxford University Press, New York.

Battese, G.E., Harter, R.M., and Fuller, W.A. (1988). An error components model for prediction of county crop areas using survey and satellite data. *Journal of American Statistical Association*, **83**, 28–36.

Belsley, D.A., Kuh, E., and Welsch, R.E. (1980). *Regression diagnostics. Identifying influential data and sources of collinearity*. Wiley Series in Probability and Mathematical Statistics. John Wiley, New York.

Berk, K. (1987). Computing for incomplete repeated measures. *Biometrics*, **43**, 385–98.

Bock, R.D. (ed.) (1989). *Multilevel analysis of educational data*. Academic Press, San Diego.

Bonney, G.E. (1987). Logistic regression for dependent binary observations. *Biometrics*, **43**, 951–73.

Boyd, L.H. and Iversen, G.R. (1979). *Contextual analysis: Concepts and statistical techniques*. Wadsworth, Belmont, CA.

Breslow, N.E. (1984). Extra-Poisson variation in log-linear models. *Applied Statistics*, **33**, 38–44.

Browne, M.W. and du Toit, S.H.C. (1992). Automated fitting of nonstandard models. *Multivariate Behavioral Research*, **27**, 269–300.

Bryk, A.S., Raudenbush, S.W., Seltzer, M., and Congdon, R.T. (1988). *An introduction to HLM: Computer program and users' guide*. University of Chicago Press.

Bryk, A.S. and Raudenbush, S.W. (1992). *Hierarchical Linear Models. Applications and Data Analysis Methods*. Sage Publications, Newbury Park, CA.

Burstein, L. (1980). The role of levels of analysis in the specification of educational effects. In *The analysis of educational productivity*, Vol. I., Issues in microanalysis (eds. Dreeben, R. and Thomas, J.A.), pp. 11–90. Ballinger Publishing Co., Cambridge, MA.

Burstein, L., Linn, R.L., and Capell, F. (1978). Analyzing multilevel data in the presence of heterogeneous within-class regressions. *Journal of Educational Statistics*, **3**, 347–83.

Chi, E.M. and Reinsel, G.C. (1987). Models for longitudinal data with random effects and AR(1) errors. *Journal of the American Statistical Association*, **84**, 452–59.

Cochran, W.G. (1947). Some consequences when the assumptions for the analysis of variance are not satisfied. *Biometrics Bulletin*, **3**, 22–38.

Connolly, M. and Liang, K.-Y. (1988). Conditional logistic regression models for correlated binary data. *Biometrika*, **75**, 501–6.

Cook, R.D. and Weisberg, S. (1982). *Residuals and influence in regression.* Chapman and Hall, New York.

Cronbach, L.J. and Webb, N. (1975). Between-class and within-class effects in a reported aptitude-by-treatment interaction: Re-analysis of a study by G.L. Anderson. *Journal of Educational Psychology*, **67**, 717–24.

Crump, S.L. (1951). Present status of variance component analysis. *Biometrics*, **24**, 527–40.

Deeley, J.J. and Lindley D.V. (1981). Bayes Empirical Bayes. *Journal of the American Statistical Association*, **76**, 833–41.

Dempster, A.P., Laird, N.M., and Rubin, D.B. (1977). Maximum likelihood from incomplete data via the EM algorithm. *Journal of the Royal Statistical Society*, Ser. B, **39**, 1–38.

Dempster, A.P., Rubin, D.B., and Tsutakawa, R.K. (1981). Estimation in covariance component models. *Journal of the American Statistical Association*, **76**, 341–53.

Dempster, A.P., Selwyn, M.R., Patel, C.M., and Roth, A.J. (1984). Statistical and computational aspects of mixed model analysis. *Applied Statistics*, **33**, 203–14.

Dobson, A. (1983). *An introduction to statistical modelling.* Chapman and Hall, London.

Donner, A. and Koval, J.J. (1980). The estimation of intraclass correlation in the analysis of family data. *Biometrics*, **36**, 19–25.

Efron, B. and Morris, C. (1975). Data analysis using Stein's estimator and its generalizations. *Journal of the American Statistical Association*, **70**, 311–19.

Eisenhart (1947). The assumptions underlying the analysis of variance. *Biometrics*, **33**, 615–28.

Fahrmeir, L. and Kaufmann, H. (1985). Consistency and asymptotic normality of the maximum likelihood estimator in generalized linear models. *The Annals of Statistics*, **13**, 342–68.

Fay, R.E. IIIrd and Herriot, R.A. (1979). Estimates of income for small places: An application of James–Stein precedures to census data. *Journal of the American Statistical Association*, **74**, 269–77.

Fieldsend, S., Longford, N.T., and McLeay, S. (1987). Industry effects and the proportionality assumption in ratio analysis: A variance component analysis. *Journal of Business Finance and Accounting*, **14**, 497–517.

Firebaugh, G. (1978). A rule for inferring individual-level relationships from aggregate data. *American Sociological Review*, **4**, 557–72.

Firth, D. and Harris, I.R. (1991). Quasi-likelihood for multiplicative random effects. *Biometrika*, **78**, 545–56.

Fisher, R.A. (1925). *Statistical methods for research workers*. Oliver and Boyd, London.

Fletcher, R. (1981). *Practical methods of optimization*. Vol. 2., Constrained maximization. John Wiley, New York.

Fuller, W.A. (1987). *Measurement error models*. John Wiley, New York.

Galton, F. (1886). Family likeness in stature. *Proceedings of the Royal Society of London*, **40**, 42–73.

Giesbrecht, F.G. and Burrows, P.M. (1978). Estimating variance components in hierarchical structures using MINQUE and restricted maximum likelihood. *Communications in Statistics*, Part A – Theory and Methods, **7**, 891–904.

Gill, P.E., Murray, W., and Wright, M.H. (1981). *Practical optimization*. Academic Press, London.

Goldstein, H. (1986a). Multilevel mixed linear model analysis using iterative generalized least squares. *Biometrika*, **73**, 43–56.

——— (1986b). Efficient statistical modelling of longitudinal data. *Annals of Human Biology*, **13**, 129–41.

——— (1987). *Multilevel models in educational and social research*. C. Griffin & Co., London and Oxford University Press, New York.

——— (1989). Restricted unbiased iterative generalized least-squares estimation. *Biometrika*, **76**, 622-623.

Goldstein, H. and McDonald, R.P. (1988). A general model for the analysis of multilevel data. *Psychometrika*, **53**, 435–67.

Graybill, F.A. (1969). *Introduction to matrices with applications in statistics*. Wadsworth, Belmont, CA.

Grizzle, J.E. and Allen, M.D. (1969). Analysis of growth and dose response curves. *Biometrics*, **25**, 357–81.

Harrison, D. and Rubinfeld, D.A. (1978). Hedonic prices and the demand for clean air. *Journal of Environmental Economics and Management*, **5**, 81–102.

Hartley, H.O. and Rao, J.N.K. (1967). Maximum likelihood estimation for the mixed analysis of variance model. *Biometrika*, **54**, 93–108.

Harville, D.A. (1974). Bayesian inference for variance components using only error contrasts. *Biometrika*, **61**, 383–5.

——— (1977). Maximum likelihood approaches to variance component estimation and to related problems. *Journal of the American Statistical Association*, **72**, 320–40.

Harville, D.A. and Mee, R.W. (1984). A mixed model procedure for analyzing ordered categorical data. *Biometrics*, **40**, 393–408.

Healy, M.J.R. (1989). *NANOSTAT user manual*. Alphabridge, London.

Hedges, L.V. and Olkin, I. (1985). *Statistical methods for meta-analysis*. Academic Press, New York.

Hemmerle, W.J. and Hartley, H.O. (1973). Computing maximum likelihood estimates for the mixed A.O.V. model using the W transformation. *Technometrics*, **15**, 819–31.

Henderson, C.R. (1953). Estimation of variance and covariance components. *Biometrics*, **9**, 226–52.

Holland, P.W. (1986). Statistics and causal inference. *Journal of the American Statistical Association*, **81**, 945–68.

Holt, D. and Scott, A.J. (1982). Regression analysis using survey data. *The Statistician*, **30**, 169–77.

Hui, S.L. (1983). Curve fitting for repeated measurements made at irregular time points. *Biometrics*, **40**, 691–7.

Hui, S.L. and Berger, J.O. (1983). Empirical Bayes estimation of rates in longitudinal studies. *Journal of the American Statistical Association*, **78**, 753–60.

Im, S. and Gianola, D. (1988). Mixed models for binomial data with an application to lamb mortality. *Applied Statistics*, **37**, 196–204.

James, W. and Stein, C. (1961). Estimation with quadratic loss. *Proceedings of the 4th Berkeley Symposium on Mathematical Statistics and Probability*, Vol. I., pp. 361–79.

Jamshidian, M. and Jennrich, R.I. (1988). *Conjugate gradient methods in confirmatory factor analysis*, UCLA Statistics Series, No. 8. University of California, Los Angeles.

Jennrich, R.I. and Sampson, P.F. (1976). Newton–Raphson and related algorithms for maximum likelihood variance component estimation. *Technometrics*, **18**, 11–17.

Jennrich, R.I. and Schluchter, M.D. (1986). Unbalanced repeated-measures models with structured covariance matrices. *Biometrics*, **42**, 805–20.

Johnson, E.G. and Zwick, R. (1990). *Focusing the new design. The NAEP technical report*. Educational Testing Service, Princeton, NJ.

Jones, K. and Moon, G. (1991). Multilevel assessment of immunisation uptake as a performance measure in general practice. *British Medical Journal*, **303**, 28-31.

Jöreskog, K.G. (1977). Factor analysis by least-squares and maximum-likelihood methods. In *Statistical methods for digital computers*, (eds. Enstein, K., Ralston, A., and Wilf, H.S.), pp. 125–53. John Wiley, New York.

Jöreskog, K.G. and Sörbom, D. (1979). *Advances in factor analysis and structural equation models*. Abt Books, Cambridge, MA.

Kackar, R.N. and Harville, D.A. (1984). Approximations for standard errors of estimation of fixed and random effects in mixed linear models. *Journal of the American Statistical Association*, **79**, 853–62.

Kass, R.E. and Steffey, D. (1989). Approximate Bayesian inference in conditionally independent hierarchical models (parametric empirical Bayes models). *Journal of the American Statistical Association*, **84**, 717–26.

Khuri, A.I. and Sahai, H. (1985). Variance components analysis: A selective literature survey. *International Statistical Review*, **53**, 279–300.

Knight, T. and Troop, N. (1988). *The Sackville illustrated dictionary of athletics*. Sackville Books, Stradbroke, England.

Kreft, I.G.G., DeLeeuw, J., and Kim, K.S. (1990). *Comparing four different statistical packages for hierarchical linear regression: GENMOD, HLM, ML2, and VARCL*, UCLA Statistics Series, No. 50. University of California, Los Angeles.

Kreft, I.G.G. and Kim, K.S. (1991). ML3. In *Statistical software reviews. Applied Statistics*, **40**, 343–7.

Laird, N.M., Lange, N., and Stram, D. (1987). Maximum likelihood computations with repeated measures: Application of the EM algorithm. *Journal of the American Statistical Association*, **82**, 97–105.

Laird, N.M. and Ware, J.H. (1982). Random-effects models for longitudinal data. *Biometrics*, **38**, 963–74.

Lange, N. and Laird, N.M. (1989). The effect of covariance structure on variance estimation in balanced growth-curve models with random parameters. *Journal of the American Statistical Association*, **84**, 241–7.

Lange, N. and Ryan, L. (1989). Assessing normality in random effects models. *Annals of Statistics*, **17**, 624–43.

Lawley, D.N. and Maxwell, A.E. (1971). *Factor analysis as a statistical method*, (2nd edn). Butterworth & Co., London.

Lee, S.Y. (1990). Multilevel analysis of structural equation models. *Biometrika*, **77**, 763–72.

Lee, S.Y. and Jennrich, R.I. (1979). A study of algorithms for covariance structure analysis with specific comparison using factor analysis. *Psychometrika*, **43**, 99–113.

Liang, K.-Y. and Zeger, S.L. (1986). Longitudinal data analysis using generalized linear models. *Biometrika*, **73**, 13–22.

Liang, K.-Y., Zeger, S.L., and Qadish, B. (1992). Multivariate regression analyses for categorical data. *Journal of the Royal Statistical Society*, Ser. B, **54**, 3–40.

Lindley, D.V. and Smith, A.M.F. (1972). Bayes estimates for the linear model. *Journal of the Royal Statistical Society*, Ser. B, **34**, 1–18.

Lindstrom, M.J. and Bates, D.M. (1989). Newton–Raphson and EM algorithms for linear mixed-effects models for repeated measures data. *Journal of the American Statistical Association*, **84**, 1014–22.

Lindstrom, M.J. and Bates, D.M. (1990). Nonlinear mixed effects models for repeated measures data. *Biometrics*, **46**, 673–87.

Lockheed, M.E. and Longford, N.T. (1989). *A multilevel model of school effectiveness in a developing country*, World Bank Discussion Papers 69, Policy Planning Research. World Bank, Washington.

Longford, N.T. (1985). Mixed linear models and an application to school effectiveness. *Computational Statistics Quarterly*, **2**, 109–17.

————— (1987). A fast scoring algorithm for maximum likelihood estimation in unbalanced mixed models with nested random effects. *Biometrika*, **74**, 817–27.

————— (1988a). A quasilikelihood adaptation for variance component analysis. Proceedings of the Section on Statistical Computing Section of the American Statistical Association. American Statistical Association, Alexandria, VA.

————— (1988b). *VARCL — software for variance component analysis of data with hierarchically nested random effects (maximum likelihood)*, Educational Testing Service, Princeton, NJ.

————— (1989). Fisher scoring algorithm for variance component analysis of data with multilevel structure. In *Multilevel analysis of educational data* (ed. R.D. Bock), pp. 297–310. Academic Press, San Diego.

————— (1990). Multivariate variance component analysis: An application in test development. *Journal of Educational Statistics*, **15**, 91–112.

————— (1991). *Negative coefficients in GRE Validity Study Service*, GRE Board Professional Report 89-05P, ETS Research Report 91-26. Educational Testing Service, Princeton, NJ.

————— (1993). Logistic regression with random coefficients. *Computational Statistics and Data Analysis*. In press.

Longford, N.T. and Muthén, B.O. (1992). Factor analysis for clustered observations. *Psychometrika* **57**, 581-97.

Lord, F.M. (1980). *Applications of item response theory to practical testing problems*. Lawrence Erlbaum Associates, Hillsdale, NJ.

Louis, T.A. (1982). Finding the observed information matrix when using the EM algorithm. *Journal of the Royal Statistical Society*, Ser. B, **44**, 226–33.

Luenberger, D.G. (1984). *Linear and nonlinear programming* (2nd edn). Addison-Wesley, Reading, MA.

Lundbye-Christensen, S. (1991). A multivariate growth curve model for pregnancy. *Biometrics*, **47**, 637–57.

Magnus, J.R. and Neudecker, H. (1986). *Matrix differential calculus and static optimization.* John Wiley, Chichester.

Malec, D. and Sedransk, J. (1985). Bayesian inference for finite population parameters in multistage cluster sampling. *Journal of the American Statistical Association*, **80**, 897–902.

Malley, J.D. (1986). *Optimal unbiased estimation of variance components*, Lecture Notes in Statistics, No. 39. Springer-Verlag, New York.

Marcuse, S. (1949). Example of clustered sampling in chemical analysis. *Biometrics*, **5**, 189–206.

Mardia, K.V., Kent, J.T., and Bibby, J.M. (1979). *Multivariate analysis.* Academic Press, London.

Maritz, J.S. (1970). *Empirical Bayes methods.* Methuen, London.

Mason, W.M., Wong, G.Y., and Entwisle, B. (1984). Contextual analysis through the multilevel linear model. In *Sociological Methodology* (ed. S. Leinhardt), Jossey Bass, 72–103.

McCullagh, P. and Nelder, J.A. (1989). *Generalized linear models* (2nd edn). Chapman and Hall, London.

McDonald, R.P. (1985). *Factor analysis and related methods.* Lawrence Erlbaum Associates, Hillsdale, NJ.

McDonald, R.P. and Goldstein, H. (1989). Balanced versus unbalanced designs for linear structural relations in two-level data. *British Journal of Mathematical and Statistical Psychology*, **42**, 215–32.

McIntosh, A. (1982). *Fitting linear models: An application of conjugate gradient algorithms*, Lecture Notes in Statistics, No. 10. Springer Verlag, New York.

Meilijson, I. (1989). A fast improvement to the EM algorithm on its own terms. *Journal of the Royal Statistical Society*, Ser. B, **51**, 127–38.

Meng, X.L. and Rubin, D.B. (1991). Using EM to obtain asymptotic variance–covariance matrices: The SEM algorithm. *Journal of the American Statistical Association*, **86**, 899–909.

Miller, J.J. (1977). Asymptotic properties of maximum likelihood estimates in the mixed model of the analysis of variance. *The Annals of Statistics*, **5**, 746–62.

Moore, D.F. (1986). Asymptotic properties of moment estimators for overdispersed counts and proportions. *Biometrika*, **73**, 583-588.

—————— (1987). Modelling the extraneous variance in the presence of extra-binomial variation. *Applied Statistics*, **36**, 8–14.

Morris, C.N. (1983). Parametric empirical Bayes inference: Theory and applications. *Journal of the American Statistical Association*, **78**, 47–65.

Morrison, D.F. (1967). *Multivariate statistical methods.* McGraw-Hill, New York.

Morton, R. (1987). A generalized linear model with nested strata of extra-Poisson variation. *Biometrika*, **74**, 247–57.

Multilevel Modelling Newsletter (1989, 1990). Institute of Education, University of London.

Muñoz, A., Rosner, B., and Carey, V. (1986). Regression analysis in the presence of heterogeneous intraclass correlations. *Biometrics*, **42**, 653–8.

Muthén, B.O. (1989). Latent variable modeling in heterogeneous populations. *Psychometrika*, **54**, 557–85.

Muthén, B.O. and Satorra, A. (1989). Multilevel aspects of varying parameters in structural models. In *Multilevel analysis of educational data* (ed. R.D. Bock), pp. 87–99. Academic Press, San Diego.

NAG (Numerical Algorithms Group) (1986). *The GLIM system. Release 3.77, manual*, Royal Statistical Society, London.

Nelder, J.A. and Wedderburn, R.W.M. (1972). Generalized linear models. *Journal of the Royal Statistical Society*, Ser. A, **135**, 370–84.

Nelder, J.A. and Pregibon, D. (1987). An extended quasi-likelihood function. *Biometrika*, **74**, 221–32.

Novick, M.R., Jackson, P.H., Thayer, D.T., and Cole, N.S. (1972). Estimating multiple regression in m groups: A cross-validation study. *British Journal of Mathematical and Statistical Psychology*, **25**, 33–50.

Patterson, H.D. and Thompson, R. (1971). Recovery of inter-block information when block sizes are unequal. *Biometrika*, **58**, 545–54.

Paul, S.R. (1990). Maximum likelihood estimation of intraclass correlation in the analysis of familial data: Estimating equation approach. *Biometrika*, **77**, 549–55.

Potthoff, R.F., Woodbury, M.A., and Manton, K.G. (1992). 'Equivalent sample size' and 'equivalent degrees of freedom' refinements for inference using survey weights under superpopulation models. *Journal of the American Statistical Association*, **87**, 383–96.

Pregibon, D. (1981). Logistic regression diagnostics. *Annals of Statistics*, **9**, 705–24.

Prentice, R.L. (1986). Binary regression using an extended beta-binomial distribution with discussion of correlation induced by covariate measurement errors. *Journal of the American Statistical Association*, **81**, 321–7.

———— (1988). Correlated binary regression with covariates specific to each binary observation. *Biometrics*, **44**, 1033–48.

Qu, Y.S., Williams, G.W. Beck, G.J., and Goormastic, M. (1987). A generalized model of logistic regression for correlated data. *Communications in Statistics*, Part A – Theory and Methods, **16**, 3447–76.

Rao, C.R. (1965). *Linear statistical inference and its applications*. John Wiley, New York.

———— (1971a). Estimation of variance and covariance components – MINQUE theory. *Journal of Multivariate Analysis*, **1**, 257–75.

———— (1971b). Minimum variance quadratic unbiased estimation of variance components. *Journal of Multivariate Analysis*, **1**, 445–56.

Rasbash, J., Prosser, R., and Goldstein, H. (1991). *Software for three-level analysis. Users' guide*. Institute of Education, University of London.

Raudenbush, S.W. (1988). Educational applications of hierarchical linear models: A review. *Journal of Educational Statistics*, **13**, 85–116.

Raudenbush, S.W. and Bryk, A.S. (1986). A hierarchical model for studying school effects. *Sociology of Education*, **59**, 1–17.

Raudenbush, S.W. and Willms, J.D. (eds.) (1991). Schools, classrooms, and pupils. International studies of schooling from a multilevel perspective. Academic Press, San Diego.

Robinson, D.L. (1987). Estimation and use of variance components. *The Statistician*, **36**, 3–14.

Robinson, G.K. (1991). The estimation of random effects. *Statistical Science*, **6**, 15–51.

Rosner, B. (1984). Multivariate methods in ophthalmology with application to other paired-data situations. *Biometrics*, **40**, 1025–35.

————— (1989). Multivariate methods for clustered binary data with more than one level of nesting. *Journal of the American Statistical Association*, **84**, 373–80.

Rotnitzky, A. and Jewell, N.P. (1990). Hypothesis testing of regression parameters in semiparametric generalized linear models for clustered correlated data. *Biometrika*, **77**, 485–97.

Rubin, D.B. (1980). Using empirical Bayes techniques in the law school validity studies. *Journal of the American Statistical Association*, **75**, 801–27.

————— (1983). Some applications of Bayesian statistics to educational research. *The Statistician*, **32**, 55–68.

Rubin, D.B., Laird, N.M., and Tsutakawa, R.K. (1981). Estimation in covariance component models. *Journal of the American Statistical Association*, **76**, 341–53.

Rudan, J.W. and Searle, S. (1971). Large sample variances of maximum likelihood estimators of variance components in the 3-way nested classification, random model, with unbalanced data. *Biometrics*, **27**, 1087–91.

Sahai, H. (1976). A comparison of estimators of variance components in the balanced three-stage nested random effects model using mean squared error criterion. *Journal of the American Statistical Association*, **71**, 435–44.

————— (1979). A bibliography on variance components. *International Statistical Review*, **47**, 177–222.

Searle, S.R. (1971). Topics in variance component estimation. *Biometrics*, **27**, 1–76.

————— (1982). *Matrix algebra useful for statistics*. John Wiley, New York.

Stanek, E.J. and Diehl, S.R. (1988). Growth curve models of repeated binary response. *Biometrics*, **44**, 973–83.

Stanek, E.J. and Koch, G.G. (1985). The equivalence of parameter estimates from growth curve models and seemingly unrelated regression models. *American Statistician*, **39**, 149–52.

Sterne, J.A.C., Johnson, N.W., Wilton, J.M.A., Joyston-Bechal, S., and Smales, F.C. (1988). Variance components analysis of data from periodontal research. *Journal of Periodontal Research*, **23**, 148-153.

Stigler, S.M., (1986). *The History of statistics. The measurement of uncertainty before 1900*. The Belknap Press of Harvard University Press, Cambridge, MA, and London.

Stiratelli, R., Laird, N.M., and Ware, J.H. (1984). Random effects models for serial observations with binary response. *Biometrics*, **40**, 961–71.

Stokes, L. (1988). Estimation of interviewer effects for categorical items in a random digit dial telephone survey. *Journal of the American Statistical Association*, **83**, 623–30.

Stram, D.O., Wei, L.J., and Ware, J.H. (1988). Analysis of repeated order categorical outcomes with possibly missing observations and time dependent covariates. *Journal of the American Statistical Association*, **83**, 631–37.

Strenio, J.F., Weisberg, H.I., and Bryk, A.S. (1983). Empirical Bayes estimation of individual growth-curve parameters and their relationship to covariates. *Biometrics*, **39**, 71–86.

Thompson, R. (1979). The estimation of variance and covariance components with an application when records are subject to culling. *Biometrics*, **29**, 527–50.

Thompson, R. and Meyer, K. (1986). Estimation of variance components: What is missing in the EM algorithm? *Journal of Statistical Computing and Simulation*, **24**, 215–30.

Ware, J.H. (1985). Linear models for the analysis of longitudinal studies. *The American Statistician*, **39**, 95–101.

Wedderburn, R.W.M. (1974). Quasi-likelihood functions, generalized linear models and the Gauss–Newton method. *Biometrika*, **61**, 439–47.

Williams, D.A. (1982). Extra-binomial variation in logistic linear models. *Applied Statistics*, **31**, 144–8.

Wilm, H.G. (1945). Notes on analysis of experiments replicated in time. *Biometrics Bulletin*, **1**, 16–20.

Wolter, K.M. (1985). *Introduction to variance estimation.* John Wiley, New York.

Wong, G.Y. and Mason,W.M. (1991). Conceptually specific effects and other generalizations of the hierarchical linear model for comparative analysis. *Journal of the American Statistical Association*, **86**, 487–503.

Wu, C.F.J. (1983). On the convergence properties of the EM algorithm. *Annals of Statistics*, **11**, 95–103.

Zeger, S.L. and Liang, K.-Y. (1986). Longitudinal data analysis for discrete and continuous outcomes. *Biometrics*, **42**, 121–30.

Zeger, S.L., Liang, K.-Y., and Albert, P.S. (1988). Models for longitudinal data: A generalized estimating equation approach. *Biometrics*, **44**, 1049–60.

Zeger, S.L. and Qadish, B. (1988). Markov regression models for time series: A quasi-likelihood approach. *Biometrics*, **44**, 1019–31.

Zelen, M. (1957). The analysis of covariance for incomplete block designs. *Biometrics*, **13**, 309–32.

Zellner, A. (1962). An efficient method for estimating seemingly unrelated regression equations and test for aggregation bias. *Journal of the American Statistical Association*, **57**, 348–68.

Zhao, L.P. and Prentice, R.L. (1990). Correlated binary regression using a quadratic exponential model. *Biometrika*, **77**, 642–8.

Index